Air Pollution in the Ural Mountains

NATO ASI Series

Advanced Science Institutes Series

A Series presenting the results of activities sponsored by the NATO Science Committee, which aims at the dissemination of advanced scientific and technological knowledge, with a view to strengthening links between scientific communities.

The Series is published by an international board of publishers in conjunction with the NATO Scientific Affairs Division

A Life Sciences **B Physics**	Plenum Publishing Corporation London and New York
C Mathematical and Physical Sciences **D Behavioural and Social Sciences** **E Applied Sciences**	Kluwer Academic Publishers Dordrecht, Boston and London
F Computer and Systems Sciences **G Ecological Sciences** **H Cell Biology** **I Global Environmental Change**	Springer-Verlag Berlin, Heidelberg, New York, London, Paris and Tokyo

PARTNERSHIP SUB-SERIES

1. Disarmament Technologies	Kluwer Academic Publishers
2. Environment	Springer-Verlag / Kluwer Academic Publishers
3. High Technology	Kluwer Academic Publishers
4. Science and Technology Policy	Kluwer Academic Publishers
5. Computer Networking	Kluwer Academic Publishers

The Partnership Sub-Series incorporates activities undertaken in collaboration with NATO's Cooperation Partners, the countries of the CIS and Central and Eastern Europe, in Priority Areas of concern to those countries.

NATO-PCO-DATA BASE

The electronic index to the NATO ASI Series provides full bibliographical references (with keywords and/or abstracts) to more than 50000 contributions from international scientists published in all sections of the NATO ASI Series.
Access to the NATO-PCO-DATA BASE is possible in two ways:

– via online FILE 128 (NATO-PCO-DATA BASE) hosted by ESRIN, Via Galileo Galilei, I-00044 Frascati, Italy.

– via CD-ROM "NATO-PCO-DATA BASE" with user-friendly retrieval software in English, French and German (© WTV GmbH and DATAWARE Technologies Inc. 1989).

The CD-ROM can be ordered through any member of the Board of Publishers or through NATO-PCO, Overijse, Belgium.

Series 2: Environment – Vol. 40

Air Pollution in the Ural Mountains

Environmental, Health and Policy Aspects

edited by

Igor Linkov

and

Richard Wilson

Department of Physics,
Harvard University,
Cambridge, U.S.A.

Springer Science+Business Media, B.V.

Proceedings of the NATO Advanced Research Workshop on
Air Pollution in the Ural Mountains
Magnitogorsk, Russia
26–30 May 1997

A C.I.P. Catalogue record for this book is available from the Library of Congress.

ISBN 978-94-010-6192-6 ISBN 978-94-011-5208-2 (eBook)
DOI 10.1007/978-94-011-5208-2

Printed on acid-free paper

The first editor dedicates this book to his grandfather, Yankel Roitman, who helped to build Magnitogorsk in 1930-1937, and the second editor to those american engineers who also helped.

TABLE OF CONTENTS

PREFACE

Mankind has created pollution, and has suffered its consequences since time immemorial. This has intesified greatly since the industrial revolution. One of the main problems in society, and a major function of government is how to cope with this pollution. 80 years ago the maxim used to be "the solution to pollution is dilution"; to dilute any pollted water supply in a large river, or to build a tall chimney stack to dilute air pollutants into the air so that concentrations of pollutants are always low. Since 1950 western countries have gone further and made major attempts to reduce the emissions of the most important pollutants.

The discussion of what is an important pollutant has changed. To SO_2 and heavy metals such as cadmium or arsenic we now add fine particles and even (when we discuss global climate change) CO_2. The experience and practice of the western countries was only partly followed in the USSR (although the switch from use of coal to natural gas in major cities around 1970 was very important). Since the collapse of the USSR it has become fashionable both in the west and inside Russia to blame all society's ills on pollution. The statistics do not bear out that conclusion, but pollution remains an important issue which can be reduced without significant detriment to other societal values.

In choosing Magnitogorsk for the site of this workshop the organizers were well aware of the three centuries of industrial pollution in the Urals, and of the fact that Magnitogorsk in particular once made half the steel of the Soviet Union. In the 1930s American engineers came to help in building this great steel complex, so it seemed appropriate for American environmental experts to come and help discuss how to make it better.

The western organizers and editors of this volume are optimistic. The air pollution by the steel works itself is clearly less than London, Pittsburgh, Gary or Tokyo of 1952. At the beautiful lake where the workshop was held little pollution was evident. With a civic will the pollution everywhere can be made to be negligible.

Igor Linkov and Richard Wilson
Harvard University
October 1997.

ACKNOWLEDGEMENTS

The editors would like to thank Dr. V.D. Cherchintsev and his colleagues of the Magnitogorsk State Academy of Mining and Metallurgy and Magnitogorsk Integrated Iron Works, Inc. for their help in the workshop organization and hospitaility. We also thank workshop participants for their contribution to the book and peer review of manuscripts. Special thanks to Elena Belinkaia for editorial and technical assisstance. Financial support for the workshop organization was provided in the main parts by NATO. We also thank the Andrei Sakharov Foundation, US EPA and Harvard University for their joint sponsorship.

Introduction

Magnitogorsk: Construction of blast furnaces in 1931.
Postcards from Y. Roitman's archive

PARTICULATE AIR POLLUTION

R. WILSON
Department of Physics, Harvard University
Cambridge, MA 02138, USA

The attention of the whole world was drawn to air pollution by the London fog of 1952. I was in that fog, although somewhat upriver at Oxford. To the best of my knowledge, anyone who has looked at the data from the official government report [1] shown here as Figure 1, agrees that about 4,500 people died within 2 weeks because of the air pollution.

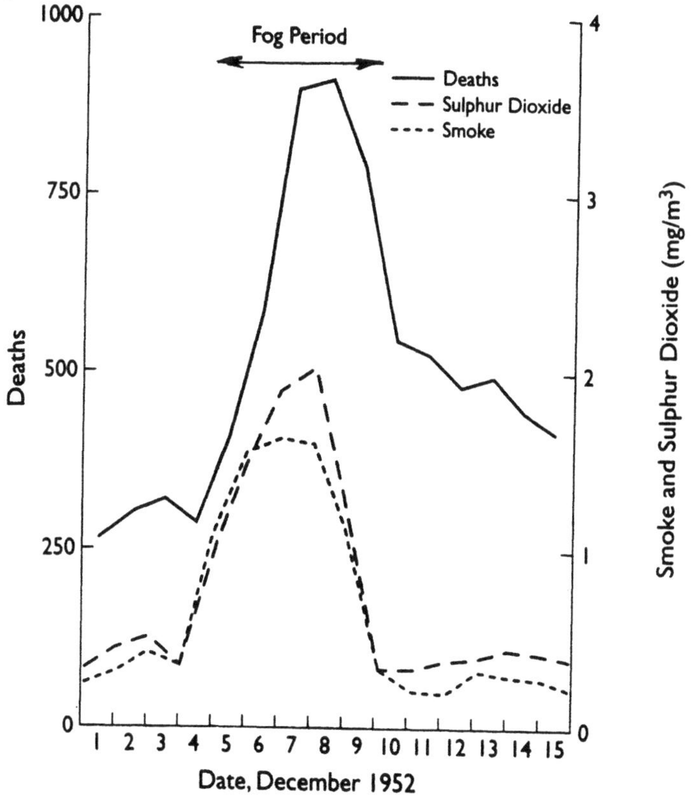

FIGURE 1. Daily mean pollution concentration and daily numbers of deaths during the London fog episode of 1952 (from [1]).

I. Linkov and R. Wilson (eds.), Air Pollution in the Ural Mountains, 3–8.

At the time, most public health experts felt that the major preventive action necessary was for the sick, aged and infirm to stay indoors and refrain from exercise during an air pollution incident. If the exposure could be reduced 5 fold, it was argued that the dose would be below an assumed threshold and that all would be well. To the best of my knowledge there was and is no data to prove a threshold but this was one of the implicit assumptions made by most experts and implied in most studies.

A second commonly made implicit assumption was that the air pollution merely accelerated the death of those who were already sick; a phenomenon sometimes called graphically "harvesting". If so, the increase in death rate in the first few weeks of December 1952 should have been followed by reduction in death rate of equal integrated amount. Yet the data of that winter showed no week in which the death rate fell below normal at least up to the middle of January 1953.

Thirdly, no one was sure what were the proximate causes of mortality. The London study showed similar correlations with SO_2, and smoke (particles). Experts thought that if one removed most SO_2 and 99% of their particles, then all would be well.

Further research has suggested that none of these three assumptions are true. The third was disproved by toxicological experiments in animals by Dr Mary Amdur, just awarded a "Merit Award" by the Society of Toxicology for her seminal work. She found in 1973 [2-3] that bronchial constriction in guinea pigs was more severe for fine particles than for coarse ones; and for sulphate particles than for SO_2 itself (Figure 2).

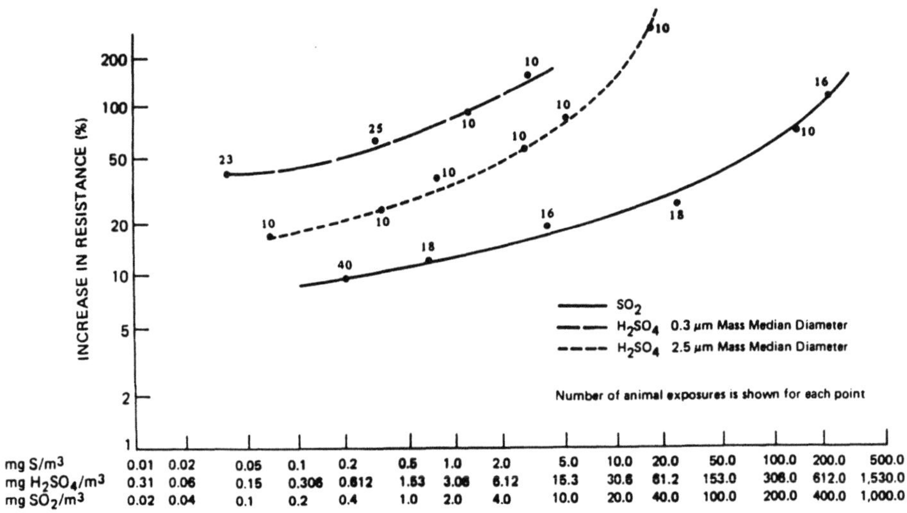

FIGURE 2. Change in bronchial resistance of guinea pigs exposed to various pollutants. From [2].

Studies confirmed that it is the fine particles that penetrate to the lung as shown in Figure 3 [4]. Moreover fine particles have two other troubling features. Particulate removal at a factory or power plant chimney removes the coarse particles more than the fine particles so that there can be appear to be considerable particulate removal without corresponding improvement in public health. And it is these fine particles that do not immediately fall to the ground but travel thousands of miles in the air. However we are still not sure of the details. Are particles of nitrates as bad as sulphates? What is the role of ozone? These tend to vary collinearly with time in the polluted cities, so that a clean epidemiological result seems impossible.

The second idea, now known to be erroneous, that air pollution only shortened life by a few days, was thrown into doubt by (ecological) epidemiological studies, begun by Lave and Seskin [7]. These showed a correlation of mortality rate with air pollution variables, and a bigger correlation with fine particles than TSP. The suggestion was clear: it is the fine particles that pass the electrostatic depressors, or are made in the air by the SO_2 precursors, that free pass indoors and easily enter the airway passages, that are the proximate problem.

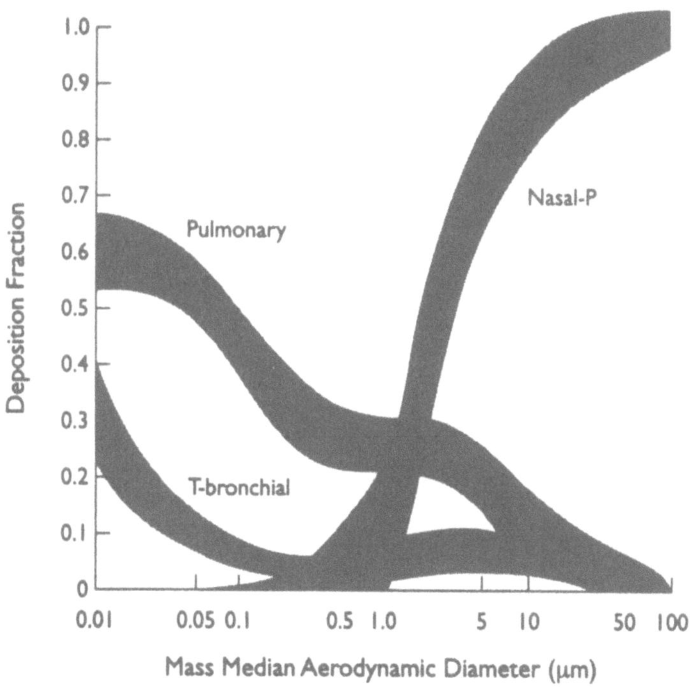

FIGURE 3. Regional deposition predictions based on the model proposed by the International Commission on Radiological Protection task group on lung dynamics [4]. Each of the shaded areas (envelopes) indicate the variable deposition for a given mass median (aerodynamic) diameter in each compartment when the distribution parameter σ_g varies from 1.2 to 4.5 and the tidal volume is 1,450 ml.

Not everyone believed these studies. They were "ecological" studies, which only compare population averages and are prone to misinterpretation. But two cohort studies, the Harvard Six City study [5], shown in figure 4, and an American Cancer Society study [6] show a variation of annual mortality between people in places with differing air pollution.

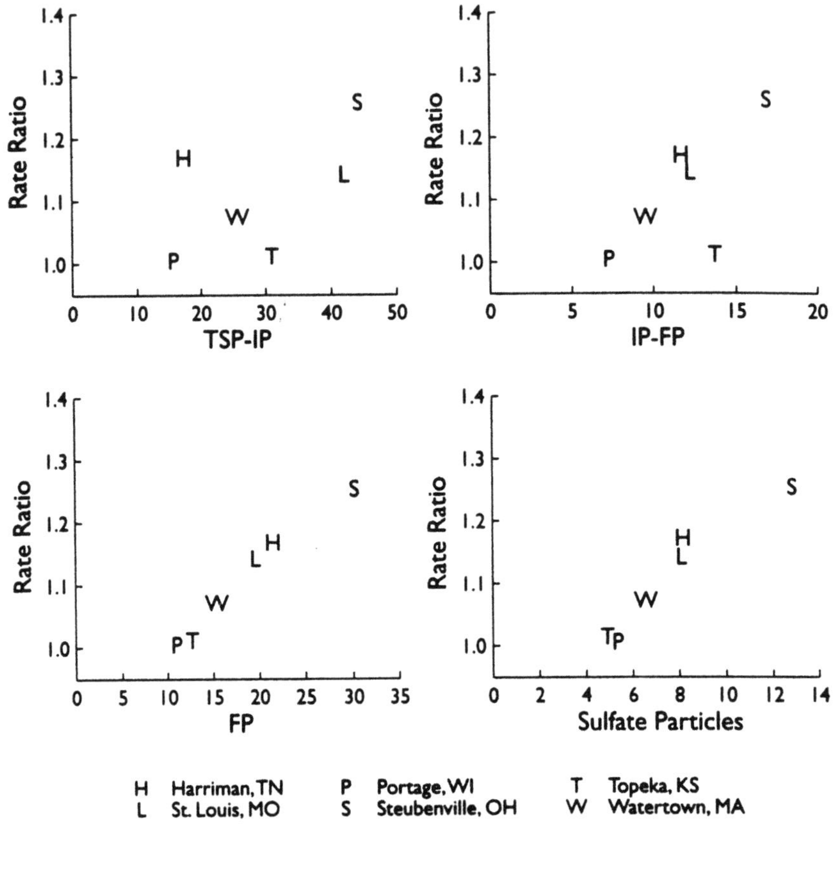

FIGURE 4. Estimated adjusted mortality rate-ratios from the six-cities study plotted against non-inhalable particles (TSP-IP), the coarse fraction of inhalable particles (IP-EP), and sulfate particles [6]. Units are $\mu g/m^3$

As late as 1980 most scientists believed either in a threshold or that particulate removal without regard for size would be adequate. Moreover, both the data of Lave and Seskin [7] were related to concentrations outdoors and not to individual exposures. Yet people spend more time indoors than outdoors! Further research has convinced many of the wavers. First the prejudice in favor of a threshold has been weakened by the failure to find one, and the realization that while some people may be completely insensitive if a threshold is not exceeded, a simple model can predict a linear dose response for a population with approximately the observed coefficients [8]. This model is that every little bit of pollution reduces the lung function a little. This permanent reduction is not important until late in life, or among the sick and infirm, whose lung function is inadequate. Figure 5 illustrates this first issue schematically. This may not be the correct model, but it does suggest that a linear dose response for a population could exist and is even likely. The second point is clarified when measurements of fine particles showed that they penetrate readily indoors. Our recent book [9] distributed at the conference, summarizes the scientific arguments.

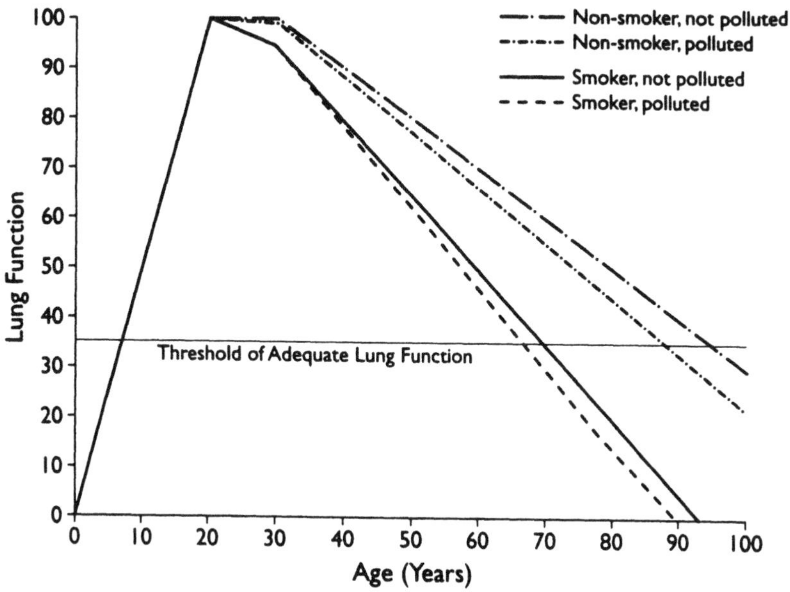

FIGURE 5. Schematic of lung function vs. age showing loss of life expectancy (LOLE) (from [8]).

It will be interesting to see whether the data from Magnitogorsk will follow the same trend as the data in the polluted cities of the USA, or whether the large pollution from the metals industry lead to additional or alternative hazards. If indeed a linear dose response relationship holds down to levels appreciably below ambient, 50,000 Americans are STILL dying each year from air pollution, with a lifetime risk of a few percent. While this is far less than the 30% risk of cigarette smoking in the USA it is large enough that it could not be ignored by the Environmental Protection Agency.

In this conference we also hope that the residents of Magnitogorsk will learn from the US experience as much as possible, and act to reduce the fine particles and their precursors. In addition to note the observations of Mary Amdur that suggest that these particles are far worse if heavy metals are attached.

References

1. Beaver H. (Chairman) (1954) "Interim Report" *Committee on Air Pollution:Cmd 9011* Her Majesty's Stationary Office, London
2. Amdur, M.O. (1973) in "Animal Studies - Proceedings of the Conference on Health effects of Air Pollutants" *U.S. National Academy of Sciences*
3. Amdur, M.O. (1989) "Sulfuric Acid: The Animals Tried to Tell US" *Appl. Ind. Hygiene* 4(8):189-197
4. International Commission on Radiological Protection (ICRP) 1966 "Task Force on Lung Dynamics" *Health Physics* 12:173-207
5. Dockery, D.W., Pope C.A. III, Xu X., Spengler, J.D., Ware, J.H., Fay, M.E., Ferris, B.G., and Speizer, F.E. (1993) "Mortality Risks of Air Pollution: a Prospective Cohort Study" *New England Journal of Medicine* 329:1753-1759
6. Pope C.A.III, Thun, M.J., Namboodiri, M.M., Dockery, D.W., Evans, J.S., Speizer, F.E., and Heath C.W. (1995) "Particulate Air Pollution as a Predictor of Mortality ina prospective study of US Adults" *American Journal of Respiratory and Critical Care Medicine* 151:669-674
7. Lave L.B., and Seskin, E.P. (1970) "Air Pollution and Health" *Science* 169:723-733
8. Evans J.S. And Wolff S.K. (1996) "Modeling of Air pollution Impacts: One Possible Explanation of Chronic Mortality" Chapter 9 of *Particles in Our Air: Concentrations and Health Effects*, Harvard University Press
9. Wilson, R. and Spengler, J.D. (Editors) (1996) *Particles in Our Air: Concentrations and Health Effects.* Harvard University Press

MINING AND ENVIRONMENT IN THE URALS DURING THE 18TH AND 19TH CENTURIES

J.O. NRIAGU
Department of Environmental & Industrial Health, School of Public Health, University of Michigan, Ann Arbor, MI 48109 USA

1. Introduction

From about the mid 1500s onwards, the massive supply of precious metals from Spanish America totally marginalized the mining activities in much of Europe. This cycle in the European mining history came to an end during the period of 1670-1760 when several events converged to stimulate the revival of intensive mining especially in central Europe. High silver prices brought about by mining crisis in South and Central America, the equally exorbitant copper prices caused by the collapse of Swedish production in 1655, the cessation of copper imports from Japan in the 1670s, and the depressed lead prices as the expansion of the British industry flooded much of Europe with cheap supplies of the metal all combined to create a price climate conducive to mining renaissance in many European countries.

The old mining districts of German states were the first to try to take the center stage but never rose above the ashes of their glorious past. The once famous Bohemian mines of Joachimstahl and Kuttenberg could only yield 644 kg of silver per year between 1670 and 1740. The legendary Schwaz and Salzburg mines in Tyrol yielded about 80 kg/year of silver on the average between 1670 and 1740 [1]. The fabled Hungarian mines did not fare much better, silver outputs from Neusohl and Smolnik mines averaged only 2500-3000 kg/year. With such marginal production from these once-famous centers, attention began to be focused on other areas. The Norwegian mine at Kongsberg, and later the argentiferous copper deposits in Oberharz and Saxony (Schneeberg, Annaberg, Marienberg, Weisenthal, Freiberger Revier and Johanngeorgenstadt) dominated silver and copper production in central and northern Europe for most of the 1670-1740 period [1]. Between 1740 and 1780, however, the mines of Hungary and Siebenbergen became the primary focus of new central European metallurgical complex [1].

The lands of the Russian Empire became a frontier for new opportunities during this revival of mining in Europe. Mining of the rich deposits of Nerchinsk in the Amur valley of southeastern Siberia began in 1704, and the Kolyvan-Voskressensk mines in the Altai region were opened in 1729. Production from these mines grew rapidly from the 1740s onwards, and with the opening of the Uralian gold mines at Beresov in 1752, production of silver in Russia reached 40,000 kg/year in the early 1770s and exceeded the productions from the rest of Europe put together [1]. The onset

I. Linkov and R. Wilson (eds.), Air Pollution in the Ural Mountains, 9–20.

of vigorous mining the Urals thus marked the first great period of Russian precious metal production, often known as "Russia's Age of Silver". During this period, which lasted from about 1745 to 1795, Russia became the leading producer of precious metals in Europe. Russian base metal industries underwent a parallel process of rapid expansion and by the 1760s, the annual outputs of copper (about 3,500 tons) and iron were second to none in the world. The Age of Silver was short lived, however. With the recovery of American silver production and increased production from Saxony and Bohemia following the introduction of patio amalgamation technique, the Russian pre-eminence in precious metal production came to an end. The next boom, or Russia's "Golden Age" began in 1845 or thereabouts and lasted until 1885. Gold production from the Urals and remote mines of northeastern Siberia reached nearly 29,000kg/year in the 1840s, which made Russia the leading gold producer in the world and second only to Mexico in combined precious metal production. The 18th and 19th centuries thus witnessed two great periods of Russian precious metal production which were determined, to a large extent, by intensification of mining activities in the Ural mountains [2-14]. The gold and silver which fueled the Russian economy also left behind an unprecedented legacy of forest destruction and massive environmental pollution.

2. Mining in the Urals

The Urals contain thousands of mineral deposits and represent one of the most important mining regions of Russia. The legendary richness of the mineral resources of the region is embellished in a number of reports and books. In a travelogue entitled *Oriental and Western Siberia*, it was noted that "on the vast estate of the Demidoff, containing 3,095,700 acres, nearly equalling Yorkshire, nature has been most bountiful. Iron and copper ore appear to be inexhaustible. Platinum and gold are in the upper valleys, and malachite is found there also in enormous quantities, with porphyry and jasper of great beauty, and various colored marbles. The forests extend over more than 10,000 square versts, and are thickly covered with timber" [15, p.121]. The fact that the iron, gold, copper and chromium resources of the Urals had attained national and world significance during the 18th and 19th centuries is well chronicled in the annals of mining and technology.

Mining for copper in the Urals probably began during the Russian Bronze Age. Exploitation of the gold deposits also began in remote antiquity, as suggested in the following excerpt from Fullartonís *Gazetteer* on the origins of gold mining in the Urals:

> "The continental tracts formerly so rich, as cited by Strabo, with the exception of the North Ural or country of the Arimaspes, whence the Scythians ores came are no longer gold-bearing districts. The Scythian or Uralian tract had, in fact, remained unknown and unattended to from the classical age until this century, and so completely ignorant were the modern Russians of the existence of gold in the Ural mountains, or that they had in their hands the country which supplied so much gold to Greece and Rome, that excellent German

> miners had long worked the iron and copper mines of that chain before any gold veins were discovered ..... the period is gradually arriving when the local depressions or basins of auriferous detritus of the region will be successively dug and washed out, and the Ural will then resemble many other countries in possessing actual mines of iron and copper but merely a history of its gold" [15, p. 68].

No attempt will be made in this report to discuss the long history of mining in the Urals [2-16]. The focus instead will be on the intensive mining activities and environmental pollution that began at the end of the 17th century and were sustained during the next two centuries.

2.1. GOLD PRODUCTION

The principal Uralian gold fields consisted of the Beresov and Chussovsk located near Ekaterinburg with lesser workings at Miask and Neviansk. The techniques used in mining and recovery of gold from the quartz host rocks differed but little from those used in central Europe during the 16th century. Once raised, the ore was crushed and carried by laborers to the washing place where it was poured into long water-filled troughs beaten with iron stampers. The hydraulic straining process resulted in a concentration of the auriferous quartz particles which were collected and heated with lead in hearths (*Stoffherden*) to extract the gold by cupellation. Separation of gold from gangue material by mercury amalgamation was introduced into Uralian workings in 1790s. This involved piling the ore into heaps in an open paved yard (patio). Water was added to the heaps until a thick slime was formed. Salt and mercury were then mixed in and the mixture allowed to "cook" for several months. The amalgam formed was removed by gravity separation in washing vats and squeezed in canvas bags to recover the free mercury. The amalgam was subsequently heated to recover some of the mercury from the hood of the furnace leaving behind an impure gold (*legatur* or *blickgold*) consisting of about 87% gold, 8.5% silver and 4.5% iron [16]. It should come as no surprise that with the introduction of mercury amalgamation, the combined output of refined gold from the Beresov and Miask mines rose sharply and by 1808/89 exceeded the Siberian production (Table 1). For well over 100 years (until the 1930s), mercury remained a common staple of gold production in the Urals and other regions of Russia.

Mining at Beresov was very extensive and relied exclusively on near-surface deposits during the 18th century [2-14]. Blanchard [1, p. 157] noted that "new mines had been opened up all over the open plain creating an infinity of conical heaps of mining rubbish overspreading its [plainís] entire extent". Because of the peculiarity of the geology of the region which led to extensive flooding in the mines, mine shafts rarely exceeded 100 ft. The flooding problem was solved in 1799 with the introduction of horse-whims with a capacity to raise over 1 million liters of water per day; the horse-whims were later replaced with steam engines. The impact of the de-watering equipment was instantaneous ñ ore production rose from about 6,600 tons/yr in the 1790s to 29,770 tons in 1806 (Table 1). In 1797, there were 81 stamps and 1025 washing tables at Beresov, and at the peak of production in 1809/10, there were 283 stamps

TABLE 1. Mine production in the Urals during the 18th century (derived from data provided in [1]).

Period	Gold/silver ore crushed and washed (tons)	Gold produced (kg)	Silver produced (kg)	Copper produced (tons)
1720-24				56
1725-29				1,686
1730-34				1,711
1735-39				2,148
1740-44				2,662
1745-49				5,460
1750-54				6,180
1755-59	5,031	1,879	252	9,295
1760-64	19,216	1,985	886	13,316
1765-69	17,947	1,772	1,772	13,963
1770-74	24,673	2,464	656	11,894
1775-79	32,921	2,552	532	11,100
1780-84	29,544	1,772	674	8,732
1785-89	33,555	2,818	567	12,723
1790-94	40,817	2.393	1,684	11,324
1795-99	60,463	2,747	1,736	9,314
1800-04	116,591	2,995	2,340	
1805-09	146,007	3,492	2,251	

and over 4000 washing tables in operation [1]. This massive complex must have exerted a heavy toll on the local environment.

Extraction of silver, as a by-product or co-product of gold recovery, likewise employed the technology developed during the Middle Ages. The ore was first reduced in a simple hearth (*plavilínaya pechí*) to produce argentiferous lead, described as "fertile", which was placed in a cupel and subjected to an oxidizing blast of air. The lead was oxidized to litharge leaving a residuum of "unclean" silver with traces of other metals. The litharge was re-smelted to obtain "sterile" or de-silvered lead. For rich ores (over 100 oz of silver per ton of ore), the overall loss of lead was about 10% but for marginal ores (about 12 oz silver per ton) the loss was over 80% [2-14].

Mining of gold and silver in the Urals during the 18th and 19th centuries thus affected the local environment in two fundamental ways: it resulted in (a) extensive exfoliation of landscape and (b) the contamination of the region with lead and mercury.

2.2. COPPER PRODUCTION

Impetus for expanded exploitation of copper deposits in the Urals came from Peter the Great, who in response to the decline in the importation of silver, set in motion a grandiose plan to create coinage from indigenous sources of metallic supply. Although the issue of copper coins in the Russian Empire began before 1700, it was the reorganization of mint activity in 1723-25 which created a mint for copper coins in Ekaterinburg that spurred the growth of copper mining in the Urals. To supply the mint, 32 hearths capable of producing over 200 tons of copper per year from the Uralian ores were set up between 1725 and 1735. Imports of precious metals into Russia rapidly dwindled and was completely eliminated by the 1740s. For the next 60 years (1735-1795), the countryís economy and monetary stocks grew on the basis of increasing supply of locally produced, mainly Uralian, metals (Table 1). Copper thus became exceptionally important to the Russia economy, being a major component of the monetary stock. The composition of monetary stock in the 18th century increased from 32 million roubles (mr) in silver and gold and 2.1 mr in copper in 1718 to 51 mr in in gold and silver and 99 mr in copper between 1763 ana 1795 [1]. In 1795, there were 37 mr of copper coins in circulation and 78 mr in reserves [5]. The Uralian copper that sustained the Russian economy during the 18th and 19th centuries [2-14] was obtained at considerable cost to the local environmental quality.

During the 18th century, the principal copper mines in Russia were located in the Altai, Olonets and Urals, with the later being the most important by far. Large number of workshops or copper factories were established in Perm, Ufa, Vyatka and Kazan provinces (Figure 1), usually by few aristocratic families such as the Chernishevs (Nzhne- and Verkhne-Yugovskii and Anninskii works), the Vorontsovs (Yagoshikhinskii, Pisorskii, Visimskii and Motovilikhinskii works), Demidovs (Viiskii, Suksunskii, Shakvinski, Bimovskii, and Ashapskii works), Osokin brothers (Yugovskii, Biaziarskii and Kurashimskii works in Kungur region) and the Strogonovs (Tamanskii, Domrianskii, Yugokamsii, Kuse-Alexandrovskii, Pozhevskii, Khoklovskii and Nytvenskii works in Kama region) [2-14]. The establishment of hegemony over the copper industry by private entrepreneurs in the Urals during this period had major environmental significance: greed, mismanagement and financial limitations brought about complete disregard for the environment in most mining endeavors.

The production of copper in the Urals increased from about 5 tons per year (t.a.) in 1720 to over 3000 t.a. in the 1760s (Table 1). Figure 1 shows the clustering of some of these works in central and southern Urals in 1762, the distribution of the number of hearths and production capacity being as follows [1]:

Region	Number of works	Production Capacity (t)
Central Urals	180	2,500
Southern Urals	83	1200

The copper factories were small, often under-capitalized and relied primarily on technology of the Middle Ages imported from Germany (Figure 2). Their impacts on the local environment could have been severe. In general, the copper ores were roasted in an open hearth. The molten mass was then subjected to a strong and continuous blast which resulted in impurities being oxidized and taken off as a slag as well as in the conversion of copper to cuprous oxide. The cuprous oxide served to further catalyze the oxidation of impurities. After oxidizing most of the impurities and copper, the cuprous oxide was reduced by introducing a billet of green wood, the dry distillation of which generated large quantities of gas for the reductive reaction [2-14].

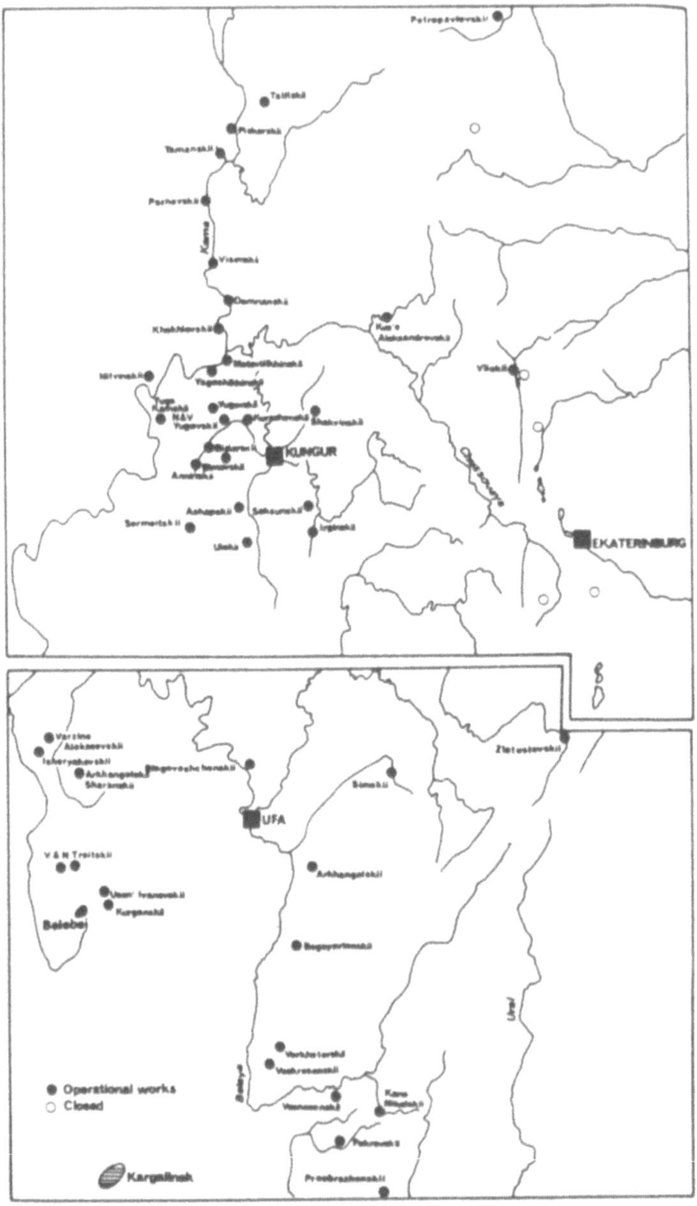

FIGURE 1. Copper mines in Central Urals (Top) and Southern Urals (Bottom) in 1762 (from [1]).

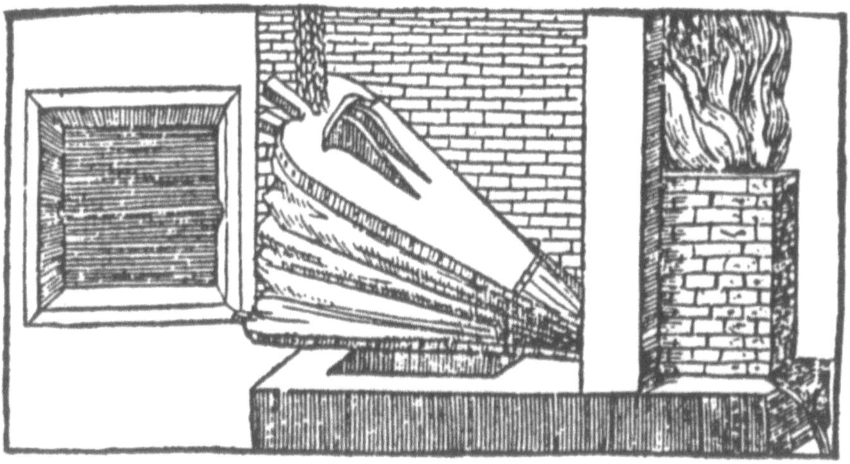

Blast furnace for smelting refractory iron ores.

Hearth for refining copper.

FIGURE 2. Refining hearth for copper (bottom) and blast furnace for the smelting of iron ores (top). Taken from [17].

It should be noted that no restraints were imposed on emissions from any of the steps used in copper roasting or refining steps. Furthermore, many of the copper ores were polymetallic and the subsidiary metals were invariably allowed to go up in smoke.

The following symptomatic description of activities in the copper mines conjures up images of dirty water and smokey air:

> "Among the remarkable things to be seen at these hives of industry were a machine for drawing water by a cord from a copper mine two miles off, a steam-hammer of seven tons weight, an iron furncace of 10,000 cubic feet dimensions, said to be the largest high furnace for wood in the world, and a machine for splitting their fuel wood, of which they burn annually 100,000 sajens, or roughly speaking, a pile of logs twice as big as St. Paulís Cathedral" [15, p. 84].

2.3 IRON PRODUCTION

The Urals contain hundreds of iron deposits, some of which have been mined out, but the remaining reserves still rank the Urals among the leading iron centers in the world. Legend has it that in his desire to develop the resources of his Empire, Peter the Great in 1680 or thereabouts commissioned two men ñ Botachoff and Demidoff ñ to inspect and report upon the facilities which were offered in the country for the establishment of centers for the manufacture of iron. The two men from Tula were said to have been drawn to the immense quantities of iron that lay, scarcely covered by soil, throughout the [Russian] Empire, to abundant forests that seemed to offer inexhaustible supply of fuel, and by the advantages offered by the Government to anyone who would take in hand the development of those stores of wealth. The names of the two blacksmiths, Demidoff and Botachoff soon became household words through the length and breadth of the Empire. "Both were men of great genius, and extraordinary energy; when one looks at the work they did, and considers the difficulties they had to contend with. They both prospered nearly alike. Demidoff branching out from his first work, which he established at Neviansk, in the mouth of the Ural, had soon availed himself of every spot where water could be collected and he found ironworks there. No authentic record exists of the precise number of foundries which he built; but Botachoff, ... left behind him twenty-five large works, extending over something like a million and half acres, and employing a population of upwards of fifty thousand souls" [15, p. 77].

Many of the Demidoff, Botachoff and subsequent establishments extended their operations over a space containing from three-quarters of a million to a million and half English acres. The works were erected at favorable sites thus creating little principalities containing numerous villages, which were entirely supported by the labor of the mines [1-14]. In surface mining operations, the ores were dug out and carted by horses or people the furnace. The roasting and subsequent smelting of the iron ores were primitive and environmentally unfriendly:

> "As a general thing, the fuel used is charcoal. Some works employ wood and charcoal, and in the south-western portion of the country, iron is smelted with mineral fuel. Generally the ores are prepared by roasting and breaking, and

only in rare instances, as at Kiselovo, are they washed. Roasting is usually done in heaps in the vicinity of the mines. Kilns are rarely used. The fuel is almost exclusively wood, the quantity being, for instance, 35 cubic feet per 2.5 metric tons at Gora Blagodat, in roasting in heaps; and 35 cubic feet per 3.14 metric tons in Raschett kilns at Visokaja-Gora. Only at some of the works, where the roasting kiln are near the furnaces, as at Malceff and Stroganoff, waste gas is used. Latterly, vertical kilns, built after an Austrian model, have been put up at Katakova for furnace gas; and at several Government furnaces for smelting brown hematites, an apparatus, designed by Moskvin, has been constructed. It consists of a series of cast-iron segments placed above the tunnel-head, from which the ore, roasted by the flame issuing from the furnace is dropped into it at red heat, by loosening the keys which hold the segments together" [15, p. 87].

"The furnaces are, as a rule old with massive masonry of brick or stone, the shaft and boshes being circular, while the hearth is rectangular (see Figure 2). The lower part of the lining is usually built of stone, which material is often used also for boshes and shaft. Brick is not common. They generally have an open top, the gases either being allowed to escape from the tunnel head or being taken off by a flue. The front too is open in most cases, only a few furnaces in central Russia, in the Ural mountains, and in Poland having a closed front. The number of tuyeres is two or three, and sometimes only one; their diameter, however, is large. The blast is, in most instances, cold and the pressure is low. The height and diameter of the furnaces vary. In the Ural mountains, the height is generally 49 feet and upwards" [15, p. 88].

With the open air roasting of ores, the low stack heights, open furnaces (Figure 2) and valley locations near water supply, severe air pollution should be expected around the zavods (or factories). Extensive contamination of surface waters should also have resulted from the mine waters, coke oven and other effluents as well as by human wastes from the mining communities.

2.4 PRODUCTION OF OTHER METALS

Iron and steel, copper and gold were by far the largest metals extracted from the Urals during the 18th and 19th centuries. Other important metals that were produced in those days included platinum, lead, zinc, chromium and manganese [1-15]. While the production of each element entailed some risk to the local ecosystem, the most pervasive and severe impacts were associated with the mining and smelting of iron, copper and gold ores.

3. Environmental Impacts

The most dramatic impact of mining in the Urals was the destruction of natural forest ecosystems. Wood was to the Ural what coal used to be for England. Large quantities of wood were required for construction, heating, smelting and melting and for steam to generate power. Iron works in the Urals required prodigious amounts of wood, estimated to be 60-95% equivalent weight (or 100-400 cubic feet) of charcoal per ton of iron produced. In 1796, the Kolyvan-Voskressensk mines and smelters required over 19,000 peasant woodcutters to meet the demand for 1.5 million cubic meters of wood (see below).

The rapid and reckless felling of trees, the ensuing erosions and elevated levels of air pollution resulted in massive devastation of the forest resources which threw the mining industry into jeopardy. In assessing the state of mining and smelting in the Urals in the early 1880s, Brown [15] noted that "the failure of mining operations here to prove permanently remunerative is largely attributable to the failure of the supply of wood required in smelting, in melting and heating the metal, and in the production of motive power in the absence of such power attained from the fall of water" (p. 111). "There has been a reckless consumption of forests in supplying fuel for smelting ore. The ore is not exhausted, the fuel which was of easy access has been consumed. .... Could forests be restored the work might be prosecuted with vigor and possibly with profit. If not, if this is to be done, other fuel must be found" (p. 110). Large-scale switching to fossil fuels, which were, in fact, available locally did not begin until the end of the 19th century; Uralian coal production exceeded 3.2 million kilograms in 1880 [15]. With the transition to coal and more modern furnaces came further increases in local and region air pollution.

The environmental impacts of the human population that mushroomed around the mining centers should not be ignored. With the limited use of machinery, a huge labor force was required to perform the most basic tasks (Table 2).

TABLE 2. Labor force at the Kolyvan-Voskressensk mine, 1771-1796 (from [1] and [15]).

Activity	1771	1782/85	1796/97
Mine workers	4,775	5,573	7,453
Ascribed peasants	n.a.	54,750	55,306
"Revision Souls"*	n.a.	67,901	69,612
Peasant woodcutters	4,156	15,882	19,002
Wood output (sazhens)**	80-82,000	108,000	129,000

* Mostly convicts and "fugitives"
** Sazhen = 11.8 m^3

In discussing the labor conditions, Storkh [11] noted that in the copper and iron mines of the Urals, they [free workers] were usually employed in the transport of ore because the attached peasants were only numerous enough to cut wood and carry charcoal. As most crown peasants in the neighborhood of the mine were attached (see below), it was very difficult, even if one paid a very considerable price, to have enough free workers and a great amount of capital was required.

From the amounts of mineral resources recovered from the Urals during the two centuries, the associated environmental impacts must have been very extensive. The zavods represented foci of severe air, water and soil contamination and in terms of their numbers and geographic locations, the environmental impacts must be considered to be regional in scale. Although the focus of this symposium is on current environmental problems, we should not loose sight of the fact that large-scale contamination of the Ural ecosystems with toxic metals has been going on for about 300 years. Current emissions from mining activities continue to add to the growing fund of toxic metals stored in various components of the local and regional ecosystems. Although the ultimate end product of this on-going experiment is uncertain, now may be the time to initiate well conceived studies to evaluate the ecosystem stresses associated with the long-term exposure to toxic metals from the sustained mining activities.

4. Reference

1. Blanchard, I. (1989) *Russia's age of silver*. London, Routledge.
2. De-Gennin, V. (1735) *Description of Ural and Siberian factories*. Washington, D.C., Smithsonian Institution
3. Kafengauz, B.B. (1949) *Istoriya khozyaistva Demidovikh v XVIII-XIX. Opit issledevoniya po istorii Uralískoi metallurgii*. Moscow.
4. Kotluiaarevskii, I. (1870) *Zamuiaetki ob Ural-skom gornom khozuiaafistvuiae Omicrofrome*. St. Peterburg, V tip V. Demakova
5. Mironov, B.N. (1979) Eksport russkogo khleba vo vtoroi polovine XVIII-nachale XIX. Istoricheskii Zapiski, vol. XCIII.
6. Nikolaevich, A.N. (1909) *Gornozavodskii Ural*. St. Peterburg, Tip. F. Valinberga i.P. Gershunina
7. Orlov, A.S. (1979) *Volneniuiaa na Urale v serdine XVIII veka: K vopr. O formirovanii proleteriata v Rossii. Moskva*, Isd-vo MGU
8. Pavlenko, N.I. (1962) *Istoriya metallurgii v Russii XVIII veka. Zovodi i zavodovladelítsi*. Moscow
9. Porfir-evanich, M. (1965) *Gornozavodskifi Ural v 1900-1917 gg*. Leningrad, Nauk, Leningradskoe otd-nie
10. Renovantz, H.M. (1788) *Mineralogische-geographische und andere vermischte Nachrichten von den Altaiischen Gerburgen Russische Kanserlichen Anteil*. Reval Publisher.
11. Storch, H. (1801) *Tableau historique et statistique de líEmpire de Russie a la du dix-huitieme siecle*. Paris.
12. Rodionov, P.F. (1971) *Elektrorazvedka metodom zaruiada*. Moskva, "Nauka"
13. Vladimirovna, L. (1993) *Gornozavodskauia promyshlennost-Urala na rubezhe XIX-XX VV: k kharakteristike proutAsessov monopolizautasii*. Ekaterinburg, Rossi. Akademica nauk, Ural-skoe otdni, in-t istorii i arkheologii

14. Voevodin, L.E. (1910) *Ural i ego gornozavodskauia promyshlennost-v preduiaelakh Permskofi gubern-ii.* Perm, Trud
15. Brown, J.C. (1994) *Forestry of the Ural Mountains.* Edinburgh, Oliver & Boyd
16. Hermann, B.F.J. (1808) *Opisanie zavodov pod vedomostvom Ekaterinburgskogo gornogo nachalistva sostoyavshikh.* Ekaterinburg.
17. Biringuccio, V. (1540) *Pirotechnia.* Translated by C.S. Smith and M.T. Gnudi. Cambridge, M.I.T. Press

Part 1

Environmental Contamination in the Ural Mountains: Measurements, Methods and Models

Magnitogorsk in early 1930s
Photos from Y. Roitman's archive

ANALYSIS OF THE ENVIRONMENTAL SITUATION IN THE URAL ECONOMIC REGION

V.N.CHUKANOV and B.A.KOROBITSIN
Institute of Industrial Ecology, Ural Branch of Russian Academy of Sciences, Sophya Kovalevskaya Str., 20-a, Ekaterinburg, 620219, RUSSIA

The Urals economic region includes 5 Oblasts (Sverdlovsk, Chelyabinsk, Perm, Orenburg and Kurgan) and 2 Republics (Bashcortostan and Udmurtia). By the level of industrial output the Urals shared the second place among eleven economic regions of Russia with the traditional industrial specialization in the field of mining, metallurgy, heavy machinery, nuclear industry. The uncontrolled growth of vast industrial enterprises, most of which are outmoded, has led to the large number of pollution problems. Many articles in the popular press and scientific literature describe the pollution problems as close to a disaster. The objective of this study has been to estimate the extent of environmental impact and environmental quality problems in the region and rank Oblasts and Republics of the Urals according to the severity of environmental conditions. The following data have been analyzed: emission of air pollutants from stationary sources from an area point of view and air quality in the largest industrial cities, discharge of contaminated waters into the rivers, toxicity of discharges and surface water quality, accumulation and production of dangerous toxic wastes.

By the total emission of air pollutants from stationary sources the Urals is at the top of eleven economic regions of Russia. In 1990-1995 the emission in the Urals

TABLE 1. Total emission of air pollutants from stationary sources, 10^3 t/y [1]

Oblast or Republic	Year					
	1990	1991	1992	1993	1994	1995
Republic Bashkortastan	909.8	854.0	794.4	711.7	696.1	781.1
Kurgan Oblast	160.0	174.0	245.8	172.8	143.7	131.3
Orenburg Oblast	1056.0	912.0	864.0	683.1	632.9	632.9
Perm Oblast	869.1	874.0	787.1	624.1	617.6	557.5
Sverdlovsk Oblast	2720.0	2439.1	2439.1	1873.2	1509.8	1474.1
Udmurt Republic	270.0	254.0	330.5	296.6	251.6	244.7
Chelyabinsk Oblast	2355.6	2061.0	1825.0	1545.3	1178.4	1128.6
Russian Federation	34705.0	31804.0	28403.0	24788.3	21929.1	21269.6

I. Linkov and R. Wilson (eds.), Air Pollution in the Ural Mountains, 23–29.

was 22-26% of the total emission in Russia with Sverdlovsk and Chelyabinsk Oblasts accounting for about 54-61% of the regional emissions (Table 1).

From 1990 to 1995 the total particulate emissions in the Urals decreased from 8.34 million metric ton per year to 4.95 million t/y. The main reason of it was industrial production decrease, not the development and implementation of pollution prevention projects.

There are no national emission standards in Russia but it is interesting to compare the emission per square kilometer (Table 2) and per capita (Table 3) in the Urals and in the European countries. Because of limitations in the availability of consistent data we use emission data for different years.

TABLE 2. Total emission of air pollutants per square kilometer, t/km^2

Country or Oblast (Republic)	Year	Emission
Great Britain (only SO_2+NO_X)	1980 [2]	27.0
Poland	1992 [3]	22.7
Chelyabinsk Oblast	1995	12.9
France (only SO_2+NO_X)	1980 [2]	11.0
Spain (only SO_2+NO_X)	1983 [2]	8.3
Sverdlovsk Oblast	1995	7.6
Republic Bashkortastan	1995	6.9
Udmurt Republic	1995	5.8
Orenburg Oblast	1995	5.1
Perm Oblast	1995	3.5
Kurgan Oblast	1995	1.8
Russian Federation	1995	1.2

TABLE 3. Emission of NO_X per capita, t/y

Country or Oblast	Year	Emission
Sweden	1991 [4]	0.046
Italy	1992 [5]	0.035
Sverdlovsk Oblast	1995	0.032

Because of high emission of pollutants, air contamination in the largest industrial cities of the Urals exceeds Russian maximum permissible limits. Every year the Russian Ministry of Environmental Protection includes such Urals cities as Berezniki, Gubaha, Perm (Perm Oblast), Ekaterinburg, Kamensk-Uralski, Nizni Tagil (Sverdlovsk Oblast), Chelyabinsk, Magnitogorsk (Chelyabinsk Oblast), Kurgan (Kurgan Oblast) in the list of cities with the highest level of air contamination in which the average annual concentrations of benz(a)pyrene, formaldehyde, or nitrogen oxides exceed Russian standards by 2-10 times (Table 4-6). It means that more than 60% of population of the Urals have living conditions which don't meet sanitary standards. But despite the serious exceedance of Russian national standards, the average annual air contamination in the Ural industrial cities is quite comparable with the same in the largest cities of the world, Table 7.

Based on the data presented in Tables 2, 3 and 7 we can conclude that many air pollution and air quality problems in the Urals seems to be less than expected by wide-spread opinion and are quite comparable with the problems of foreign (including West European) countries which usually are considered to be environmentally successful.

TABLE 4. Average annual concentration of benz(a)pyrene in the air of the largest Ural cities, 10^{-6} mg/m^3 (national standard is 1.0×10^{-6} mg/m^3) [6]

City	Year				
	1991	1992	1993	1994	1995
Ekaterinburg	6.6	4.0	2.9	1.6	1.7
Kurgan	7.8	6.1	4.6	4.7	5.2
Magnitogorsk	12.9	11.8	10.0	4.8	8.0
Nizni Tagil	6.1	7.8	7.4	1.9	2.6
Perm	2.6	2.4	2.1	1.6	3.1
Chelyabinsk	7.2	7.4	6.6	2.9	3.2

TABLE 5. Average annual concentration of formaldehyde in the air of the largest Ural cities, mg/m^3 (national standard is 0.03 mg/m^3) [6]

City	Year				
	1991	1992	1993	1994	1995
Ekaterinburg	0.008	0.007	0.014	0.014	0.016
Kurgan	0.013	0.015	0.009	0.004	0.004
Nizni Tagil	0.016	0.012	0.013	0.013	0.014
Perm	0.007	0.007	0.007	0.008	0.009
Chelyabinsk	0.006	0.007	0.007	0.006	0.006

TABLE 6. Average annual concentration of NO_2 in the air of the largest Ural cities, mg/m^3 (national standard is 0.04 mg/m^3) [6]

City	Year				
	1991	1992	1993	1994	1995
Ekaterinburg	0.04	0.04	0.04	0.05	0.06
Kurgan	0.02	0.02	0.02	0.02	0.02
Magnitogorsk	0.10	0.14	0.10	0.13	0.22
Nizni Tagil	0.02	0.02	0.02	0.02	0.02
Perm	0.04	0.04	0.04	0.04	0.04
Chelyabinsk	0.05	0.05	0.04	0.04	0.04

TABLE 7. Average annual contamination of the urban air in 1991, mg/m^3

Pollutant	Tokyo [7]	Beijing [7]	London [8]	Magnitogorsk
CO	no data	no data	2.0	2.0
SO_2	0.047	0.233	0.017	0.050
benz(a)piren (10^{-6} mg/m^3)	2.25	33.7	no data	15.0
NO	0.037	no data	0.100	0.150
NO_2	0.111	0.068	0.041	0.100
TSP	no data	0.400	no data	0.300

The degradation of water quality is a far more dramatic environmental problem. In 1995 the wastewater discharge in the Urals was 12.0% of the total waste water discharge in Russia. In 1995, 31.2% of that was from Sverdlovsk Oblast and 26.7% from Chelyabinsk Oblast (Table 8). In 1995 in the Republic of Bashkortastan about 99.5% of all contaminated water was dumped without sufficient purification, in Chelabinsk Oblast about 98.3% of all contaminated water was not purified. The problem is that wastewater treatment plants have inadequate capacity, collecting more sewage than they could treat.

TABLE 8. Total wastewater discharge from industrial, municipalities and agricultural sources, million m^3/y [1]

Oblast or Republic	Year					
	1990	1991	1992	1993	1994	1995
Republic Bashkortastan	562.6	578.2	560.9	550.4	557.4	510.4
Kurgan Oblast	21.6	25.5	22.9	22.6	20.0	16.6
Orenburg Oblast	60.8	147.7	151.4	74.2	80.9	76.5
Perm Oblast	615.8	610.8	577.2	515.4	503.1	591.5
Sverdlovsk Oblast	766.1	835.2	830.0	864.2	864.0	911.3
Udmurt Republic	67.6	69.1	59.3	46.5	46.1	36.0
Chelyabinsk Oblast	632.1	662.0	646.1	707.9	704.8	779.3

To compare territories of the Urals by the impact of wastewater discharge on river waters a nondimensional reduced index of sewage was used:

$$J = \Sigma(D_i/C_i)/S,$$

Where:

D_i - annual discharge of i-th pollutant into surface water from the territory of Oblast or Republic, t/y;

C_i - maximum permissible concentration of i-th pollutant in the river water, mg/l;

S - average annual 'territory's own' (excluding transit water flaw) runoff, m^3/y.

This index indicates how much water over the 'territory's own' runoff must be increased to dilute discharged sewage to make it safety. The results are presented in Table 9.

TABLE 9. Non-dimensional reduced index of sewage (1995)

Oblast or Republic	J
Republic Bashkortastan	147
Kurgan Oblast	448
Orenburg Oblast	107
Perm Oblast	174
Sverdlovsk Oblast	1079
Udmurt Republic	356
Chelyabinsk Oblast	1090

Because of increasingly high pollution loads in many water bodies the resultant water quality problems include low dissolved oxygen, higher nutrient levels (i.e. nitrogen and phosphorous), eutrophication of lakes and rivers and the most dangerous - extremely high concentration of heavy metals. The most contaminated are rivers Chusovaya, Salda, Tagil, Neiva, Iset, Pyshma, Table 10.

TABLE 10. Average annual concentration of heavy metals in water, mg/l (1995) [6] (Figures in brackets are national standards)

River	Concentration			
	Copper (0.001)	Zinc (0.01)	Manganese (0.01)	Iron (0.1)
Chusovaya	0.103	0.109	0.32	0.95
Salda	0.240	0.181	0.39	no data
Tagil	0.095	0.104	0.23	0.72
Neiva	0.069	0.222	0.58	0.65
Iset	0.063	0.084	0.22	1.07
Pyshma	0.086	0.066	0.21	0.96

It can be seen that the exceeding of water quality standards is more serious than the same of air. But the most critical is the fact that in 1990-1995 despite a sharp decrease in industrial output, water consumption in the Ural industry remained stable, and sewage emission even increased by 7.2% and natural water quality became worse during that period.

Hazardous industrial wastes are one more continuing problem in the Urals, increasing in both quantity and toxicity. Mining and heavy industry such as steel and copper production is the largest sours of solid wastes. In the process of production of 1 metric ton of metal about 100 tons of solid wastes are produced at the stage of ore mining and processing and up to 8 tons of slugs and other wastes at the stage of smelting. Average annual production of such wastes in the Urals is about 2 billion cubic meters per year. The most part of wastes produced in the process of ore mining and processing are not extremely hazardous and so are not officially classified as toxic ones but they take up a lot of territory (at present more than 2 thousand square kilometers) and for decades will remain a source of water and air contamination in the result of leaking and wind dispersion. The most hazardous solid and liquid industrial wastes which are classified as toxic in accordance with Russian regulation practice (this classification is based on the set of factors which take into account both the impact of waste on the environment and toxicity for man) are usually stored at the territory of those enterprises were they were produced. The data on production and processing of toxic solid wastes in the Urals are presented in Table 11. We think that the industrial toxic waste problem will not be solved for many decades because of the absence of wasteless technologies in mining and metallurgy and economic ineffectiveness of their processing at the present state of the Russian economy.

A direct comparison of industrial toxic wastes production between Ural and foreign countries is impossible because of different approaches to their classification. But the municipal solid waste production in the Urals does not seem to be more dramatic problem than in the most of developed countries, Table 12.

TABLE 11. Toxic solid wastes production and processing, thousand metric tons [1]

Oblast or Republic	Accumulated in storage sites on 1 January1995	Produced in 1995	Reused, processed or rendered in 1995
Republic Bashkortastan	63850.72	3933.96	344.73
Kurgan Oblast	152.49	59.09	3.61
Orenburg Oblast	968.88	1.14	0.23
Perm Oblast	1734.53	226.46	48.10
Sverdlovsk Oblast	108844.70	5238.10	3925.40
Udmurt Republic	4161.77	228.12	159.88
Chelyabinsk Oblast	249299.08	16740.93	10990.26

TABLE 12. Municipal solid waste production per capita, kg (1990) [9]

Country	Waste production	Country	Waste production
USA	803	Sverdlovsk Oblast (1994)	428
Canada	601	Japan	408
Netherlands	497	Great Britain	398
Switzerland	441	Sweden	374

From the analyses of the data presented above the following can be identified:

1) Oblasts and Republics of the Ural economic region are distinguished greatly by environmental impact and environmental quality so it pointless to discuss the environmental problems of the Ural region as the whole. The most severe environmental situation is in the Chelyabinsk and Sverdlovsk Oblasts. The Kurgan Oblast and Udmurt Republic are relatively safe.

2) A comparison of statistics on industrial production output, environmental impact and environmental quality data identifies the fact that political and economic changes in 1990-1995 have not induced emission decreases and consequently improved environmental quality

3) The environmental situation in the Urals is one of the most important in Russia. But in spite of wide-spread opinion many environment impact and environmental quality indices in the Oblasts and Republics of the Urals seems to be less than expected and are quite comparable with the same of foreign (including West European) countries which usually are considered to be environmentally successful. So there is no reason for excessive pessimism and declaration of the whole Ural region as the zone of environmental disaster.

4) A comparison of statistics on industrial production output, environmental impact and environmental quality data identify that political and economic changes in 1990-1995 have not induced emission decreases and consequently environmental quality The reason may be that the existing statistical data on releasing of pollutants to the environment are not adequate.

References

1. Unpublished report of the Sverdlovsk Oblast's Statistics Department (1996).
2. Hordijk L., Shaw R. and Alcamo J. (1990) Background to acidification in Europe, in J.Alcamo, R.Shaw and L.Hordijk (eds.), *The rains model of acidification. Science and Strategies in Europe,* Kluwer Academic Publishers, Dordrecht, pp. 31-60.
3. Wierzbicka M., Michalak E. (1993) Environmental pollution in Poland, *AMBIO* **4**, 248-249.
4. Cristen Agren (1993) Atgarder mot utslapp av luftfororeningar i europa, *Vatten* **3**, 202-206.
5. Pasquon Italo (1994) Technologie emergentie protezione delliambiente, *Chim. e ind.* **6**, 439-447.
6. Unpublished report of the Ural Regional Hydrometheorology and Environmental Contamination Monitoring Department (1996).
7. Ando M. et al. (1996) Indoor and outdoor air pollution in Tokyo and Beijing supercities, *Atmospheric Environment* **5**, 695-702.
8. Derwent R.G. et al. (1995) Analysis and interpretation of air quality data from an urban roadside location in Central London over the period from July 1991 to July 1992, *Atmospheric Environment* **8**, 923-946.
9. World resource foundation (1995), *Warmer Bull.* **44**, i-iv.

POLLUTION STRESS OF SIBERIAN FORESTS

S. NILSSON, K. BLAUBERG, E. SAMARSKAIA
*International Institute for Applied Systems Analysis,
A-2361 Laxenburg, AUSTRIA*

V. KHARUK
*Institute of Forest, Siberian Branch, Russian Academy of Sciences,
660036 Krasnoyarsk, Academgorodok, RUSSIA*

Abstract

Many articles in the popular press and western scientific literature stress the severe environmental conditions in Russia and Siberia. The general pollution problem is described in many cases to be close to a disaster. The availability of consistent data for anthropogenic stress factors to the Siberian forests is limited, both from a registration point of view, and from a methodological one. The objective of the overall study has been to attempt to estimate the extent of anthropogenic stress factors to the Siberian forests, based on existing data and studies in Russia. The following activities have been carried out and are summarized in this paper; estimations of emissions and depositions of major pollutants, estimations of the radio nuclide contamination, analyses of the current forest decline in Siberia, and estimates on the areas and growing stock under risk of future forest decline.

From these results the following can be identified.

- The highest depositions of heavy metals are in Norilsk, the Irkutsk region, and the southern part of the Far East. Based on Russian estimates of the critical loads for heavy metals and forests, it can be concluded that even the highest depositions of heavy metals (30-41 kg $km^{-2}yr^{-1}$) are below the critical loads. However, in the vicinity of the emitters, the critical loads are substantially exceeded.
- Based on the current critical load calculations for ecosystems, it can be seen that critical loads for sulfur are seriously exceeded in the Urals and Altai Mountains, West Siberia, western East Siberia, in the boundary regions of Kazakhstan, the Norilsk region, southern Far East, Sakhalin, and the southern Kurilean islands.
- The exceedance of critical loads for nitrogen are most serious in the Urals, West Siberia, in the boundaries to Kazakhstan, Norilsk, and in the Far East.
- Based on the existing data for radioactivity, we can tentatively conclude that soil contamination and contaminated forest ecosystems do not seem to be a dramatic

I. Linkov and R. Wilson (eds.), Air Pollution in the Ural Mountains, 31–54.

problem from an area point of view. However, there are concerns which must be resolved for the Irkutsk region and the Yenisey river basin.

- The storage of nuclear waste is a high potential risk for future large-scale radioactive contamination. This high risk condition requires further investigation.

1. Introduction

According to the 1993 inventory, Siberia has 597 million ha of Forested Areas (closed forests) and a growing stock of 69.6 billion m^3 [1]. Stretching from the Ural Mountains to the Pacific, Siberia is home to about 20 percent of the world's forested areas and half of all coniferous forests. There are many reports witnessing the environmental problems in Russia and how these influence the development of the Russian forests (e.g., [2] [3] [4] [5] [6] [7] [8] [9]).

The objective of this paper is to attempt to quantify how serious the threat of air pollutants and radio nuclide contamination are to the Siberian forests. Most of the basic data used in this paper stems from work carried out by IIASA's Forest Study [10]. In the quantification we are carrying out, estimations are made for each individual ecoregion. In the IIASA Forest Study we have defined ecoregions for all of Siberia and Russia [11] [12].

2. Emissions of Sulfur and Nitrogen

Between 1988 and 1993 the total emissions of air pollutants from the industry declined by 41% in West Siberia, by 20% in East Siberia, and by 37% in the Far East [13]. This decline was a result of decreased industrial production. The impact of these declined emissions on the ecology has not yet been realized due to: 1) decreased efficiency of scrubbing of emissions (e.g. Norilsk industrial combine cleaned ~20% of the SO_2 emissions in 1988 and in 1994 - 3.5%); 2) increased automobile-related air pollution (e.g. in Krasnoyarsk region air pollution due to traffic between 1992 and 1994 increased by 64%) [10]. It should be noted that not all emissions are included in official statistics, for example, emissions caused by rail, air- and water transportation, military operations, and waste burning. Based on official statistics for the individual ecoregions of Siberia for 1993, SO_2 and NO_x emissions are presented in *Figures 1* and *2*.

The worst emissions of SO_2 occur in the western part of East Siberia and West Siberia. The dominating NO_x emissions are located in West Siberia.

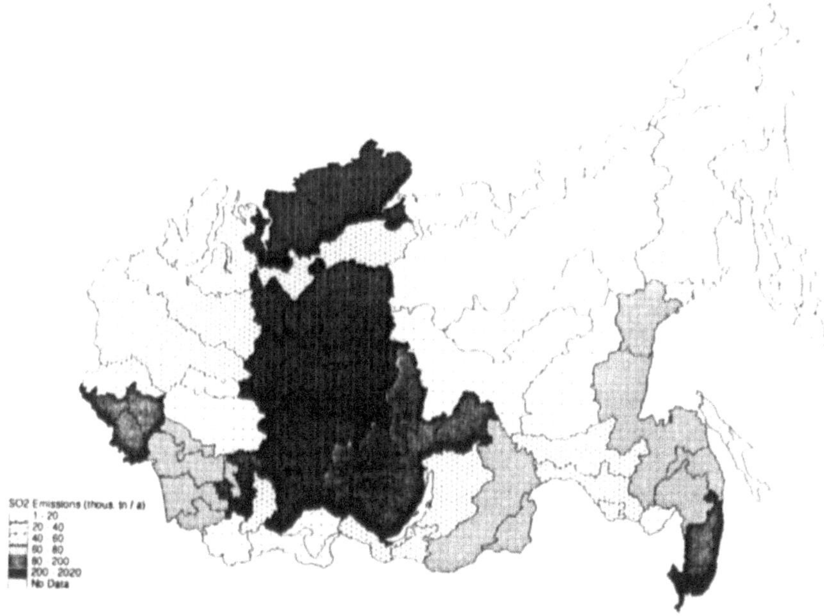

FIGURE 1 SO_2 Emissions in 1993 in 1000 tons per year. After Kharuk *et al.* [10].

FIGURE 2 No_x Emissions in 1993 in 1000 tons per year. After Kharuk *et al.*[10].

3. Depositions of Sulfur and Nitrogen

Official data on depositions include only sulfates (SO_4) and total nitrogen (nitrate + ammonia). There is some doubt on the accuracy of the official deposition data [10] and a lack of official data on the total sulfur depositions.

The depositions of sulfates and total nitrogen (minimum, maximum, average) in 1993 for each individual ecoregion based on official data [13] are presented in *Figures 3* and *8*. The largest sulfate depositions occur in southern West Siberia, Irkutsk region and the Norilsk area, where the depositions reach 1,000-3,000, and in some locations >3,000 kg $km^{-2}yr^{-1}$. The distribution of nitrogen depositions has a similar pattern with maximum loads in southern West Siberia and in the Norilsk zone (500-1,000 respectively >1,000 kg $km^{-2}yr^{-1}$). Feshbach [4] has presented aggregated maps of the sulfur sulfate burden and the total nitrogen loads for the early 1990s in Russia that show somewhat lesser depositions in comparison with the data set presented here.

Generally, sulfur and nitrogen depositions in Siberia are less than in the European part of Russia.

4. Emissions and Depositions of Heavy Metals

The major heavy metal (HM) pollutants are considered to be Pb, Cd, Cu, Zu, Ni, V, and As, which originate from many sources and are concentrated to urbanized territories. The sources of As, Cu, Zn and Cd are not abundant and have fixed geographic locations. The main sources of HM emissions are concentrated in the latitudinal belt of 52 and 60 degrees. They are copper, nickel, cobalt, aluminum, lead, steel, and cement industries, and a large number of coal power stations. The main sources of HM emissions are listed in *Table 1*. The world's main source of Ni, Cu, V and Co emissions is at the Norilsk Nickel Industry in the northern Krasnoyarsk region. Emitters of Pg are Komsomolsk-on-Amur, Novosibirsk and Kemerovo; for Zu it is Krasnoyarsk; for Cr it is Barnaul, Krasnoyarsk and Komsomolsk-on-Amur; and for V it is Omsk and Khabarovsk. All of these sources are high temperature emitters of HM in the form of aerosols or vapors. In some cases, the vapor fraction can be up to 50% of the HM emissions.

The process of HM dispersion depends on emission-source parameters (stack height, volume and temperature of emissions) and meteorology (wind speed and precipitation). Around point sources of emissions; local (0-40 km), medium (50-100 km) and distant (>100 km) zones of depositions are formed. In the local zone, HM depositions could be as high as 10-30% of the total emissions. The depositions strongly depend on the precipitation. For example, with a radius of 40 km around a copper industry the extent of Ni, Pg, Zn and Cd depositions were estimated to ~3% on dry days and reached 10-100% of the emissions on days with precipitation [10]. A considerable amount of HM emissions are involved in long-distance transportation,

FIGURE 3 Sulfate depositions (minimum) in 1993 in kg/km^2/yr. After Kharuk *et al.* [10].

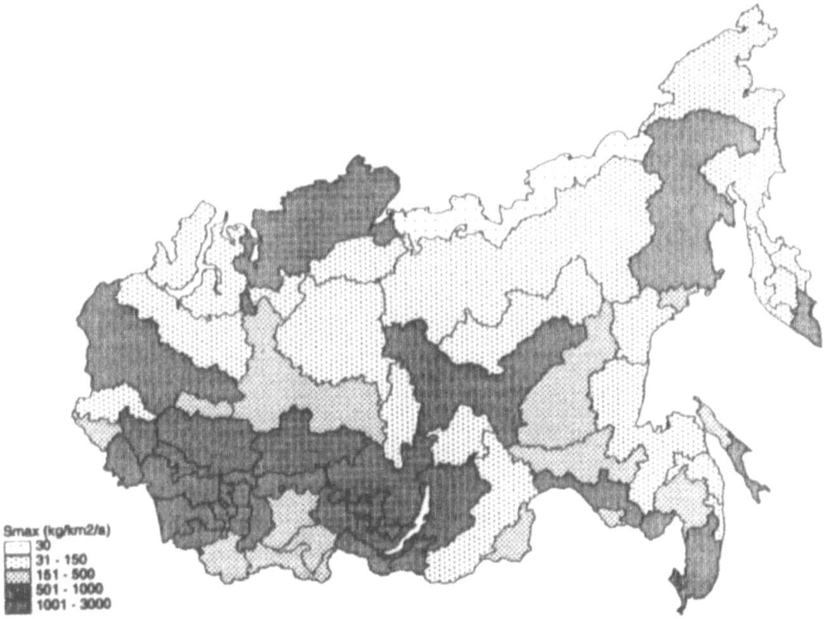

FIGURE 4 Sulfate depositions (maximum) in 1993 in kg/km^2/yr. After Kharuk *et al.* [10].

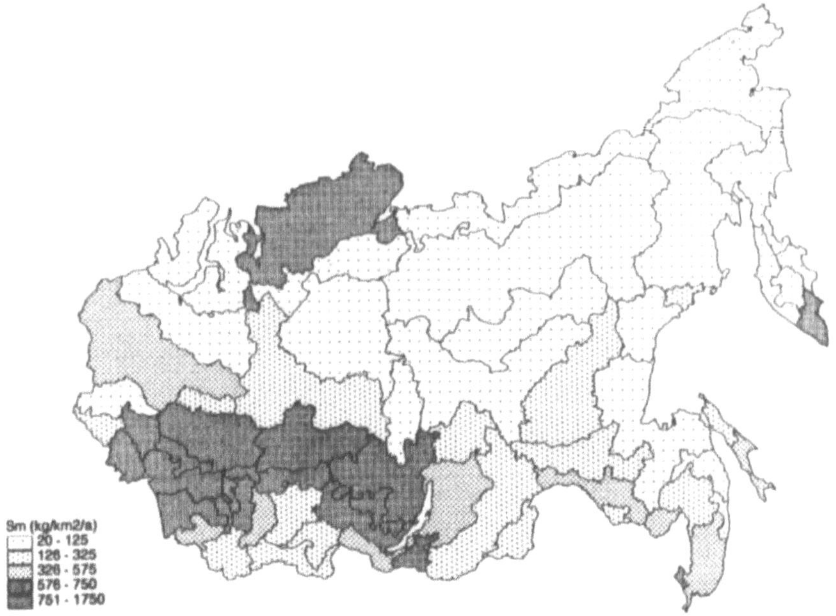

FIGURE 5 Sulfate depositions (average) in 1993 kg/km^2/yr. After Kharuk *et al.* [10].

FIGURE 6 Nitrogen depositions (minimum) in 1993 in kg/km^2/yr.
After Kharuk *et al.* [10].

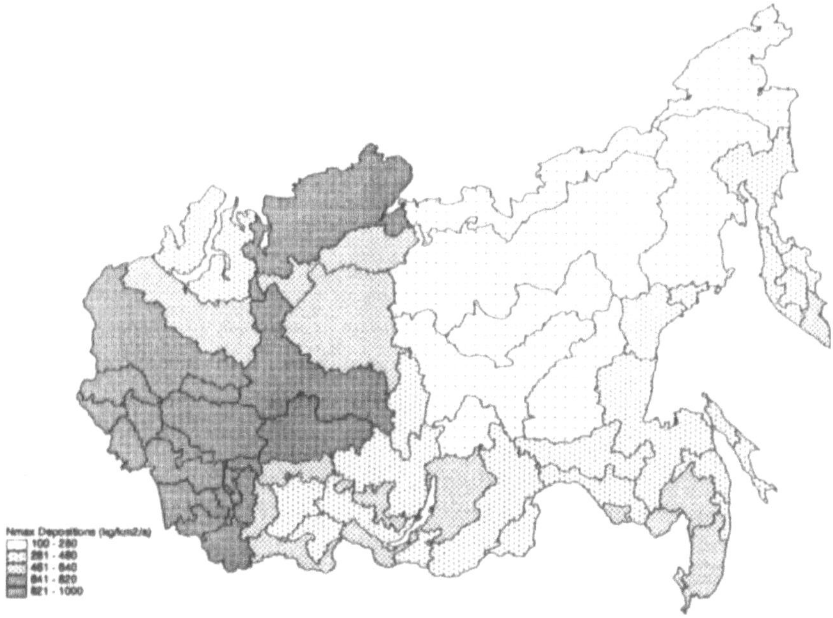

FIGURE 7 Nitrogen depositions (maximum) in 1993 in kg/km^2/yr. After Kharuk *et al.* [10].

FIGURE 8 Nitrogen depositions (average) in 1993 in kg/km^2/yr. After Kharuk *et al.*, [10]

TABLE 1. Emissions of Heavy Metals in 1993, tons/yr. After: [10].

Region/city	Pb	Cd	Ni	Cu	Zn	Cr	V	Mn	Fe
Omsk	.08				0.6	0.49	211.3	11.50	
Novosibirsk	1.62								
Tomsk	0.2								
Kemerovsk	1.30								18.0
Irkutsk	0.12								
Chitinsk							1.1		
Barnaul	0.10		543.00			1.10	2.0	0.01	
Krasnoyarsk	0.75		0.32		17.9	5.30	3.9	2.00	2.5
Norilsk			4008.00	1813			65.0		
Khabarovsk	0.18						37.0	1.14	
Amursk	0.50						3.3	0.40	
Komsomolsk	13.80				0.1	1.10	5.2	1.80	
Vladivostok	0.03	0.1		0.02		0.06		0.22	

and are subject to wet and dry depositions. The total HM wet depositions in 1993 for each individual ecoregion is presented in *Figure 9*. The wet depositions of Pg, Cd, Ni, Cu and Zn for 1993 are presented in Figure 10 (The width of the circles indicate the size of the depositions). Wet depositions are dominant in areas where the annual precipitation is >600 mm. For these areas the dry depositions are estimated to be ~one-third of the total depositions. This is valid for the dominating part of Siberia, except for Yakutia republic (~300 mm) and Tuva Republic (~250 mm). The highest depositions of HM in Siberia are in Norilsk and in the Siberian Far East. The wet depositions of Pg, Cd, and Ni in Siberia correspond to those in the USA, Canada and Japan. Cu depositions seem to be lower and Zn depositions seem to be higher in Siberia in comparison with the countries mentioned [10].

5. Critical Loads

In the international literature critical loads (CL) are defined as quantitative estimates of an exposure (depositions) to one or more pollutants, below where significant harmful effects on specific sensitive elements of the environment do not occur, according to our present knowledge. Target loads in most cases are less restrictive with respect to deposition loads in that consideration for other pollution control factors, such as economic ones are incorporated (e.g. [14]). Target loads (or levels) are based on critical loads and are developed taking into consideration possible legal, technical, ecological, economic, and political concerns. Target loads may be set lower or higher than critical loads depending on the actual conditions. The basic idea in setting target levels is that they will form the basis for negotiating acceptable emission reduction strategies (e.g. [14]).

Most models for critical loads of sulfur and nitrogen developed in the west are based on Simple Mass Balance (SMB) equations [15]. These models have gone

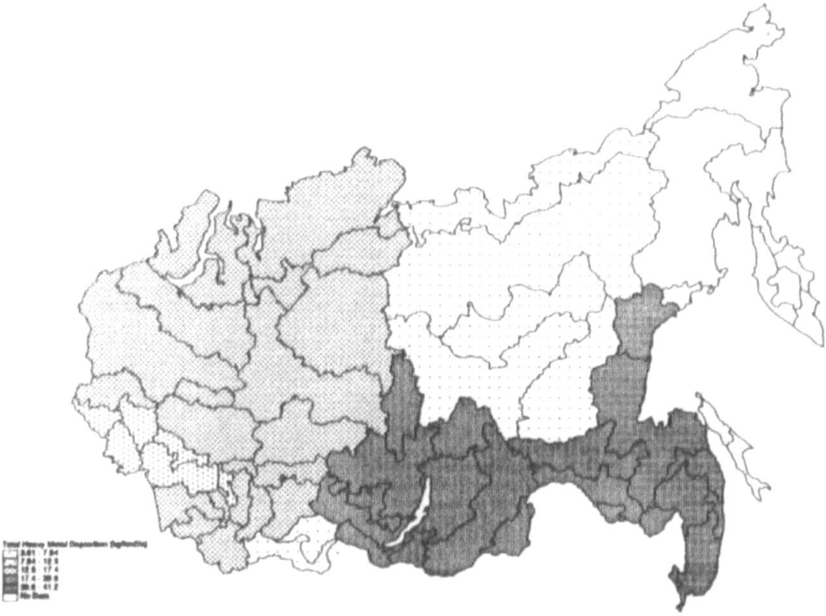

FIGURE 9 Total heavy metal deposition in 1993 in kg/km^2/yr. After Kharuk *et al.* [10].

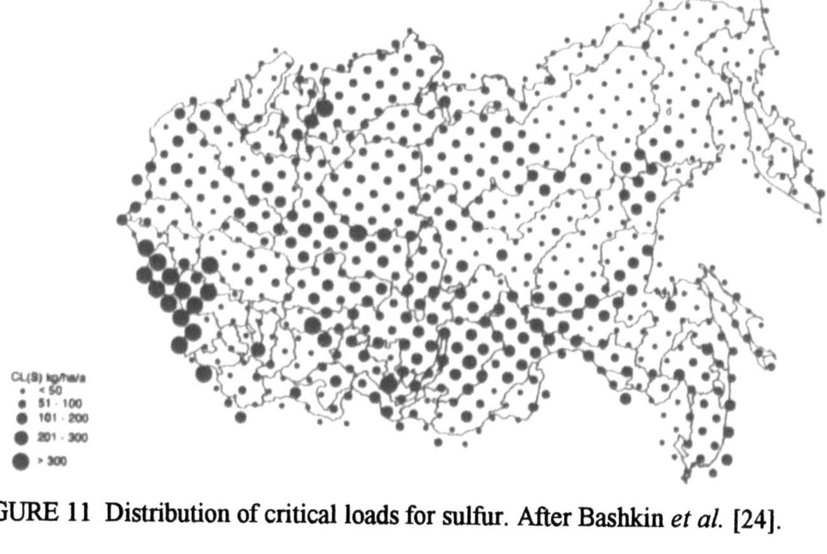

FIGURE 11 Distribution of critical loads for sulfur. After Bashkin *et al.* [24].

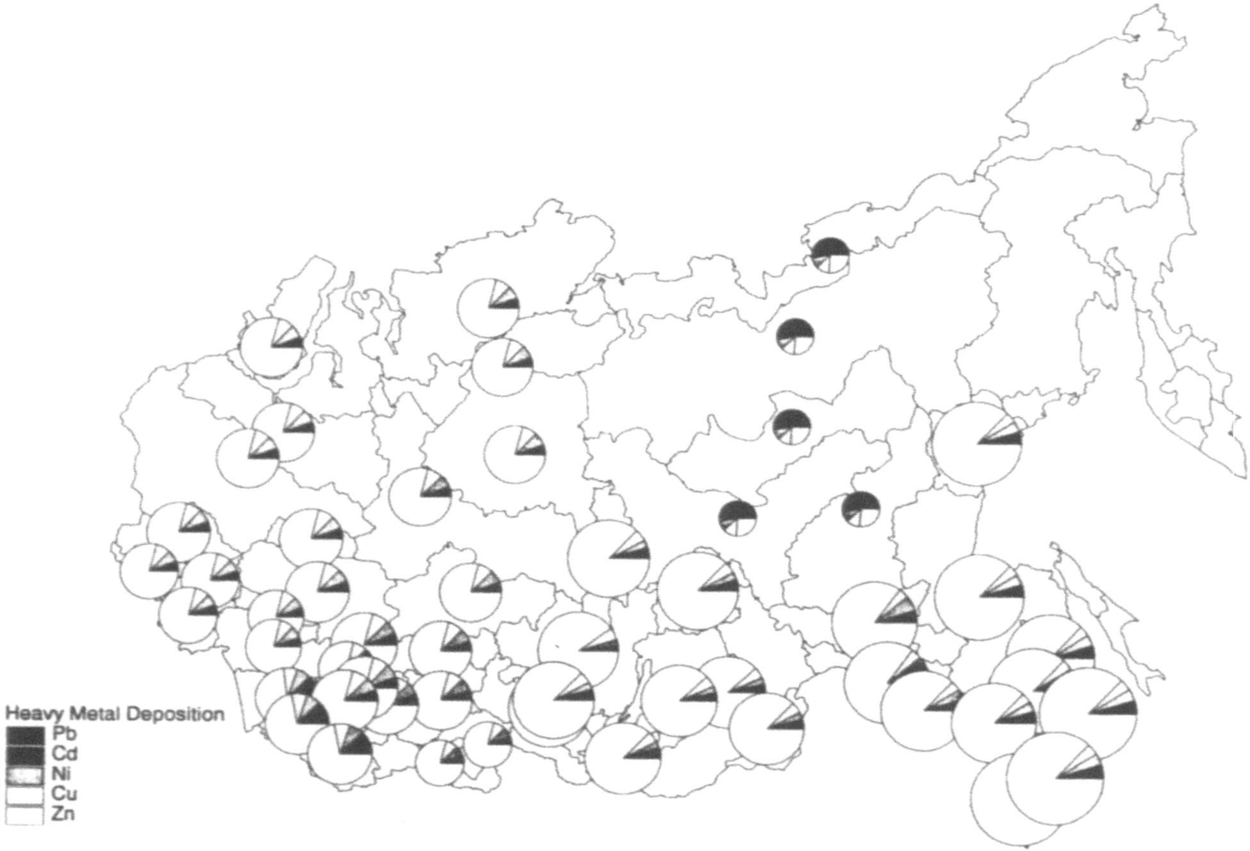

FIGURE 10 Wet deposition of some specific heavy metals in 1993 kg/km^2/yr. After Kharuk *et al.* [10].

through continuous development since the late 1980s (e.g., [16] [17] [18] [19] [20] [21]).

The current development is to move from single critical loads (for individual pollutants) to multiple critical thresholds. This has resulted in the generation of protection isolines, which illustrate combinations of sulfur and nitrogen depositions where damage is ensured [22].

Nilsson *et al.* [23] used the following single target loads with respect to forests in European Russia: sulfur depositions in coniferous forests 0.5 g $m^{-2}yr^{-1}$ (high sensitivity), 1.0 (medium), and 2.0 (low). For sulfur depositions in deciduous forests the values were: 1.0 (high), 2.0 (medium) and 4.0 (low). For nitrogen depositions the corresponding values were: 0.3, 1.0 and 1.5 (coniferous); 0.5, 1.2 and 2.0 (deciduous). Later, within the Convention on Long-Range Transboundary Air Pollution Program, Russia carried out single critical load estimates for European Russia based on the "Guidelines for the Computation and Mapping of Critical Loads and Exceedances of Sulfur and Nitrogen in Europe" [15] [20]. The distribution of critical loads achieved for European Russia using this calculation is presented in *Table 2.* These latter estimates show less endangered forests than presented by Nilsson *et al.* [23].

TABLE 2. Distribution of values of critical deposition of sulfur and nitrogen in the European part of Russia. Figures are given as percentage of total area. After: [20].

Ranges	Critical deposition		Exceedances	
(eq ha^{-1} yr^{-1})	CD(S)	CD(N)	CD(S)exc	CD(N)exc
<200	4.8	7.0	87.1	99.4
200-500	36.0	22.0	5.3	0
500-1000	26.9	15.1	5.3	0
1000-2000	19.9	22.0	2.3	0
>2000	12.4	33.9	0	0.6

Hettelingh *et al.* [20] show that by a multiple, critical threshold approach for sulfur and nitrogen depositions in 1990, the dominating part of the European Russian ecosystems are protected against harmful effects due to air pollutants.

Bashkin *et al.* [24] have made a first attempt to produce critical load estimates for Siberia with respect to nutrient and acidifying nitrogen, and for sulfur and acidity based on the international approaches discussed above. The Coordination Center for Effects, RIVM has evaluated Russian critical load calculations and conclude that the approach is scientifically sound (Posch, personal communication: May 15, 1997). There may be an overestimate in the nitrogen uptake by the soils, resulting in an overestimate of the critical loads for nitrogen. In contrast to this, Bashkin *et al.* [24] stress that the ecosystems of the arctic, subarctic and permafrost areas are very sensitive to excessive input of nitrogen which might result in the critical loads being underestimated. The calculations of critical loads for sulfur and nitrogen are carried out for Asian EMEP grids of 150x150 km. Professor V.N. Bashkin, Pushkino has kindly given us permission to use the original calculations of the critical loads for the grid data. The distribution of critical loads for sulfur and nitrogen are presented in *Figures 11* and *12*.

Russian scientists [25] [26] [27] have tried to introduce ecologically-based normatives for critical load estimates. These estimates have only been presented for a few heavy metals for specific regions. The critical loads for trees and forests are presented in *Table 3*. de Vries and Bakker [28] have tried to estimate critical loads of heavy metals for the soils of Western Europe. Their estimates are substantially lower than the Russian estimates on critical loads for forests.

TABLE 3. Critical loads (CL) of some heavy metals for trees, $kg/km^{-2}yr^{-1}$. After: [27].

	CL, current	CL, prospective
Cu	91.1-187.7	59.4-122.4
Pb	20.2-45.3	13.2-30.9
Cd	1.4-3.7	0.9-2.4
Zn	94.8-213.9	61.8-139.5

6. Forests with Depositions Exceeding Critical Loads

To estimate the exceedance of the critical loads to determine the ecological risks the rate of depositions is crucial. As stated earlier, official deposition data have some doubts and the total sulfur depositions are not available. Therefore, in this case the calculations of sulfur and nitrogen depositions were made by a model using meteorological and emission data for 1991 and the Asian EMEP grid system [29]. The distribution of exceedance of critical loads for sulfur and nitrogen are presented in *Figures 13* and *14* (expressed in e.q. $ha^{-1}yr^{-1}$)[1]. The most dramatic exceedance of critical loads for sulfur is in West Siberia, the western part of East Siberia, the Altai Mountains, boundary regions of Kazakhstan, and southern Far East. A serious exceedance of nitrogen occurs in western Siberia and southern Far East.

But not all of these areas are covered by closed forests (Forested Areas according to Russian classification). To estimate the forests at risk due to harmful effects of air pollution we have overlaid the forest resources information from IIASA's Forest Study database [11] [23] with the grid data on exceedance of critical loads.

The Forested Areas with serious exceedance of critical loads for sulfur is illustrated in *Figure 15*, and corresponding information for growing stock is illustrated in *Figure 16*. The size of the circle indicates the size of Forested Areas respectively growing stock in the different ecoregions. The corresponding information for nitrogen is presented in *Figures 17* and *18*. The distribution of the exceedances of the critical loads is presented in *Table 4*.

---

[1] equivalents per hectare and year

FIGURE 12 Distibution of critical loads for nitrogen. After Bashkin *et al.* [24].

FIGURE 13 Distribution of exceedance of critical loads for sulfur. After Bashkin *et al.* [24].

FIGURE 14 Distribution of exceedance of critical loads for nitrogen. After Bashkin *et al.* [24].

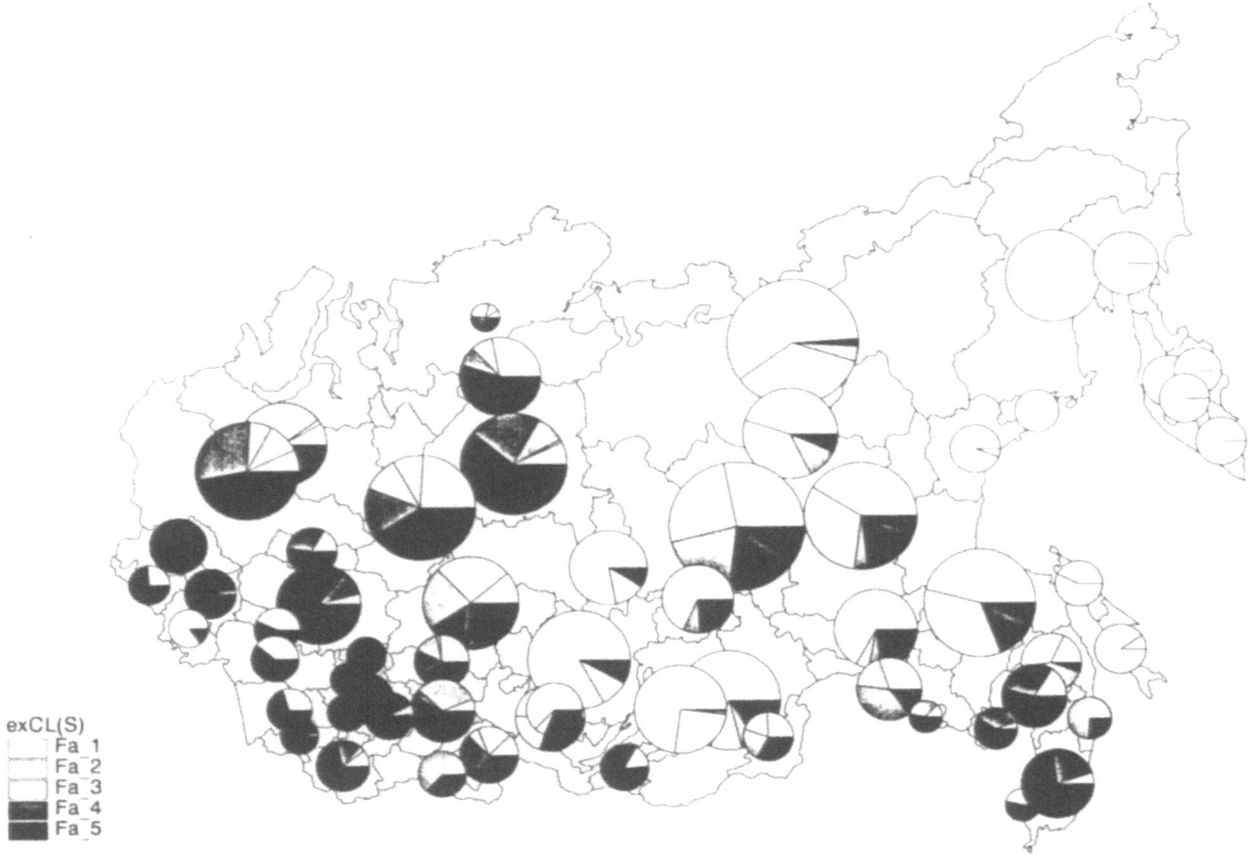

FIGURE 15 Forested areas with exceedance of the critical loads for sulfur. The same classification as in FIGURE 13.

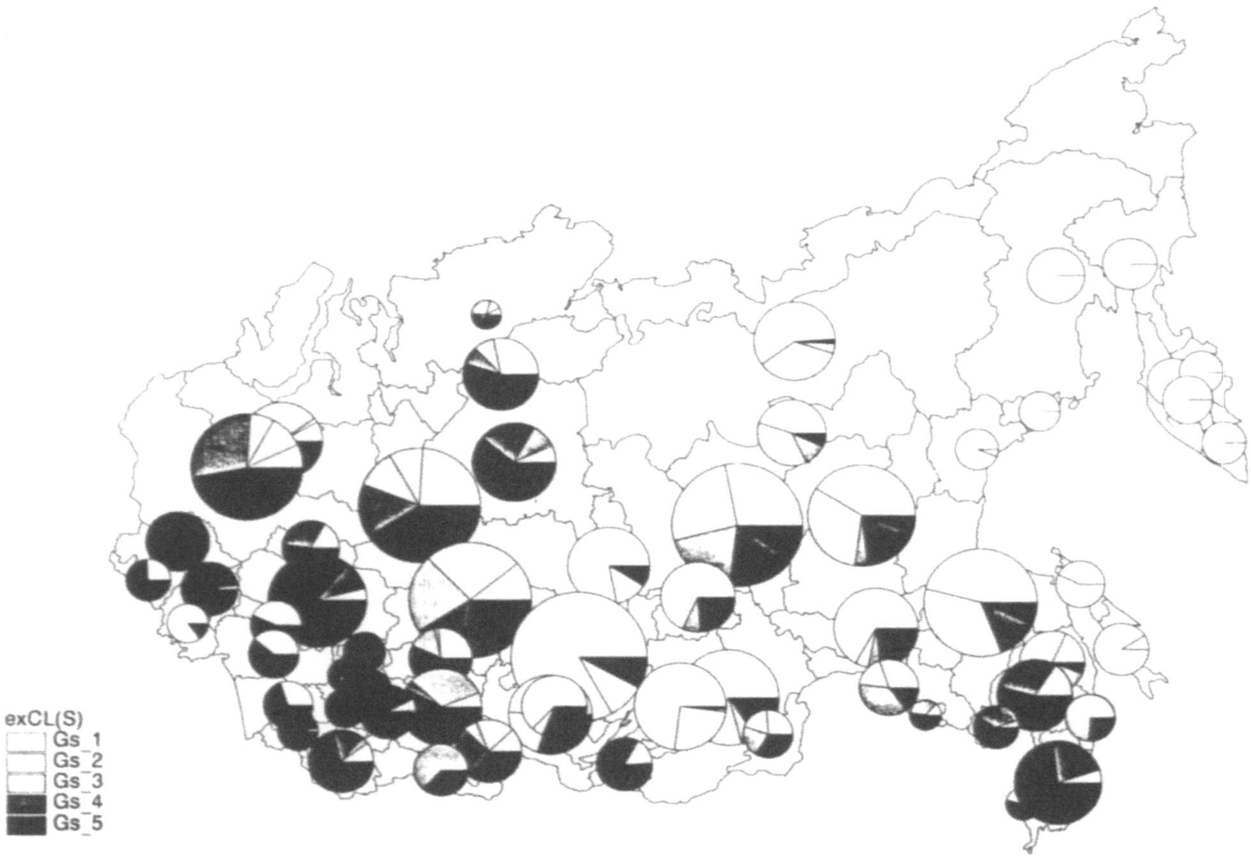

FIGURE 16 Growing stock of forested areas with exceedance of the critical loads for sulfur. The same classification as in FIGURE 13.

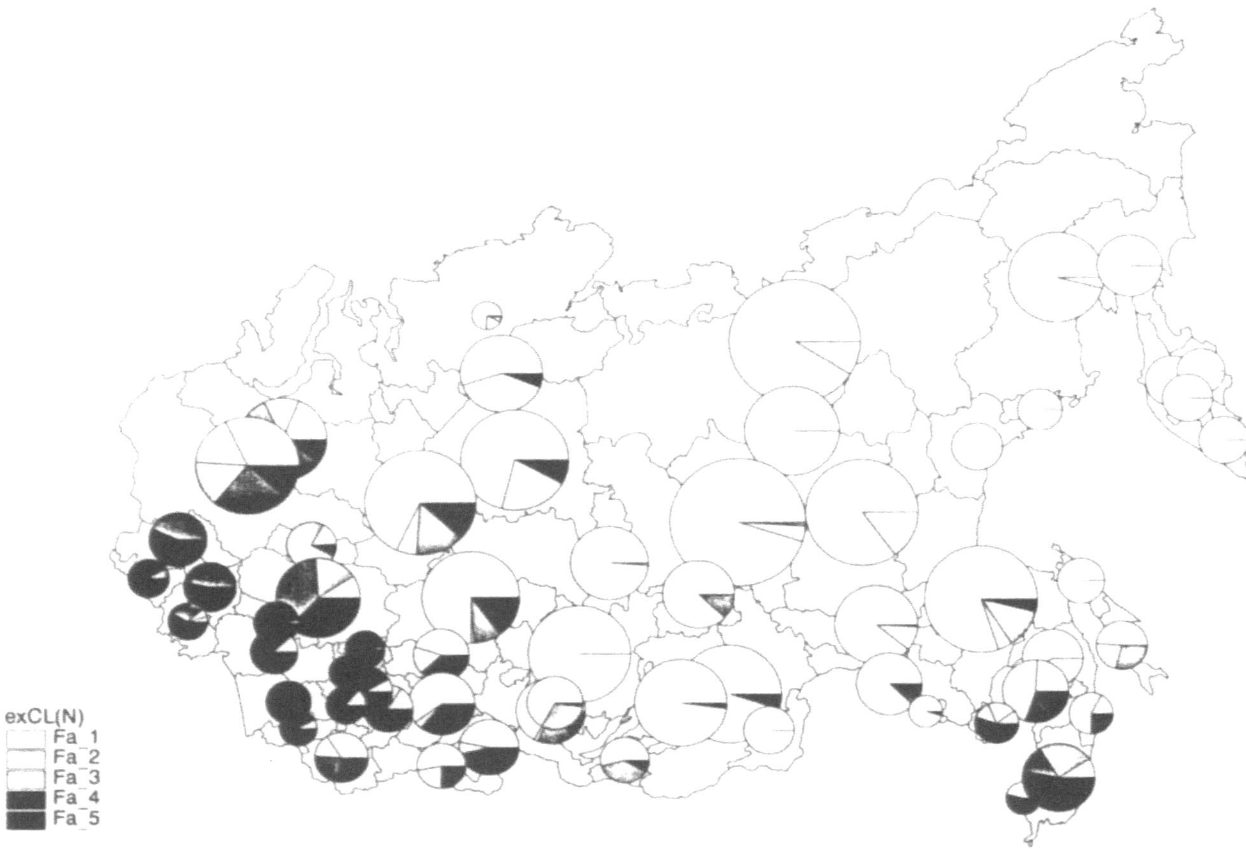

FIGURE 17 Forested areas with exceedance of the critical loads for nitrogen. The same classification as in FIGURE 14.

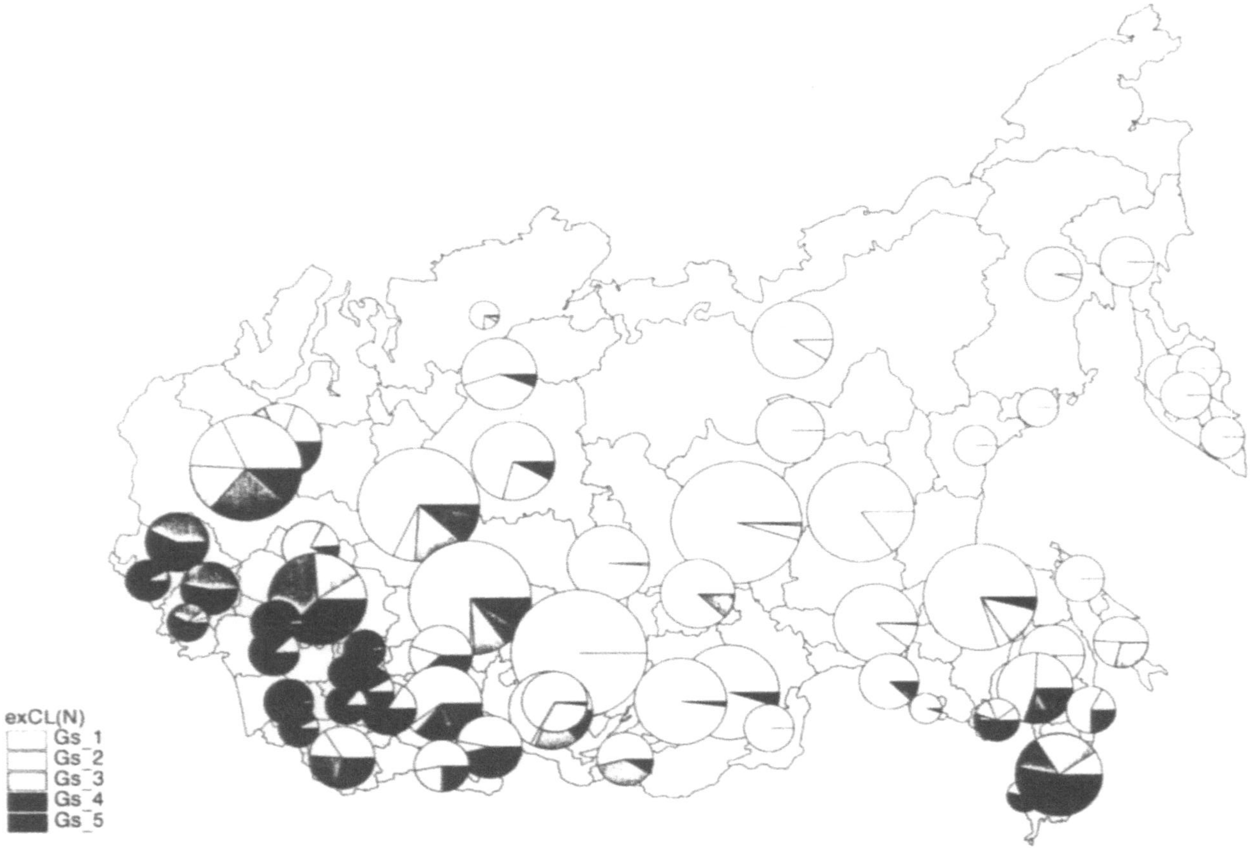

FIGURE 18 Growing stock of forested areas with exceedance of the critical loads for nitrogen. The same classification as in FIGURE 14.

TABLE 4. Distribution of exceedance in percentage.

Exceedance	Sulfur		Nitrogen	
e.q. $ha^{-1}yr^{-1}$	Forested Area	Growing Stock	Forested Area	Growing Stock
< 0	39.4	35.0	71.6	66.2
0-50	16.5	15.0	8.0	7.5
51-100	8.5	8.9	5.8	7.1
101-200	10.9	11.6	6.5	8.1
> 200	24.6	29.5	8.1	11.1

If we consider the two last exceedance classes (> 101 e.q. $ha^{-1}yr^{-1}$) there are more than 210 million ha of Forested Areas at risk of sulfur pollution and decline. The growing stock at risk of decline due to sulfur pollution is 24.5 billion m^3. The corresponding figures for nitrogen are 87 million ha of Forested Areas and 11.4 billion m^3 of growing stock.

The Forested Area at risk of sulfur pollution in Siberia is larger than all exploitable forests in the USA (196 million ha). The growing stock at risk is similar to the growing stock of the U.S. exploitable forests (23.1 billion m^3). The Forested Area and growing stock at risk in Siberia due to nitrogen pollution is some 65% of the total exploitable forests in Europe (133 million ha respectively 18.5 billion m^3).

With respect to heavy metals, it can be concluded that in Siberia HM depositions are below critical loads for forests set by Russian experts (see *Table 3*) except in the vicinity of emitters of heavy metals. Hence, we currently estimate that heavy metal pollution does not seem to be an overwhelming problem in Siberia. But further development of the critical loads for heavy metals is necessary.

7. Radionuclide Contamination

Official data on radio nuclide contamination in Siberia derive from different sources and have different reliability. Data from official reports and unofficial measurements do not always coincide. The radioactive conditions in Siberia are determined by: global radioactive background; natural radioactivity; by radio nuclides due to radioactive outbreaks and underground nuclear explosions; and by the nuclear industry and nuclear waste storage. The main sources for nuclear contamination are listed in *Table 5*.

We have been through all available information (to the best of our knowledge) on radio nuclide contamination in Siberia. It would be too extensive to report all details here and, hence, we make reference to Kharuk *et al.* [10] for further information. The information available is sparse, but based on our investigations we can only conclude that there seems to be only limited contaminated areas that exceed the critical loads (levels) in Siberia today. From the Kharuk *et al.* [10] investigation the following can be stated:

TABLE 5 Main Sources of Nuclear Contamination in Russia and Siberia. After: [30]. Number of sources.

	Russia	Siberia
1. Nuclear materials mining	12	8
2. Nuclear fuel and nuclear weapon industry	15	8
3. Nuclear power stations (NPS)	12	4
4. Power units and reactors, which are equivalent to NPS (no longer in operation)	38(5)	10(2)
5. Research nuclear reactors	39	4
6. Underground nuclear explosions	93	35
7. Radionuclide storage ("cemeteries")	21	7

- Atmospheric deposition in Siberia is not regarded as a major source of nuclear contamination due to the fact that existing contamination is comprised mainly of natural isotopes.
- The soil's natural radioactivity in Siberia is considered to be within the background level except for areas with naturally increased levels of radioactivity and sites of nuclear minerals mining.
- Soil contamination by ^{137}Cs and ^{90}Sr through natural radioactivity is highest in the latitudinal belt of 50-60 degrees north. In addition, there are areas of radioactive contamination within urban areas caused by nuclear waste, and by building materials with naturally high levels of radioactivity. Areas of contamination are identified in the vicinity of nuclear waste storages and along nuclear waste transportation tubes. Underground explosions have caused local soil contamination in at least two cases.
- Currently, the nuclear industry is not considered a significant source of nuclear contamination in Siberia. But until 1992, the so-called direct-current reactors caused problems with nuclear contamination in the Yenisey river valley. Concerns remain about the severity of the problems in this river basin and in the Irkutsk region.
- The current nuclear waste storage is a high potential risk for future radioactive accumulation in the Siberian forest ecosystems. This high risk condition requires urgent further investigation.

8. Impacts of Air Pollution and Radio Nuclide Contamination on Forest Ecosystems

Kharuk *et al.* [10] have made an overview of the forest decline in Siberia caused by present air pollutants. They conclude that there may be forest decline in Siberia caused by air pollutants on 3-3.5 million ha, which is 3-4 times more than that officially reported for total Russia.

Investigations in the Norilsk region did not reveal any sign of nitrogen eutrophying in spite of substantial nitrogen oxide emissions since 1944 [31]. Also, in spite of extremely heavy S pressure in that region (~2 million/tons/year), there is no evidence of any considerable soil acidification. From the Norilsk investigations it can be concluded that stand vigor correlates better with needle sulfate-ion concentrations than with S depositions. Menzhikov *et al.* [32] also reports a poor correlation between soil sulfur content and stand vigor in other areas. Hence, it seems that the primary cause of forest decline is not sulfur depositions, but SO_2 or its derivatives in gaseous or aerosol forms and fluoride emissions (Irkutsk region). The impact of nitrogen emissions on the current extent of forest decline is negligible [10]. Negative impacts of heavy metals is limited, but even in these cases the damage is caused mainly by SO_2 or its derivatives.

A decreased increment has been discovered in damaged areas, but also in areas with no signs of decline. Estimates by Russian scientists (see [10]) show that with the current rate of emissions of primarily sulfur and nitrogen, forest decline will proceed and probably accelerate in the future.

We have studied the investigations available in Russia on the migration of radionuclides within forest ecosystems [10]. The conclusion of our investigation is that there are no data supporting the idea that forest ecosystems in Siberia are currently at risk of radioactive damage. There also seems to be a negligible amount of radioactive accumulation in the woody parts of the Siberian forests. But again, this issue requires further investigation.

9. Policy Recommendations

Based on the results presented, the most severe problem to the forest ecosystem and anthropogenic stress seems to be SO_2 emissions. More than 210 million ha of Forested Areas and 24.5 billion m^3 of growing stock are at risk of decline due to sulfur emissions.

The following policy recommendations require the immediate attention of Russian authorities [33]:

- Establishment of new abatement strategies to reduce primarily SO_2 emissions, but also heavy metals and nitrogen.
- The development of a complete critical load calculation system for all major pollutants in Siberia is required.
- A reliable system for measuring radioactive contamination in Siberia needs to be established.
- A risk evaluation of current nuclear waste storage and possible migration of radio nuclides from these storages must be carried out in the very near future.
- A reliable monitoring system of the forest health in Siberia should be established immediately. Reliable estimates on the health status will influence policy-setting on forest management, annual allowable cuts, abatements of greenhouse gases and environmental policies.

10. References

1. Shvidenko, A. and Nilsson, S. (1997) Are the Russian forests disappearing? *Unasylva*, **48,** 57-64, 1997/1.
2. Dudley, N. (1992) Forests in trouble, WWF International, Gland, Switzerland.
3. European Parliament (1995) Siberian forests and timber trade, STOA, European Parliament, Brussels, Belgium.
4. Feshbach, M. (ed.) (1995) *Environmental and Health Atlas of Russia*, PAIMS Publishing House, Moscow.
5. Gusewelle, C.W. (1992) World forests: Siberia on the brink, *American Forests*, May/June 1992.
6. Knight, R. (1992) Northern exposure: the bills are due for 30 years of communist exploitation, *US News World Report*, March 30, 1992.
7. Newell, J. and Wilson, E. (1996) The Russian Far East, forests, biodiversity hotspots, and industrial developments, Friends of the Earth-Japan, Tokyo, Japan.
8. Stanglin, D. (1992) Toxic wasteland, *US News World Report*, April 13, 1992.
9. Tracy, L.A. (1994) The deforestation of Siberia: economic and environmental problems in Russian forest management, Forest Industry Lecture No. 32, Faculty of Agriculture, Forestry and Home Economics, University of Alberta, Edmonton, Canada.
10. Kharuk, V.I., Nilsson, S. and Samarskaia, E. (1996) Anthropogenic and Technogenic Stress Factors to Forests in Siberia, WP-96-104, International Institute for Applied Systems Analysis, Laxenburg, Austria.
11. Blauberg, K. (1996) Siberian Forest Study Data Dictionary, Unpublished Manuscript, International Institute for Applied Systems Analysis, Laxenburg, Austria.
12. Shvidenko, A. and Raile, G. (1996) Description of Data Elements in the Ecoregion Database of the Siberian Forest Study. Unpublished Manuscript, International Institute for Applied Systems Analysis, Laxenburg, Austria.
13. Okruzayuschaya prirodnaya sreda Rossii (Russian Environment) (1995), *Ecos*, Moscow, pp. 20-99, (In Russian).
14. Nilsson, S. (ed.) (1991) *European Forest: The Effect of Air Pollutants and Suggested Remedial Policies*, IIASA, The Royal Swedish Academy of Agriculture and Forestry, and InterAction Council, Stockholm.
15. Posch, M., Hettelingh, J.-P., Sverdrup, H., and de Wries, W. (1993) Guidelines for the Computation and Mapping of Critical Loads and Exceedances of Sulphur and Nitrogen in Europe, in R.J. Downing, J.-P. Hettelingh, and P.A.M. de Smet, (eds.) *Calculation and Mapping of Critical Thresholds in Europe*, CCE Status Report, RIVM, Bilthoven, The Netherlands.
16. Nilsson, J. (ed.) (1986) Critical Loads for Sulphur and Nitrogen, Report from Nordic Working Group, Nordic Ministerrod Milj^ Rapport, pp. 232.
17. Nilsson, J. and Grennfelt, P. (eds.) (1988) Critical Loads for Sulphur and Nitrogen, *Nord* 1988:97, Nordic Council of Ministers, Copenhagen, Denmark.

18. Sverdrup, H., de Wries, W., and Henriksson, A. (1990) Mapping Critical Loads, *NORD* 1990:**98**. Nordic Council of Ministers, Copenhagen, Denmark.
19. Grennfelt, P. and Th^rnel^f, E. (eds.) (1992) Critical Loads for Nitrogen, *NORD* 1992:**41**, Nordic Council of Ministers, Copenhagen, Denmark.
20. Downing, R.J., Hettelingh, J.-P., and de Smet, P.A.M. (eds.) (1993) Calculation and Mapping of Critical Loads in Europe, CCE Status Report, RIVM, Bilthoven, The Netherlands.
21. Hornung, M., Sutton, M.A., and Wilson, R.B. (eds.) (1995) Mapping and Modelling of Critical Loads for Nitrogen: a workshop report, Proceedings of the Grange-over-Sands Workshop, 24-26 October 1994, Institute of Terrestrial Ecology, Edinburgh, United Kingdom.
22. Hettelingh, J.-P., Posch, M., and de Smet, P.A.M. (1995), in M. Posch, P.A.M. de Smet, J.-P. Hettelingh, and R.J. Downing (eds.), Calculation and Mapping of Critical Thresholds in Europe, Status Report 1995, Coordination Center for Effects, RIVM Report N. 259101004, RIVM, Bilthoven, The Netherlands.
23. Nilsson, S., Salln‰s, O., Hugosson, M., and Shvidenko, A. (1992) *The Forest Resources of the Former European USSR*, The Parthenon Publishing Group, Carnforth, UK.
24. Bashkin V.N., Kozlov, M.Ya., Priputina, I.V., Abramychev, A.Yu., and Dedkova, I.S. (1995) Calculation and mapping of critical loads of S, N and acidity on ecosystems of the northern Asia. *Water, Air and Soil Pollution*, **85**, 2395-2400.
25. Sadykov, O.F. (1991) Today's problems and prospectives of applied ecology, in *Development of S.S. Schwarz ideas on Modern Ecology*, Russian Academy of Sciences, Moscow, pp. 143-213.
26. Stepanov, A.M. (1991) Bioindikaziya na urovne ecosystem (Bioindication at Ecosystem Level), in Bioindikaziya and Biomonitoring, Russian Academy of Sciences, Moskow, pp.56-64. (In Russian)
27. Vorobeichik E.L., Sadikov O.F., and Farafontov M.G. (1994) *Ecologicheskoe normirovanie tehnogennikh zagryaznenii nazemnikh ecosystem* (Ecologically Based Normative for Tehnogenic Pollution for Terrestral Ecosystems), Russian Academy of Sciences, Ekatrinburg, Nauka, pp. 280. (In Russian).
28. de Vries, W. and Bakker, D.J. (1996) Manual for Calculating Critical Loads of Heavy Metals for Soils Surface Waters, Report 114, DLO Winand Staring Centre, Wageningen, The Netherlands.
29. Galperin, M.V., Erdman, L.K., and Subbotin, S.P. (1994) Modelling of the Pollution of the Arctic by S and N Compounds and Heavy Metals from Sources in the Northern Hemisphere, MSC-E Report, Moscow, Russia.
30. Bulatov, V.I. and Chirkov, V.A. (1994) *Tomskaya avariya: bil li sibirskii Chernobyl?* (Tomsk Accident: If there was a Siberian Chernobyl?), Novosibirsk, CERIS, (In Russian).
31. Kharuk, V.I. (ed.) (1992) Monitoring drevostoev i pochv Norilskogo regiona (Monitoring of Forest Stands and Soils of Norilsk Region, Annual report), Institute of Forests, Krasnoyarsk, pp. 142. (In Russian).

32. Menzchikov, S.L., Mahnev, A.K., and Vlasenko, V.E. (1990) Osobennosty aerotehnogennogo zagryazneniya lesotundrovikh biogeocenozov (Air pollution specificity in forest-tundra ecosystems) in *Problemy Lesovedeniya i Lesnoy Ecologii* (Problems of Forest Ecology), Minsk, pp. 593-594. (In Russian).
33. Nilsson, S. (ed.) (1997) Dialogue on Sustainable Development of the Russian Forest Sector - Volume II, Interim Report 97-010/March, International Institute for Applied Systems Analysis, Laxenburg, Austria.

DRY DEPOSITION OF PARTICLES FROM THE ATMOSPHERE

M.J. ZUFALL and C.I. DAVIDSON
Departments of Civil & Environmental Engineering and Engineering & Public Policy, Carnegie Mellon University Pittsburgh, Pennsylvania 15213, USA

1. Introduction

Atmospheric inputs of particulate pollutants to a region may cause environmental damage. For example, acidic particles reaching sensitive ecosystems can harm vegetation and degrade water quality. Particles depositing in urban areas can damage structures, such as buildings and monuments.

Airborne particles may be delivered to surfaces by wet or dry deposition. Wet deposition is from precipitation, while dry deposition refers to the set of mechanisms that deliver airborne constituents to surfaces in the absence of precipitation. Despite its importance, much uncertainty remains in our understanding of the dry deposition process. Efforts have been made to quantify the amount of dry deposition by modeling and experimental methods. In general, the models are limited in their predictive ability due to our incomplete understanding of the many factors influencing deposition. Methods of measurement are complicated by difficulties in analyzing trace quantities of the depositing species, and by variability in deposition from one location to another.

In this paper, we present a brief review of particle dry deposition. We begin with an overview of the process and describe some of the key mechanisms involved. Then we discuss a few of the mathematical models used to calculate deposition. Next we summarize currently available methods for measuring dry deposition, covering both surface accumulation methods as well as techniques to estimate the flux from airborne data. Finally, we discuss some special considerations regarding the role of coarse particles in dry deposition.

2. Overview of the Dry Deposition Process

The deposition process can be idealized as taking place in three separate steps: aerodynamic transport, boundary layer transport, and surface interactions. Aerodynamic transport refers to particles carried from any height in the free atmosphere to the viscous sublayer. This transport is characterized mainly by sedimentation and turbulent diffusion. The viscous sublayer is a thin (several mm or

I. Linkov and R. Wilson (eds.), Air Pollution in the Ural Mountains, 55–73.

smaller) layer between a surface and the free atmosphere which contains a steep velocity gradient. The sublayer consists of mainly laminar flow with intermittent bursts of turbulence. Transport across this layer, the second step, is governed by Brownian diffusion for small particles, and by interception, inertial forces, and sedimentation for larger particles. Electrical migration, thermophoresis, and diffusiophoresis may also influence small particles under certain conditions. The third and final step in the process involves surface interactions. Once at the surface, particles may adhere or bounce off, depending on the characteristics of the contaminant and the surface. Any of these three steps may be rate-limiting, depending on the specific conditions of the atmosphere, surface, and contaminant [1].

It is of interest to consider in more detail the various mechanisms involved in these steps. Turbulent diffusion utilizes turbulent energy to transport particles along a concentration gradient. As particles are continually deposited there is a concentration gradient between the atmosphere and a perfect sink surface. This mechanism of transport is most important in the free atmosphere where turbulence is greatest.

Transport by sedimentation is due to the force of gravity. A drag force, due to the viscosity of air, opposes the acceleration of gravity and the two forces on the particle reach equilibrium relatively quickly ($<<$1s for a 1 μm spherical particle). At equilibrium the particle will fall at a constant velocity. Sedimentation transport is greatest for large ($>$ 1 μm) particles since the settling velocity strongly depends on particle size [2]. The equation for sedimentation also includes a buoyant force which acts in the same direction as the drag force, but this is usually very small for airborne particles.

Brownian diffusion is similar to turbulent diffusion, as it also depends on the concentration gradient. With Brownian diffusion, the driving force is the random thermal energy of air and particles. Brownian diffusion is much weaker than turbulent diffusion and only becomes significant when turbulent diffusion is weak, such as in the viscous sublayer. This mechanism mainly applies to gases and submicron particles, as transport by Brownian diffusion decreases with increasing particle size [2].

Interception and inertial forces may transport large ($>$ 1 μm) particles across the viscous sublayer. A particle with a diameter greater than the sublayer height moving with the mean motion of the air will be deposited by interception when it collides with an obstacle. This occurs when the particle is traveling on an air streamline that passes within one particle radius of the obstacle. Inertial forces may lead to impaction and turbulent inertial deposition. Impaction occurs when the particle leaves the streamline since it cannot follow rapid changes in the air flow, and collides with an obstacle. Turbulent inertial deposition occurs when inertial energy derived from the component of air flow perpendicular to the surface (a turbulent eddy) carries a particle close to the surface.

Once a particle has traversed the viscous sublayer, it will interact with the surface. Depending on the characteristics of the contaminant and surface, a particle may stick or bounce off. There may also be subsequent chemical reactions.

Particle deposition by any combination of these transport mechanisms largely depends on atmospheric, surface, and particle characteristics. Examples of atmospheric parameters include wind speed, humidity, stability, and temperature. In addition, micrometeorological parameters related to the surface such as friction velocity, roughness height and zero-plane displacement will affect the deposition. Surface properties which influence deposition include chemical and biological reactivity, geometry of roughness elements, terrain characteristics, and wetness. Characteristics of the depositing particles also affect deposition; the most important of these include size, shape, density, reactivity, hygroscopicity, and solubility [3]. We now consider some of the formulations developed for modeling the dry deposition process.

3. Mathematical Models for Dry Deposition

We normally quantify dry deposition in terms of a flux F(g/cm^2 s):

$$F = v_d(z)\, C(z) \qquad (1)$$

where v_d is the deposition velocity (cm/s), and C is the airborne concentration of particles (g/cm^3). Note that v_d and C are functions of height z, although F is assumed to be invariant with z. Many models develop formulations for v_d based on either theoretical or empirical considerations. In this section, we first summarize mathematical modeling that has been conducted for each of the three steps in the dry deposition process as described above. We then discuss the role of turbulent bursts in affecting dry deposition. Finally, we present alternative models for the various resistances to transport that incorporate empirical expressions.

3.1 AERODYNAMIC TRANSPORT

Transport of particles in the free atmosphere occurs by turbulent diffusion and gravitational sedimentation. The minor influence of Brownian diffusion may also be included. The flux due to either diffusion process is modeled as a coefficient times the concentration gradient. The turbulent diffusion coefficient K (cm^2/s) of a contaminant above the vegetation canopy is assumed to be the same as the kinematic eddy viscosity of the air, whose expression is derived by analogy to transport of air momentum:

$$K = k u_*(z-d) \qquad (2)$$

where k is von Karman's constant and is equal to 0.4, u_* is the friction velocity (cm/s), and d is the zero plane displacement (cm). Within the canopy, the turbulent transfer coefficient is taken to be

$$K = khu_* \exp\left[-n\left(1-\frac{z}{h}\right)\right] \qquad (3)$$

where h is the height of the canopy, and n is an empirical coefficient dependent on characteristics of the canopy.

In a nonadiabatic atmosphere, the expression on the right side of each equation is divided by a correction factor ϕ_c to account for instability. Correction factors are determined experimentally, and are a function of ζ= (z-d)/L, where L is the Monin-Obukhov length. Expressions for stability factors have been determined by Businger et al. [4], Wesely and Hicks [5], and Hicks et al. [6].

The Brownian diffusivity D of a contaminant (cm^2/s) is defined as :

$$D = \frac{kTc}{3\pi\mu d_p} \tag{4}$$

where k is the Boltzmann constant (g cm^2/s^2 K), T is the temperature (K), c is the Cunningham correction factor, μ is the dynamic viscosity of air (g/cm s), and d_p is the physical diameter of the particle (cm).

For particles of sizes typically encountered in the atmosphere, the transport due to sedimentation by gravity is usually taken as the Stokes settling velocity:

$$v_g = \frac{\rho_p g c d_p^2}{18\mu} \tag{5}$$

where ρ_p is the particle density (g/cm^3). This formulation assumes that the density of air is negligible compared to the particle density so that the buoyant force can be neglected. The flux due to settling is calculated as the sedimentation velocity times the concentration.

Since model calculations assume spherical, unit density particles, the aerodynamic diameter is introduced to apply these models to real particles. The aerodynamic diameter is defined as the diameter of a spherical, unit density particle with the same motion characteristics as the actual particle [7].

The total flux to the surface can be represented as (where downward flux is positive):

$$F = (D + K)\frac{dC}{dz} + v_g C(z) \tag{6}$$

Use of Equation 6 requires several empirical parameters, since it is difficult to determine the boundary condition for C(z) and there are complexities in estimating values of some parameters in the equation. For example, determining C(z) requires estimates of particle removal by leaves, stems, and other parts of the canopy. Collection efficiencies of canopy elements are discussed in section 3.2.

To estimate dry deposition, it is sometimes necessary to determine the velocity profile. The velocity profile above the canopy can be calculated by analyzing momentum transport:

$$u(z) = \frac{u_*}{k}\ \ln\left(\frac{z-d}{z_o}\right) \text{ for } z > h \tag{7}$$

where z_o is the roughness height (cm). Within a canopy, the wind speed profile is:

$$u(z) = u(h)\exp\left[-n\left(1-\frac{z}{h}\right)\right] \text{ for } z < h \tag{8}$$

where u(h) is the wind speed at the height of the canopy.

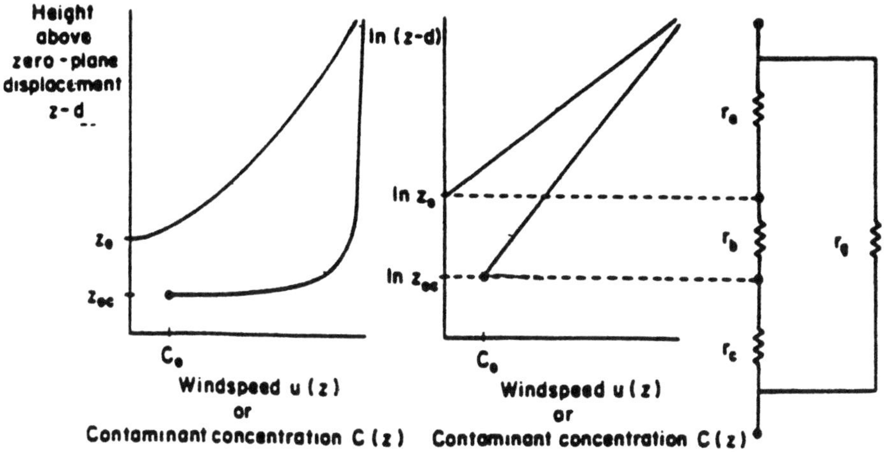

FIGURE 1. Wind speed and contaminant concentration as functions of height above the zero-plane displacement of a rough surface. Hypothetical sinks for air momentum and contaminant concentration are denoted by $(z\text{-}d)=z_o$ and $(z\text{-}d)=z_{oc}$, respectively. (From Davidson, C. I. and Wu, Y.-L., In Lindberg, S. E., Page, A. L., and Norton, S. A., eds., *Acidic Precipitation*, Springer-Verlag, 1990. With permission.)

There are distinct differences between the shapes of the wind profile u(z) and the contaminant concentration profile C(z) above the surface, as shown in figure 1. The shape of the wind and concentration profiles can be used to identify the region where there is the greatest resistance to air momentum flux and contaminant flux, respectively. Under most conditions, there is a continual flux of momentum downward toward the surface by turbulent eddies; the wind speed decreases starting well above the surface, as momentum is rapidly destroyed at the surface by friction. However, the concentration profile remains relatively constant until very near the surface because most particles are only slowly transported through the viscous sublayer. Most of the gradient in momentum therefore occurs well away from the surface where resistance to momentum transport is greatest, while most of the gradient in particle concentration occurs in the viscous sublayer where resistance to particle mass transport is greatest.

3.2 BOUNDARY LAYER TRANSPORT: COLLECTION BY CANOPY ELEMENTS

To fully represent the particle dry deposition flux, equation 6 is combined with a formulation representing the flux of contaminants removed at the surface. This is determined from the rate at which particles are transported parallel to the ground integrated over the height of interest, combined with the efficiency of collection. The particle transport parallel to the ground is determined from the wind speed profile multiplied by the concentration profile.

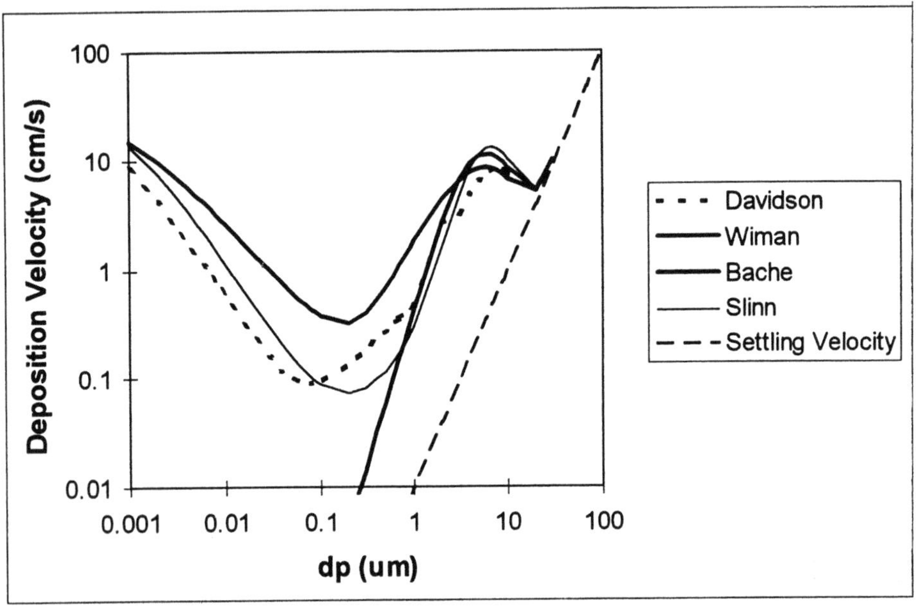

FIGURE 2. Deposition velocities from the models of Davidson et al. [8], Slinn [12], Bache [10], and Wiman and Agren [13]. (Redrawn from Ruijgrok, W., Davidson, C., and Nicholson, K., *Tellus*, **47B**, 587, 1995).

The efficiency of collection depends on the area available for collection and the efficiency of the mechanisms which deposit the particle on the receptor. A "receptor" is a portion of the surface that collects a contaminant, either a horizontal area or a vertical protrusion from a horizontal plane. Due to the complexity of any natural surface, this efficiency is extremely difficult to model. Mathematical models have been created to represent the area available for collection. One such model describes the surface as a field made up of blades of grass, where each blade is considered as an individual receptor.[8] Another model, in this case for a pine forest, uses a Gaussian distribution of foliage to determine the area available for collection [9][10][11]. Slinn [12] created a more general model in which a leaf area index was used to describe the surface area of vegetation per unit volume at a specific height. Finally, Wiman and Agren [13] created a model for transport and deposition through an open field to a forest where the particles were deposited. The ranges of deposition velocities resulting from these models are presented in figure 2.

The collection efficiency depends on the surface and particle characteristics as well as the specific mechanisms. Brownian diffusion, interception, and impaction are the main mechanisms by which particles are transported across the viscous sublayer. For a vegetative canopy, each leaf, stem or other canopy element may have a viscous sublayer. The collection efficiency of particles transported by Brownian diffusion can be represented by analogy to heat flow and is dependent on the diffusion coefficient D. The efficiency of interception depends on the ratio of particle to receptor size. Impaction efficiency is a function of the Stokes number and depends on the wind speed and specific characteristics of the particle and surface. All of the individual efficiencies can be combined to give an overall efficiency [1].

3.3 SURFACE INTERACTIONS: PARTICLE REBOUND

Some contaminants may come into contact with a surface but do not stick to it. This bounce off is usually characterized by a fractional value representative of the surface and type of contaminant.

Bounce off occurs when the kinetic energy of the particle after impact is large enough to overcome surface attraction forces. Wu et al. [14] found that bounce off is dominated by energetic particles driven by turbulent fluctuations, and that the rebound fraction increases as wind speed or particle size becomes greater. This analysis assumed that the attractive forces between the particle and surface are due only to van der Waals forces. The fraction of rebound is determined from the probability that the particle velocity v_i just before reaching the surface is greater than the critical velocity v_i^*. Based on conservation of energy and the energy of van der Waals attraction, v_i^* is calculated to be:

$$v_i^* = \frac{u_*[A(1-e^2)/18\pi\mu\nu h_o)]}{(\tau_p^+)^{0.5}} \tag{9}$$

where A is the Hamaker constant (g cm^2/s^2), e is coefficient of restitution, h_o is the smallest distance between the particle and the surface (cm), and τ_p^+ is the non-dimensionalized particle relaxation time:

$$\tau_p^+ = \frac{\tau u_*^2}{\nu} \tag{10}$$

$$\tau = \frac{\rho_p c d_p^2}{18\mu} \tag{11}$$

where ν is the kinematic viscosity of air (cm^2/s). Predictions of the rebound fraction for a Monte Carlo simulation (50 runs) of the probability of $v_i > v_i^*$ are shown in Figure 3.

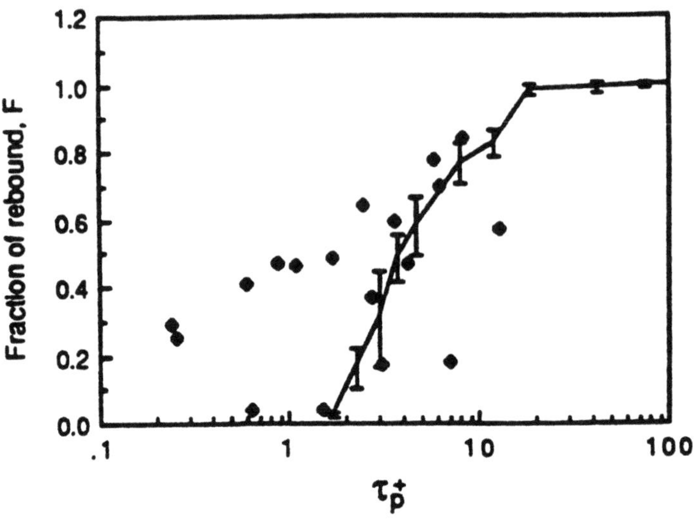

FIGURE 3. Model predictions for rebound fraction (solid line with one standard deviation error bars) compared with results measured in a wind tunnel. (From Wu, Y.-L., Davidson, C. I., and Russell, A. G., *Aerosol Sci. Technol.*, **17**, 231, 1992. With permission.)

Wind tunnel experiments were also conducted to determine bounce off fractions experimentally. Uranine particles were generated and discharged into the wind tunnel. Rebound fractions were calculated from the ratio of deposition rates to greased and ungreased Teflon plates. The results of the experimental fraction of rebound are presented with calculated results in figure 3.

Paw U [15] suggests that the critical rebound velocity should be calculated from the vertical component of the particle velocity, not from the particle speed. Through wind tunnel studies with Lycopodium spores depositing on leaves, it was found that rebound began to occur at wind speeds of several m/s for flow parallel to the surface. However, for flow perpendicular to the surface, particle bounce off occurred at speeds of 0.7 m/s. These results are consistent with earlier theories which used the kinetic energy based on the vertical component of the particle velocity [16][17][18].

3.4 TURBULENT BURSTS

Although many models assume that the viscous sublayer consists of entirely laminar flow, this is not the case. There are bursts of turbulent air which move swiftly through the viscous sublayer, carrying particles with them. Turbulent bursts may be eddies from the free atmosphere that extend down to the surface [14] or eddies created due to surface roughness [19]. Particles transported by these bursts are not affected by the viscous sublayer and move at approximately the same rate as in the aerodynamic transport region. Recently, developments have been made in quantifying this phenomenon.

Wu et al. [14] developed a model for particle dry deposition which includes the effects of turbulent bursts. In this model, particles near the center of a turbulent burst deposit, while those far from the centerline are transported back into the main flow regime. The average lateral distance between bursts λ determines the probability of a particle depositing by this method. The limiting distance between the particle and center of the burst within which deposition occurs, y_{lim}, is determined by the trajectory of a particle which deposits $\lambda/4$ from the centerline. The fraction of particles in the viscous sublayer that deposits is $2y_{lim}/\lambda$, and the particle deposition velocity is $2w_o y_{lim}/\lambda$, where w_o is the instantaneous vertical velocity at the viscous sublayer edge.

Cleaver and Yates[19] developed estimates of turbulent burst size, spacing and rate of occurrence. On average, turbulent bursts have a diameter of $20(\nu/u_*)$, are laterally spaced $135(\nu/u_*)$ apart, and occur every $75(\nu/u_*^2)$.

The effect of turbulent bursts on deposition rates depends on particle size. Particles in the viscous sublayer greater than 10 μm are deposited swiftly due to gravity sedimentation and are not affected by turbulent bursts. Smaller particles carried by turbulent bursts will move at a speed comparable to that in the aerodynamic transport region. These small particles will have a deposition velocity up to 20 times greater than if they were only influenced by Brownian diffusion in the viscous sublayer.

3.5 ALTERNATIVE MODELS USING RESISTANCE TO TRANSPORT

Resistance models for dry deposition are used to calculate the dry deposition velocity based on the resistance to transport (s/cm) for each step in the process:

$$v_d = \frac{1}{r_a + r_b + r_c} + v_g \qquad (12)$$

where r_a is the resistance to aerodynamic transport, r_b is the resistance to transport through the boundary layer, and r_c is resistance to collection by the surface. Note that r_a does not include the effects of gravity.

The expression for aerodynamic resistance is based on equation 2, combined with the equation for flux in the absence of gravity ($F = K\, dC/dz$):

$$r_a = \frac{\ln\left(\frac{z-d}{z_o}\right) - \Psi_c}{k u_*} \qquad (13)$$

where Ψ_c is the stability correction factor for contaminants. The correction factor is introduced after integration of the equation for flux. Note that for an adiabatic atmosphere ($\Psi_c = 0$), combining equation 13 with equation 7 yields $r_a = u(z)/u_*^2$.

The boundary layer resistance incorporates many mechanisms including eddy and Brownian diffusion, interception, inertial motion and sedimentation. The total effect of all of these mechanisms, when combined with aerodynamic transport, can be represented by a contaminant sink z_{oc}. The contaminant sink is the height at which the concentration is effectively zero for cases where the particles stick to the surface without bounce off:

$$r_b = \frac{\ln(z_o / z_{oc})}{ku_*} \tag{14}$$

z_{oc} is usually very difficult to determine, since it depends on the various collection efficiencies of the different mechanisms. Instead of Equation 14, a simple empirical equation has been developed:[6]

$$r_b = \frac{2\left(Sc/Pr\right)}{ku_*} \tag{15}$$

where the Prandtl number Pr accounts for the difference between heat and momentum transport:

$$Pr = \nu/\kappa \tag{16}$$

where κ is the thermal diffusivity (cm^2/s).

The Schmidt number Sc accounts for the difference between momentum and mass transport [6]:

$$Sc = \nu/D \tag{17}$$

The canopy resistance depends on the type of vegetation at the surface. Resistances for individual parts of the canopy may be summed according to the "big leaf" model [6].

$$\frac{1}{r_c} = \frac{1}{r_s(Sc / Pr) + r_m} + \frac{1}{r_{cut}} + \frac{1}{r_{soil}} \tag{18}$$

where r_s is the stomata resistance, r_m is the mesophyll resistance, r_{cut} is the cuticle resistance, and r_{soil} is the resistance of the soil. The model of Kramm et al.[20] provides a similar formulation but also includes the resistance of nearly dry soil, and resistances of water due to leaf and soil wetness.

4. Measurements of Dry Deposition

The dry deposition flux may be measured using either surface analysis methods or atmospheric flux methods. Surface analysis methods are those which analyze material accumulated over time on a surface. Examples of this type of measurement include foliar extraction, throughfall and stemflow methods, use of surrogate surfaces, and snow sampling. Atmospheric flux methods are those in which the flux is estimated from measurements of contaminants in the atmosphere. Examples include eddy correlation, the profile method, the tracer method, aerometric balance studies, eddy accumulation, and the variance method. In this section, we describe these categories and cite recently published results. Additional descriptions and data for earlier measurements are provided by Davidson and Wu [1].

4.1 SURFACE ACCUMULATION MEASUREMENTS

These methods provide a direct measurement of the amount of contaminant delivered to a surface. However, making accurate measurements on natural surfaces such as vegetation is extremely difficult due to the low rates of accumulation. Furthermore, it is difficult to extrapolate the results at one location to an entire region.

4.1.1 *Foliar Extraction*

This involves "washing" of a plant surface to obtain material that has accumulated by dry deposition over an exposure period. The measurements are calibrated against leaves where the amount of contaminant deposited prior to the exposure period is known. One potential problem with the method is that some of the chemical species extracted using this method may be due to plant leaching, not dry deposition. It is also difficult to extrapolate the flux to an entire canopy since leaves in different positions in a canopy will collect different amounts of atmospheric material. Therefore, large amounts of data are needed for reliable results.

Shanley [21] made measurements of dry deposition of Ca^{2+}, Mg^{2+}, Mn^{2+}, Pb^{2+}, and SO_4^{2-} to spruce foliage in the Black Forest, Germany. This experiment used branches of high elevation Norway spruce near the top of the canopy. The branches were rinsed with deionized water and the resulting water samples were analyzed. Corrections for leaching were made in two ways. One method used correction factors from other experiments, while the second method assumed that loadings during a clean air period were all from leaching. These two methods provided similar results. The results were compared to measurements made in nearby petri dishes. The comparisons showed wide variations and significant differences over the course of the experiment. Deposition velocities determined from foliar washing ranged from 0.034 - 0.78 cm/s, with a median of 0.33 cm/s.

4.1.2 *Throughfall*

Throughfall is the water which falls to the ground beneath plant canopies during precipitation [22]. Chemical species concentrations in precipitation are subtracted from concentrations in throughfall to determine the dry deposition. The method assumes that material accumulated on surfaces by dry deposition since the previous storm is completely washed off by the precipitation. Throughfall can thus provide an estimate of the flux to an entire canopy. The measurements depend on the amount and strength of the rain, and the length of the preceding dry period [23]. Because the method relies on precipitation to remove contaminants, those particles irreversibly absorbed by the plant surface are not measured. In addition, leaching from foliage may contribute contaminants to the collected rain water.

In a study in the Hubbard Brook Experimental Forest, throughfall was used to estimate the dry deposition of sulfur [24]. Twenty-four collectors, each consisting of two 21 cm diameter funnels, were mounted 1m above the ground. Throughfall results were compared with inferential estimates made from average measured airborne concentrations and deposition velocities calculated from the Hicks et al. [6] resistance model. The throughfall study estimated a flux of 3.3 kg S/ha yr, while the inferential estimate yielded 2 kg S/ha yr.

Although there is much uncertainty associated with this method, researchers continue to use it since it provides an estimate of dry deposition flux over complex canopies [22][23][25]. It is also often used in conjunction with other methods for comparison and corroboration [26][27].

4.1.3 *Surrogate Surfaces*

Surrogate surface measurements involve the use of artificial collectors. These surfaces may be aerodynamically designed so that flow characteristics can be analyzed. Other surfaces with less well known airflow characteristics such as petri dishes, buckets, and Teflon plates have also be used. Surrogate surfaces offer a convenient method of measurement, but results of experiments using them are difficult to extrapolate to natural surfaces.

In a study of deposition of polychlorinated dibenzo-p-dioxins and dibenzofurans (PCDD/F), Koester and Hites [28] used inverted frisbees and square flat plates with glass collection surfaces to measure dry deposition at two sites in Indiana. PCDD/F are toxic compounds which enter the atmosphere as by-products of certain combustion sources. The aerodynamically designed inverted frisbee samplers created minimal wind flow disturbances. The depth of the sampler and a coating of mineral oil minimize bounce off. This study found that the dry deposition flux was inversely correlated with the temperature. At cooler temperatures more PCDD/F are adsorbed to particles which dominate their deposition flux. The average dry deposition velocity was found to be 0.2 cm/s.

4.1.4 *Snow Sampling*

This method uses increases in contaminant concentration in surface snow over time to estimate accumulation by dry deposition. The method attempts direct measurement of dry deposition to a natural surface, although processes other than deposition can affect the concentrations. Sublimation, migration within the snowpack, and chemical changes may also affect the data. Furthermore, the results may be difficult to interpret due to heterogeneity in concentrations and the inability to distinguish between gaseous and particulate forms of chemical compounds.

Snow samples were collected at Summit, Greenland after a three month period of accumulation in July 1993 and July 1994 [29]. The samples were sliced into two cm vertical intervals and then analyzed by ion chromatography for major anions and cations. The contaminant concentrations measured in the snow were compared with those predicted using a model for snow, fog, and dry deposition. The ratios of modeled to measured concentrations are 0.85 and 0.95 for SO_4^{2-} for the 1993 and 1994 field seasons respectively, and are 0.45 and 0.90 for Ca^{2+} for the 1993 and 1994 field seasons respectively.

4.2 ATMOSPHERIC FLUX METHODS

Atmospheric measurements can be used to infer dry deposition fluxes. Generally short sampling times allow for better time resolution than surface accumulation methods. Since atmospheric methods estimate flux from turbulent transfer only, they work best when applied to small particles which are less affected by sedimentation [30]. The methods require adequate fetch and uniform canopy characteristics for accurate results.

4.2.1 *Eddy Correlation*

Eddy correlation measurements may be taken from a tower or an aircraft. This technique involves measurement of the vertical wind velocity w and the contaminant concentration C. Each of these two properties is divided into a time-averaged component ($\overline{w}, \overline{C}$) and a deviation from the time-averaged component (w', C'). The deviations are caused by random turbulent fluctuations. The dry deposition flux is calculated as the time averaged value of the product, $\overline{w'C'}$. The vertical wind fluctuations are measured with a fast-response wind sensor, and the particle concentration fluctuations are measured with a sensor for the contaminant. Although the method provides a wealth of high resolution flux data, sensors are expensive, are subject to noise and detection limit problems, and are not available for all contaminant species.

Eddy correlation measurements were made for small particles at high wind speeds at the Boulder Atmospheric Conservatory [31]. This study used an active-scattering aerosol spectrometer probe to measure particle concentration, and a sonic anemometer to measure vertical wind speed from a tower. For particles with a mean diameter of 0.2 μm, the deposition velocity was found to be 0.6 - 1.0 cm/s for an unstable period with a mean horizontal wind speed of 8.7 m/s, and 0.1 - 0.4 cm/s for a stable period with mean wind speed of 6.4 m/s. These deposition velocities were compared to those estimated from the model of Slinn [12]. The measured results were nearly a factor of two greater than the modeled results.

4.2.2 *Profile Method*

The dry deposition flux of a contaminant is determined in this method from F = K(dC/dz), where K is the estimated eddy diffusivity of the contaminant, and dC/dz is the measured concentration gradient. The eddy diffusivity can be estimated by analogy to sensible heat. The concentration gradient is obtained by measuring the concentration at more than one height. Correction factors are included for nonadiabatic atmospheric conditions. This method does not require fast response sensors as with eddy correlation. However, the results are very sensitive to the precision of the concentration data. In addition, number concentration and size-specific mass concentration measurements may not accurately represent deposition flux because deposition of small particles onto large particles will be measured as deposition to the surface. Furthermore, it is desirable to keep sampling times limited to periods of unchanging meteorology.

The profile technique was used to measure dry deposition of small particles to a short grassy surface [32]. Wind speed and temperature profiles were obtained from measurements taken at four heights between 0.15 m and 2 m above the surface. Aerosol concentrations were measured at six heights over this same range using a Nuclepore polycarbonate prefilter and a PTFE membrane at each height. The prefilter was used to exclude particles greater than 2 μm. Sampling occurred discontinuously over fifty separate intervals with a mean duration of 162 minutes, during which atmospheric conditions were neutral 29 times, unstable 17 times, and stable 4 times. The gradients were determined from the least squares regression method, and the deposition velocity was calculated to be 0.08 cm/s for neutral and unstable conditions and 0.25 cm/s for stable conditions.

4.2.3 *Tracers*

When a non-reacting, non-depositing tracer is released simultaneously with a contaminant of interest, the amount of dry deposition of the contaminant may be obtained. The method is useful over a wide range of distances (100 m - 100 km), and sampling times may range from one hour to one day. It is assumed that dispersion of the non-depositing species is identical to that of the depositing species. This technique is primarily used for intensive field studies rather than routine dry deposition monitoring.

Lead deposition over Los Angeles was estimated by the dual tracer method using CO as the conservative species [33]. Vehicle emissions in a tunnel were used as the base Pb/CO ratio. The fraction of aerosol remaining airborne was calculated as were deposition velocities over different size ranges. An overall dry deposition velocity of 0.26 cm/sec was found for the lead.

4.2.4 *Aerometric Balance Method*

This method consists of taking measurements of contaminant concentrations by aircraft around the perimeter of a region. The data are used to estimate inflow and outflow from the region. When combined with estimates of contaminant loss by chemical reactions, the amount lost to dry deposition can be determined. This method works best over large regions and for contaminants with slow reaction rates. However, extensive aircraft sampling is necessary to construct the mass balance.

4.2.5 *Eddy Accumulation*

This technique measures the difference in contaminant concentration in updrafts and downdrafts to determine the flux from dry deposition. Two samplers are used: one operates only when the vertical wind velocity is downward, while the other sampler operates only when the wind is directed upward. The method obtains direct flux data over long averaging times for species without fast-response sensors. However, accurate flow controllers and precise chemical analyses are required since differences in concentrations in updrafts and downdrafts are likely to be small. Total averaging times of several days or longer may be needed.

4.2.6 *Variance Method*

Since turbulent fluctuations in temperature, humidity, and contaminant mass concentration are highly correlated, relationships can be developed to determine contaminant flux based on measurements of the other variables. When the sensible heat flux F_s is known, the contaminant flux is estimated as $F=F_s\sigma_c/\sigma_T$, where σ_c is the standard deviation of the contaminant concentration, and σ_T is the standard deviation of the temperature. The particle dry deposition flux may also be determined from the water vapor flux F_w, where $F=F_w\sigma_c/\sigma_w$. Here σ_w is the standard deviation of the water vapor concentration.

Aerometric balance studies, eddy accumulation, and variance methods are not yet widely used for measuring atmospheric particle dry deposition, so no recent results are presented for these methods.

5. Special Considerations for Coarse Particles

Investigators have provided evidence of the importance of large particles in affecting overall mass dry deposition of a chemical species. For example, Davidson and Friedlander [34] estimated that the bulk of the mass deposition of certain trace metals and SO_4^{2-} was due to a small fraction of airborne particles. Coe and Lindberg [35] analyzed particle dry deposition using a scanning electron microscope. Although 40% of the deposited number of particles were less than 2 μm in diameter, the deposited particle mass was dominated by particles greater than 10 μm.

Recently several deposition studies have focused on larger (d_p > 5 μm) particles. These studies found that despite the small number of these large airborne particles, their deposition accounts for a substantial portion of total dry deposition (Holsen et al., [36] Holsen and Noll, [37] Baeyens et al., [38] Dulac et al., [39] Zufall and Davidson [40]). Dry deposition modeling formulations which use total particle mass concentrations in all size ranges and a single deposition velocity tend to underestimate the contribution of coarse particles and thus the dry deposition [37]. The importance of the large size fraction has prompted the development of new sampling techniques for large particles [41] and new coarse particle models [42].

6. Summary

While many advances have been made in the study of dry deposition, uncertainties still remain. This chapter has considered the current state of research on particle dry deposition, focusing on vegetative canopies.

Particles are transported through the free atmosphere mainly by turbulent transfer and gravitational settling. The particles are then transferred by interception, impaction, or Brownian diffusion through a relatively quiescent viscous sublayer to a surface. Once at the surface the particles may stick or bounce off, depending on the atmosphere, particle, and surface characteristics. At a later date, the particle may also be resuspended into the atmosphere.

Dry deposition may be quantified by modeling or experiments. The latter may include field tests or laboratory experiments such as wind tunnel studies. The deposition velocity, defined as the contaminant mass flux divided by its airborne concentration, depends largely on particle size, wind speed and a variety of other characteristics of the particle, surface and atmosphere.

The dry deposition flux can be measured using surface accumulation methods or by atmospheric flux techniques. The former refers to experiments which measure accumulation of contaminants at a particular location and extrapolate this flux to an entire area. Atmospheric flux methods use airborne concentrations and atmospheric conditions to infer deposition. While certain models and experimental methods provide reasonable estimates of dry deposition for specific circumstances, none have been proven to provide consistently accurate results over a wide range of conditions.

Future research on dry deposition is needed in many areas, including development of improved models that incorporate details of atmospheric turbulence and improved measurement methods for vegetative canopies and for other surfaces such as water and building materials.

The paper is a shortened version of the more complete description of dry deposition given by Zufall and Davidson [43]; readers can consult this reference for more detailed information on the topic.

7. Acknowledgments

The authors would like to thank the organizers of the NATO Advanced Research Workshop on Air Pollution in the Ural Mountains. This work was funded in part by EPA contract CR 82-2054-01-1 through a subcontract with the University of Michigan.

8. References

1. Davidson, C. I. and Wu, Y.-L. (1990). Dry deposition of particles and vapors, *In* S. E. Lindberg, A. L. Page, and S. A. Norton, eds. *Acidic precipitation*, Springer-Verlag, New York.
2. Friedlander, S. K. (1977). *Smoke, Dust and Haze*, John Wiley and Sons, New York.
3. Sehmel, G. A. (1980). Particle and gas dry deposition: a review, *Atmos. Environ.*, **14**, 983.

4. Businger, J. A., Wyngaard, J. C., Izumi, Y., and Bradley, E. F. (1971). Flux-profile relationships in the atmospheric surface layer, *J. Atmos. Sci.,* **28**, 181.
5. Wesely, M. L. and Hicks, B. B. (1977). Some factors that affect the deposition rate of sulfur dioxide and similar gases on vegetation, *J. APCA,* **27**, 1110.
6. Hicks, B. B., Baldocchi, D. D., Hosker, R. P. Jr., Hutchison, B. A., Matt, D. R., McMillen, R. T., and Satterfield, L. C. (1985). *On the use of monitored air concentrations to infer dry deposition*, NOAA Technical Memorandum ERL ARL-141, Silver Springs, MD.
7. Flagan, R. C. and Seinfeld, J. H. (1988). *Fundamentals of Air Pollution Engineering*, Prentice-Hall, Inc. Englewood Cliffs, NJ, 307.
8. Davidson, C. I., Miller, J. M., and Pleskow, M. A. (1982). The influence of surface structure on predicted particle dry deposition to natural grass canopies, *Water Air Soil Pollut.,* **18**, 25.
9. Bache, D. H. (1979). Particle transport within plant canopies -I. A framework for analysis, *Atmos. Environ.*, **13**, 1257.
10. Bache, D. H. (1979). Particulate transport within plant canopies - II. Prediction of deposition velocities, *Atmos. Environ.,* **13**, 1681.
11. Peters, K. and Eiden, R. (1992). Modeling the dry deposition velocity of aerosol particles to a spruce forest, *Atmos. Environ.,* **26A**, 2555.
12. Slinn, W. G. N. (1982). Predictions for particle deposition to vegetative canopies, *Atmos. Environ.,* **16**, 1785.
13. Wiman, B. L. B. and Agren, G. I. (1985). Aerosol depletion and deposition in forests - a model analysis, *Atmos. Environ.,* **19**, 335.
14. Wu, Y.-L., Davidson, C. I., and Russell, A. G. (1992). A stochastic model for particle deposition and bounceoff, *Aerosol Sci. Technol.,* **17**, 231.
15. Paw U, K. T. (1992). Rebound and reentrainment of large particles, *In* S. E. Schwartz and W. G. N. Slinn, eds. *Precipitation scavenging and atmosphere exchange processes*, Proceedings, 5th International Conference on Precipitation Scavenging and Atmosphere-Surface Exchange Processes, Richland, WA, July 15-19, 1153.
16. Dahneke, B. (1971). The capture of aerosol particles by surfaces, *J. Coll. Interface Sci.,* **37**, 342.
17. Dahneke, B. (1975). Further measurements of the bouncing of small latex spheres, *J. Coll. Interface Sci.,* **51**, 58.
18. Paw U, K. T. (1983). The rebound of particles from natural surfaces, *J. Coll. Interface Sci.,* **93**, 442.
19. Cleaver, J. W. and Yates, B. (1973). Mechanism of detachment of colloidal particles from a flat substrate in a turbulent flow, *J. Colloid Interface Sci.,* **44**, 464.
20. Kramm, G., Dlugi, R., Dollard, G. J., Foken, T., Molders, N., Muller, H., Seiler, W., and Sievering, H. (1995). On the dry deposition of ozone and reactive nitrogen species, *Atmos. Environ.,* **29**, 3209.

21. Shanley, J. B.(1989). Field measurements of dry deposition to spruce foliage and petri dishes in the Black Forest, F. R. G, *Atmos. Environ.*, **23**, 403.
22. Lindberg, S. E., Cape, J. N., Garten, C. T. Jr., and Ivens, W. (1992). Can sulfate fluxes in forest canopy throughfall be used to estimate atmospheric sulfur deposition?-A summary of recent results, *In* S. E. Schwartz and W. G. N. Slinn, eds. *Precipitation scavenging and atmosphere exchange processes*, Proceedings, 5th International Conference on Precipitation Scavenging and Atmosphere-Surface Exchange Processes, Richland, WA, July 15-19, 1367.
23. Erisman, J. W. (1993). Acid deposition onto nature areas in the Netherlands: part II. throughfall measurements compared to deposition estimates *Water Air Soil Pollut.*, **71**, 81.
24. Lovett, G. M., Likens, G. E., Nolan, S. S. (1992). Dry deposition of sulfur to the Hubbard Brook Experimental Forest: A preliminary comparison of methods, *In* S. E. Schwartz and W. G. N. Slinn, eds. *Precipitation scavenging and atmosphere exchange processes*, Proceedings, 5th International Conference on Precipitation Scavenging and Atmosphere-Surface Exchange Processes, Richland, WA, July 15-19,1391.
25. Beier, C., Gundersen, P., and Rasmussen, L. (1992). A new method for estimation of dry deposition of particles based on throughfall measurements in a forest edge, *Atmos. Environ.*, **26A**, 1553.
26. Miller, E. K., Panek, J. A., Friedland, A. J., Kadlecek, J., and Mohnen, V. A. (1993). Atmospheric deposition to a high-elevation forest at Whiteface Mountain, New York, USA, *Tellus*, **45B**, 209.
27. Butler, T. J. and Likens, G. E. (1995). A direct comparison of throughfall plus stemflow to estimates of dry and total deposition for sulfur and nitrogen, *Atmos. Environ.*, **29**, 1253.
28. Koester, C. J., and Hites, R. A. (1992). Wet and dry deposition of Chlorinated Dioxins and Furans, *Environ. Sci. Technol.*, **26**. 1375.
29. Bergin, M. H., Davidson, C. I., Dibb, J. E., Jaffrezo, J. L., Kuhns, H. D., and Pandis, S. N. (1995). A simple model to estimate atmospheric concentrations of irreversibly deposited aerosol chemical species based on snow core chemistry at Summit, Greenland, *Geophys. Res. Letters.*, **22**, 3517.
30. Nicholson, K. W. (1988). The dry deposition of small particles: a review of experimental measurements, *Atmos. Environ.*, **22**, 2653.
31. Sievering, H. (1987). Small-particle dry deposition under high wind speed conditions: eddy flux measurements at the Boulder Atmospheric Observatory, *Atmos. Environ.*, **21**, 2179.
32. Allen, A. G., Harrison, R. M., and Nicholson, K. W. (1991). Dry deposition of fine aerosol to a short grass surface, *Atmos. Environ.*, **25A**, 2671.
33. Main, H. H., and Friedlander, S. K. (1990). Dry deposition of atmospheric aerosols by dual tracer method-I. area source, *Atmos. Environ.*, **24A**, 103.

34. Davidson, C. I. and Friedlander, S. K. (1978). A filtration model for aerosol dry deposition: application to trace metal deposition from the atmosphere, *J. Geophys. Res.*, **83**, 2343.
35. Coe, J. M. and Lindberg, S. E. (1987). The morphology and size distribution of atmospheric particles deposited on foliage and inert surfaces, *J. APCA*, **37**, 237.
36. Holsen, T. M., Noll, K., Fang, G., Lee, W., Lin, J., and Keeler, G. J. (1993). Dry deposition and particle size distributions measured during the Lake Michigan urban air toxics study, *Environ. Sci. Technol.*, **27**, 1327.
37. Holsen, T. M. and Noll, K. E. (1992). Dry deposition of atmospheric particles: Application of current models to ambient data, *Environ Sci. Technol.* **26**, 1807.
38. Baeyens, W., Dehairs, F., and Dedeurwaerder, H. (1990). Wet and dry deposition fluxes above the North Sea, *Atmos. Environ.*, **24A**, 1693.
39. Dulac, F., Bergametti, G., Lasno, R., Remoudaki, E, Gomes, L., Ezat, U. and Buat-Menard, P. (1992). Dry deposition of mineral aerosol particles in the marine atmosphere: significance of the large size fraction, *In* S. E. Schwartz and W. G. N. Slinn, eds. *Precipitation scavenging and atmosphere exchange processes*, Proceedings, 5th International Conference on Precipitation Scavenging and Atmosphere-Surface Exchange Processes, Richland, WA, July 15-19, 841.
40. Zufall, M. J. and Davidson, C. I. (1995). Dry deposition of atmospheric contaminants to Lake Michigan, *In* R. D. Vidic and F. G. Pohland, eds., *Innovative Technologies for Site Remediation and Hazardous Waste Management.* Proceedings of the National Conference of the American Society of Civil Engineers. Pittsburgh, Pa, 628.
41. Noll, K. E. and Fang, K. Y.-P. (1986). A rotary impactor of size selective sampling of atmospheric coarse particles, Presented at the 79th annual meeting of the air pollution control association, Minneapolis, MN, June 22-27.
42. Noll, K. E. and Fang, K. Y.-P. (1989). Development of a dry deposition model for atmospheric coarse particles, *Atmos. Environ.*, **23**, 585.
43. Zufall, M. J. and Davidson, C. I. (1997) Dry deposition of particles. *In* R. M. Harrison and R. Van Grieken, eds. *Atmospheric Particles*. IUPAC. John Wiley and Sons. New York.

RUSSIAN REGULATORY DIFFUSION MODELS: STATUS, RESULTS OF VALIDATION AND INTERNATIONAL INTERCOMPARISONS

E.L. GENIKHOVICH
Main Geophysical Observatory
7 Karbysheva Street
194021 St. Petersburg, RUSSIA

1. Introduction

Since 1962 Russian regulatory diffusion models have been based on techniques developed by a team of specialists from the Main Geophysical Observatory (MGO) in cooperation with other scientific institutes of the former USSR. A description of these techniques was presented in publications [1,2] and in the last version of the national regulatory document [3] which will be referred here as OND-86 (an abbreviation which stays for "National Regulatory Document introduced in 1986"). Unlike the majority of western regulatory dispersion models, the MGO model [3] does not use Gaussian formulae and is based on analytical approximations of the numerical solution of the advection-diffusion equation (ADE). This solution, which in fact is the Green function of the ADE, is also used to describe concentration fields from multiple, line, and area sources as a superposition of concentration fields from single sources. Coefficients in ADE are parametrized using surface meteorological data (wind speed, wind direction and temperature gradient). The solution obtained was generalized to the cases of complex terrain and building environments using results of numerical, field, and laboratory experiments.

Concentration fields calculated with Russian dispersion models should be compared with ambient air quality standards called in Russia "maximum permissible concentrations" (MPCs). Two lists of MPCs were established (each one containing several hundred pollutants) which correspond to short-term (20-30 min) and long-term (up to one year) averaging times. Their ratios, however, are small in comparison with ratios of maximum short-term concentrations to long-term ones measured in field conditions. Thus, Russian short-term MPCs are more restrictive than long-term ones. That is why OND-86 is based on calculations of short-term worst-case concentration fields. In such a sense, OND-86 can be compared with SCREEN2, SCREEN3, or other American screening models [4] which also predict short-term worst-case concentrations and cannot be used for estimations of the long-term averages (those

I. Linkov and R. Wilson (eds.), Air Pollution in the Ural Mountains, 75–80.

should be estimated using different models).

At present, about ten packages of modeling software for IBM and IBM-compatible personal computers are on the market in Russia, which were developed by different institutions and endorsed by MGO. All of them are based on the same OND-86 but have different coding. These models are widely used in Russia and other republics of the former USSR. The main areas of applications are: regulatory practice; designing new and modifying existing industrial sources; calculating background air pollution in cities; and other topics related to decision-making in the reduction of non-radioactive air pollution. Input information for the models includes the climatological characteristics of the critical dispersion conditions and parameters of the emission sources.

2. Validation and Intercomparisons

The MGO regulatory dispersion model was participating in several intercomparisons of the models (see for example [5]). It was found that ground-level concentrations generated by different atmospheric dispersion models, even regulatory ones, are characterized by significant scatter. As an example, maximum (more exactly, 98th percentile) ground-level concentrations (MGLC) calculated with nine different European regulatory dispersion models are shown on Fig. 1; models developed in the Netherlands, Austria, Denmark, Greece, Belgium, Germany, Russia, Hungary, and

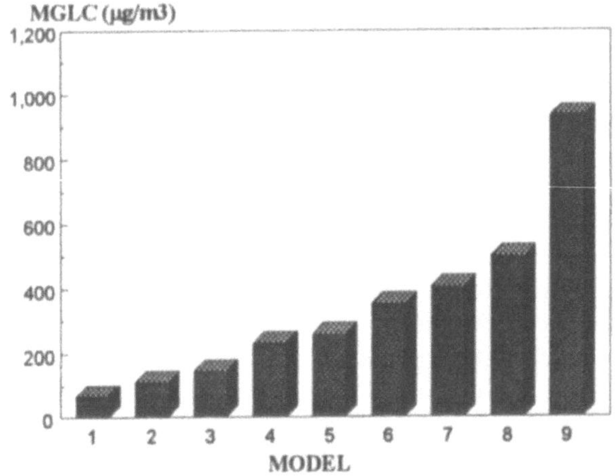

FIGURE 1. Maximum ground-level concentrations calculated with nine European regulatory dispersion models.

England are marked with the numbers from 1 to 9, respectively (number 7 corresponds to OND-86). These data published by Cosemans at al. [6] were calculated for two closely located stacks of the 600 MW electric power plant using the same input information. All the differences in MGLCs on this figure are due to the errors generated by the models. Actually, it means that only careful testing of the models against experimental data can prove that they are applicable.

Numerous data of field experiments carried out in different regions of the

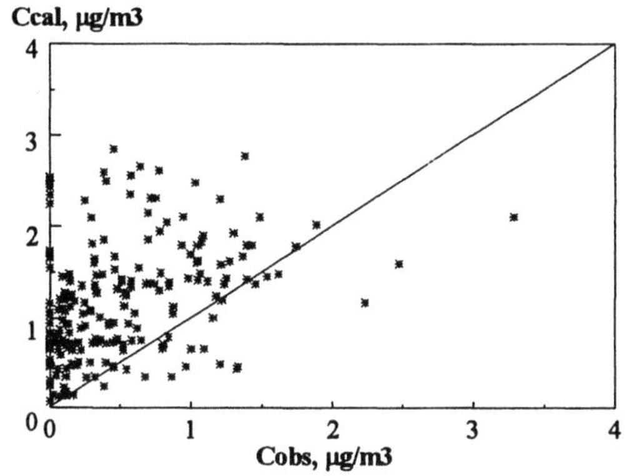

FIGURE 2. Calculated (OND-86) versus observed (Kincaid) concentrations.

former USSR were used to verify the OND-86 model (see for example [1, 2, 7]). They confirmed that the model predicted the upper 98th to 99th percentile of the PDF of concentrations with an error of about 25%. Comparisons with results of the laboratory and field tests carried out in the USA, EU countries, and Japan confirmed this

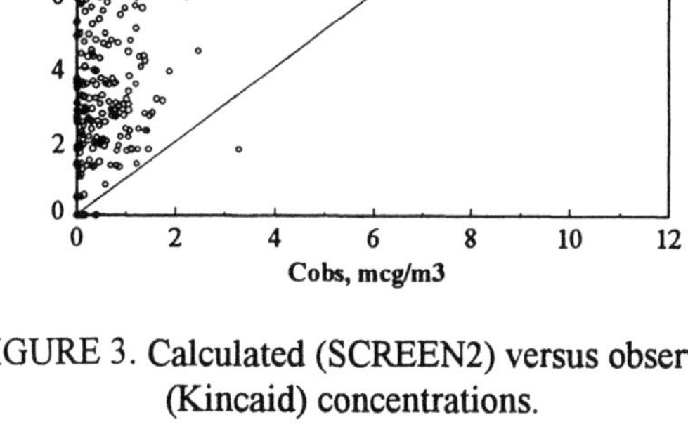

FIGURE 3. Calculated (SCREEN2) versus observed (Kincaid) concentrations.

conclusion too. An example of validation ([2]) is given on Figures 2 and 3. A correlation graph of concentrations, observed in the Kincaid experiments (C_{obs}) and calculated (C_{cal}) with OND-86, is shown on Fig. 2; a similar correlation graph for the SCREEN2 model is given on Fig. 3. Straight lines (diagonals) on these figures correspond to 1 to 1 agreement, and, if $C_{cal} > C_{obs}$ ($C_{cal} < C_{obs}$), the points are located above (below) these lines. One can see from these figures that points predicted with OND-86 lay mainly above the diagonal, but more-or-less close to this, and a small fraction of all points is located below the diagonal. The predicted with this model MGLC (2.86 $\mu g/m^3$) is close to the measured first (3.29 $\mu g/m^3$) and second (2.48 $\mu g/m^3$) maximum values. As for the SCREEN2, its predictions seem to be too conservative (in particular, their maximum value overestimates the measured one by 3.6 times). Similar conclusions were drawn in [2] using data of the Copenhagen experiment.

3. A New Long-term Dispersion Model

A new long-term dispersion model (MGOLT) developed at MGO for regulatory purposes is based on calculation of the long-term averaged concentration $<C(\underline{r})>$ from the point source at the receptor point $\underline{r}$ using the following expression:

$$< C(\underline{r}) > \; = \int c(\underline{r}, \omega) p(\omega) d\omega ,$$

where $c(\underline{r}, \omega)$ is short-term concentration at this receptor point corresponding to meteorological conditions ω (it is parametrized using the aforementioned analytical approximation of the numerical solution of the ADE), and $p(\omega)$ is the probability density. Here, ω is a multi-vector of the governing (first of all, describing the advection and diffusion processes) parameters, and integration in (1) should be carried over its phase space. Formulae for concentrations from line, area, and multiple-point sources can be obtained from (1) using the superposition principle.

As a part of the MGOLT model, a meteorological preprocessor has been developed which uses data of standard meteorological observation to determine the governing parameters. Corresponding formulae were introduced in [8] (see also [9]). When calculating climatological averages, "climatological parametrizations" for $p(\omega)$ should be used in Eq. (1). Such parametrizations were discussed in [10] and [11].

The MGOLT model was validated using emission, meteorological, and monitoring data for the city of Raahe, Finland, collected by the specialists from the Finnish Meteorological Institute. Results of this validation are presented in Table 1.

TABLE 1. Observed (C_{obs}) and calculated (C_{cal}) mean annual SO_2 concentrations in Raahe, Finland

Monitoring station	Coordinates X, m	Coordinates Y, m	C_{obs} ($\mu g/m^3$)	C_{cal} ($\mu g/m^3$)
1	1860	2860	11.6	10.2
2	4140	5540	13	13.1
3	2700	5000	22	14.3

This model was also tested upon experimental data (PM_{10} concentrations), collected in Volgograd, Russia, in the framework of the RAMP project during the Sample Saturation Study, and showed a reasonably good performance. Results of this validation will be published separately.

4. Conclusion

From the user's point of view, a dispersion model can be considered as a black box which is characterized by its input and output. Every national regulatory dispersion is tuned to the existing structure of the input information which could be obtained from the national meteorological network. At the moment the Russian National Weather Service cannot provide users with the input information required for running the western Gaussian-type models. That is why at present there are only rather limited opportunities for application of Gaussian dispersion models in Russia. As for the MGO models, their development always was oriented toward the use of the input data available in Russia. OND-86 and its predecessors were in use in Russia for more than thirty years and performed reasonably well. OND-86, however, provides only short-term worst-case concentration fields and cannot be used for estimation of long-term (*e.g.*, annual) concentrations. This drawback is now eliminated with the newly developed MGOLT model which is intended, in particular, for applications in risk assessment.

5. Acknowledgments

Development of the MGOLT model was supported in part by the US Agency for International Development via the Institute for Sustainable Communities and carried out in close contacts with Drs. T. Pace, S. Perry, and W. Beal from EPA and Mr. F. Renner from SAIC International. Special thanks to the collaborators of this work. They include: Prof. M. Berlyand, Drs. I. Gracheva, L. Khurshudyan, R. Onikul, A. Ziv, S. Chicherin, V. Eliseyev, and other colleagues from MGO.

6. References

1 Berlyand, M.E. (1982) *Moderne Problemen der atmosph‰rischen Diffusion und der Verschmutzung der Atmosph‰re*, Academie-Verlag, Berlin

2 Genikhovich, E.L. (1995) Application of regulatory diffusion models in Russia, *International Journal of Environment and Pollution*, **5**, 530-537

3 Berlyand, M.E., Gasilina, N.K., Genikhovich, E.L., Onikul, R.I., and Glukharev, V.A. (1987) *Method for Calculation of Concentrations of Air Pollutants that Industrial Emissions Contain*, USSR Regulatory Document OND-86, Gidrometeoizdat, Moscow (in Russian)

4 Brode, R.W. (1988) Screening procedures for estimating the air quality impact of stationary sources - draft for public comments, EPA-450/4-88-010, US Environmental Protection Agency, Research Triangle Park

5 Hanna, S.R. (1982) Review of atmospheric diffusion models for regulatory applications. WMO Technical Note No 177, WMO, Geneva

6 Cosemans, G., Kretzschmar, J., Janssen, L., and Maes, G. (1994) The Third Workshop's environmental impact assessment model intercomparison exercise, in J. Kretzschmar, G. Maes, and G. Cosemans (eds.), *Operational Short-Range Atmospheric Dispersion Models for Environmental Impact Assessment in Europe*, VITO, Mol, pp.405 - 418

7 Genikhovich, E.L. and Schiermeier, F.A. (1995) Comparison of United States and Russian complex terrain diffusion models developed for regulatory applications, *Atmospheric Environment*, **29**, 2375 - 2385

8 Genikhovich, E.L. and Osipova, G.I. (1984) On determination of turbulent diffusivity using standard meteorological observation data, *Trudy GGO (Transactions of the Main Geophysical Observatory)*, **479**, 62 - 69 (in Russian)

9 Groisman, P.Ya. and Genikhovich, E.L. (1997) Assessing surface-atmosphere interactions using former Soviet Union standard meteorological network data. Part 1: Method. *Journal of Climate* (in press)

10 Berlyand, M.E., Genikhovich, E.L., Kanchan, Ya. S., Onikul, R.I., and Chicherin, S.A. (1979) On estimation of mean annual pollutant concentrations from industrial sources, *Trudy GGO (Transactions of the Main Geophysical Observatory)*, **417**, 3 - 18 (in Russian)

11 Berlyand, M.E., Genikhovich, E.L., and Chicherin, S.S. (1984) Theoretical basis and calculation technique for estimation of the field of annual concentration of pollutant from industrial sources. *Trudy GGO (Transactions of the Main Geophysical Observatory)*, **479** , 3 - 17 (in Russian)

NUMERICAL SIMULATION MODELS FOR AIRBORNE HEAVY METALS IN EUROPE: A REVIEW

G. PETERSEN
GKSS-Research Centre, Institute of Hydrophysics
Max-Planck-Strasse, D-21502 Geesthacht, GERMANY

Abstract

In view of the decision of the United Nations-Economic Commission Europe (UN-ECE) to prepare a protocol for heavy metals and persistent organic pollutants under its Convention on Long-range Transboundary Air Pollution an attempt is made to summarize European activities on modeling the atmospheric long-range transport of heavy metals. Special emphasis is placed on mercury, lead and cadmium which have been defined by the UN-ECE to be the priority heavy metals of concern. In this paper, numerical models for atmospheric transport, transformations and deposition of heavy metals are discussed chronologically with respect to their development and first applications. Basic features of the models together with input data and main results are summarized in a table. Limitations of present models and the improvements that are needed to address effectively the issues presently facing the atmospheric scientific and policy-making communities are briefly outlined.

1. Introduction

Over the past two decades, significant efforts have been devoted in Europe to the monitoring of heavy metals as a result of the general awareness of the potential impacts of atmospheric long-range transport on sensitive ecosystems such as European marginal seas [17], [1], [15], [71], [24], [65], [67], [19], inland lakes and forest areas [66], [75], [39] and the Norwegian Arctic [45], [43], [44]. Relatively little attention has been given to numerical modeling of the long-range transport and deposition of heavy metals over Europe so far. However, in recent years it has been widely recognized, that the use of numerical models is important in formulating effective control strategies for the reduction of atmospheric deposition fluxes of heavy metals to the above mentioned ecosystems, because the only way of delineating their atmospheric transport pathways and hence the emitter-receiver relationship is through numerical modeling [51], [26], [27], [38].

In a broad sense, two basic types of models have been developed to date, wherein the turbulent transport of heavy metals is analyzed through either a Lagrangian or an Eulerian approach. Lagrangian models developed for heavy metals and currently

I. Linkov and R. Wilson (eds.), Air Pollution in the Ural Mountains, 81–97.

TABLE 1. A Comparative Summary of Models Applied to Heavy Metals in Europe

model							reference
name	type	application					
		institution	input		output		
			meteorological input	calculated substances	receptor area	typical time scale	
EMEP (European Monitoring and Evaluation Program)	Lagrangian 1-layer trajectory model	NILU Norsk Institutt for LUftforskning Kjeller, Norway	6-hourly from the Norwegian Numerical Weather Prediction Model	Pb,Cd,As V,Mn,Sb,Se	Birkenes, Southern Norway	months	*Pacyna et al.,1984*
TREND	statistical approach of a Gaussian plume model and a trajectory model	RIVM National Institute of Public Health and Environmental Protection Bilthoven, the Netherlands	long-term averages from the Dutch Meteorological Office and from the European Centre for Medium-Range Weather Forecast (ECMWF).	As,Sb,Cd,Cr, Cu,Ni,Zn	North Sea	years	*van Jaarsveld et al.,1986*
OPS (Operational Priority Substances)		TNO Netherlands Organization for Applied Research Delft, the Netherlands		Cd,Cr,Cu,Pb Hg,Ni,Zn	North Sea, Dutch Wadden Sea		*Warmenhoven et al.,1989*
					River Rhine drainage basin		*Baart and Diederen,1991*
EUTREND				Pb,Cd	Europe North Sea		*Baart et al.,1995*
HHLRT (Hamburg Long-Range Transport Model)	three-dimensional stochastic model	Meteorological Institute,Unversity of Hamburg Germany	6-hourly analyses from the ECMWF	Pb,Cd	North Sea	months	*Krell and Roeckner,1988*

TABLE 1 (continued)

model							reference
name	type	application					
		institution	input		output		
			meteorological input	calculated substances	receptor area	typical time scale	
HMET (Heavy Metals Eulerian Transport)	1-layer (plus 1 residence layer) Eulerian model	EMEP Meteorological Synthesizing Centre West Oslo, Norway	6-hourly from the Norwegian Numerical Weather Prediction Model	Pb,Cd,Zn,As	Europe	months years	*Bartnicki et al.*, *1993* *Bartnicki* , *1994* *Bartnicki* , *1996*
ASIMD (asymmetric advection scheme)	4-layer Eulerian model	EMEP Meteorological Synthesizing Centre East Moscow, Russia	6-hourly from the Russian Hydro-meteorological Centre	Pb,Cd	Europe		*Pekar* , *1996*
LPMOD (Large Particle MODel)	4-layer Eulerian model			Pb,Cd	Europe		*Pekar* , *1996*
MSC-E (Meteorological Synthesizing Centre East)	1-layer Eulerian model with a vertical Gaussian pollutant profile			Pb,Cd,Zn,As Hg	Europe Northern Hemisphere		*Galperin et al.*, *1995* *Galperin et al.*, *1996*
CAM (Chemistry of Atmospheric Mercury)	1-layer chemical process (box) model	Swedish Environmental Research Institute (IVL) Göteborg, Sweden		Hg		hours	*Pleijel and Munthe*, *1995a* *Pleijel and Munthe*, *1995b*
ADOM (Acid Deposition and Oxidant Model)	12-layer Eulerian model	GKSS Research Centre Geesthacht Germany	hourly from the High Resolution Limited Area Weather Prediction Model (HIRLAM)	Hg	Europe	days	*Petersen et al.*, *1996a*

TABLE 1 (continued)

model							reference
name	type	application					
		institution	input		output		
			meteorological input	calculated substances	receptor area	typical time scale	
GKSS-EMEP	Lagrangian 1-layer trajectory model	GKSS Research Centre Geesthacht Germany	6-hourly from the Norwegian Numerical Weather Prediction Model	Pb	North Sea Baltic Sea	months	*Grassl et al., 1989* *Petersen et al., 1989* *Krüger, 1996*
				Pb,Cd,Zn,As			*Petersen and Krüger, 1993*
				Hg			*Petersen et al., 1995* *Petersen, 1992a*
GESIMA (Geesthacht Simulation Model of the Atmosphere)	high-resolution non-hydrostatic mesoscale model		numerical weather prediction models e.g. 'Deutschland' model of the German Weather Service	Pb	North Sea coastal areas in Germany	days	*Eppel et al., 1992* *Kapitza and Eppel, 1995*
TRACE (TRace Metal Atmospheric Concentrations in Europe)	climatological model	IIASA International Institute for Applied System Analysis Laxenburg Austria	climatological input	Pb,Cd,Zn,As	Europe	years	*Alcamo et al., 1991* *Alcamo et al., 1992*

in use are variants of the so-called trajectory models. These models are usually formulated under assumptions of simplified turbulent diffusion, no convergent or divergent flows, and no wind shear. In these approaches only first-order chemical reactions can be treated rigorously. However, the Lagrangian approach avoids many of the computational complexities associated with the simultaneous solution of many differential equations; this generally results in requiring significantly less computational resources and can facilitate an understanding of problems that do not require descriptions of interactive nonlinear processes. Further progress in understanding the atmospheric cycling of heavy metals has emphasized the need for direct modeling of complex nonlinear processes (e. g. the physico-chemical transformations of atmospheric mercury species) by comprehensive Eulerian models. These approaches employ extensive gas- and aqueous-phase chemical mechanisms and explicitly track numerous species concentrations. Also, a more detailed numerical formulation for physical and chemical processes occurring within and below precipitating clouds is included. Typically, these models contain modules designed to calculate explicitly the chemical interactions that move gas-phase species into and among the various aqueous phases within clouds as well as calculate the aqueous-phase chemical transformations that occur within cloud and precipitation droplets. The scope and the general structure of such a comprehensive Eulerian modeling framework has been extensively discussed by Peters *et al.* [54].

Despite intense interest in using both types of models for heavy metals only a few number of operational models, which are validated through comparison with experimental data, are currently available in Europe. These models have been designed to simulate long-range and long-term transport of heavy metals assumed to be chemically inert and in association with particulate matter using a combined trajectory-climatologic approach [4], statistical approaches of a Gaussian plume model and a trajectory model [30] and an EMEP type Lagrangian trajectory model [55], [59], [36]. More recently, three-dimensional models using an Eulerian reference frame have been introduced simulating various inert species on regional geographic scales in Europe [11], [22], [53] and for high-resolution nonhydrostatic mesoscale applications [21]. Moreover, the development and application of dispersion models for heavy metals has progressed from formulations for inert species to models accounting for the physico-chemical transformations of atmospheric mercury species [56], [57], [58]. An extensive treatment of the chemistry scheme employed with the model calculations can be found in Petersen *et al.* [60]. Similar schemes are used in several other models previously developed at the EMEP Meteorological Synthesizing Center East [23], the U. S. Environmental Protection Agency [16] and at the U. S. Argon National Laboratory [68].

The major sources of information for the subsequent review on results from the above mentioned types of models are those listed in the reference section. A number of sources cited in these publications were also used in preparing this review. In addition, information was obtained from internal reports and personal communication with scientists from European and North American research institutes and environmental protection agencies.

2. Results From Models Applied in Europe

The first model for long-range transport of heavy metals [45] similar to that developed in the OECD program on long-range transport of sulphur compounds in Europe [42] was presented soon after the first European emission survey was completed [46]. The model was used to verify emission data by comparison of measured and calculated concentrations of heavy metals in ambient air. The results show that measured concentrations at a location in southern Norway can be related to calculated anthropogenic emissions for a number of elements, e. g. vanadium, lead, cadmium, manganese and chromium. It was concluded from this work that future emphasis should be placed on the deposition of heavy metals in addition to estimating air concentrations to create the opportunity to assess the contribution of emissions from various source regions to a certain receptor site in a more complete way.

A first overview of results from atmospheric transport and deposition models for heavy metals in Europe was given in the framework of a multidisciplinary study on the status of the North Sea [3]. There, attempts to calculate atmospheric inputs to the North Sea by means of transport models developed in the Netherlands are summarized [2], [30]. With these models, whose latest versions are called TREND [31] and OPS [32], annually or seasonally averaged concentrations and depositions of heavy metals over the North sea are calculated from emission data bases prepared for cadmium, zinc, lead, chromium, copper and nickel for the years 1979 and1980 [46].

TREND and its more operational versions OPS and EUTREND use statistical approaches of a Gaussian plume model for source distances within 20km and of a trajectory model for greater distances. Meteorological input data are provided by the Dutch national air quality monitoring network with additional data from the Dutch meteorological office and the European Centre for Medium-Range Weather Forecast (ECMWF). Dry deposition of particle associated metals is taken into account by deriving dry deposition velocities from different particle size and atmospheric stability classes. Wet deposition includes washout processes parameterized by a scavenging coefficient and a raindrop size distribution function and rainout is modeled on the basis of empirical scavenging ratios.

In more recent research projects commissioned by the Dutch Ministry of Housing, Physical Planning and Environment different TREND versions were used to estimate atmospheric input to the North Sea and the Dutch Wadden Sea focussing on the contributions of the surrounding countries to the total input [74]. In this study, recent updates to the above mentioned emission inventories, including mercury, are used [49], [50]. The model results concerning the contributions of the surrounding countries are considered to be realistic, because the model has been successfully validated through comparison with data from the Dutch National Air Quality Monitoring Network and because the emission data from the surrounding countries are relatively reliable. The EUTREND version covers the entire European continent including its marginal seas and calculates deposition as a function of surface characteristics. This version has been applied within the European Soil and Sea QUality and Atmospheric Deposition of Selected Substances (ESQUAD) project to assess the deposition fluxes of heavy metals to the European continent and the North Sea [28], [9].

In the framework of research contracts with the German Umweltbundesamt (UBA), the three-dimensional stochastic model HHLRT (Hamburg Long Range Transport Model) has been developed and applied at the University of Hamburg for episode studies of atmospheric sulphur transport in Europe [37] and subsequently modified for heavy metal transport and deposition [34]. The meteorological input data for the model, such as windfields and precipitation rates, are prescribed according to 6-hourly analyses of the ECMWF. Mixed-layer height analyses are from the Norwegian Weather Service. Emission data are adopted from the first European emission survey on heavy metals [46]. Results for two species, i. e. lead and cadmium, have been presented for the months of July 1984 and January 1985. From these explicit simulations, annual input of lead and cadmium into the North Sea has been extrapolated assuming no significant month-to-month variability of deposition. The annual values are considerably smaller than estimates based on extrapolated measurements at coastal sites but they are comparable to the results from the above mentioned Dutch models.

In 1985 the long-period model for sulphur [20] used under the European Monitoring and Evaluation Programme (EMEP) has been modified at the GKSS-Research Centre Geesthacht for transport and deposition of heavy metals. The model is a one-layer trajectory model. Columns of air in the atmospheric boundary layer are followed along specified 96-hours trajectories picking up emissions of heavy metals from the underlying grid. The mass balance equations are integrated along each trajectory, taking into account emission inputs, possible chemical reactions, dry and wet removal, and the influence of relevant meteorological parameters. The 6-hourly meteorological data are taken from the output of the Norwegian Numerical Weather Prediction (NWP) model.

In the first phase of a comprehensive research project funded by the German Umweltbundesamt, the model was used to calculate atmospheric lead input into the North Sea and the Baltic Sea [25], [55]. Anthropogenic lead emissions in Europe used with the model were based on emission surveys available by 1986 and published by 1988 [49], [50]. Model results in terms of weekly averages of lead concentrations compare well with observations at various sites at the North Sea coast. The calculated total annual atmospheric lead input into the North Sea is in agreement with previous model results [2], [30]. The conclusion drawn in earlier studies [2], [34], that extrapolations from measurements at coastal sites tend to overpredict the total input, is supported by the model results obtained at GKSS. These findings initiated the development of more advanced extrapolation methods [67] whose results in terms of total lead input the Baltic Sea compare well with model calculations [27].

More recent updates to emission databases for lead, cadmium, zinc and arsenic [7] are used with the GKSS model in the second phase of the project [59]. Model validation has been extended to the Baltic Sea area since data from the HELCOM monitoring network have been available. The results show that the model again performs well in predicting annual lead input into the North Sea and the Baltic Sea. Emission reduction scenario runs concerning introduction of best available flue gas cleaning technology for the entire European non-ferrous metal industry and use of lead free gasoline in all European countries have been performed resulting in 90% lead input reduction for both sea areas approximately, if both scenarios are applied [36].

In comparison with observations, calculated concentrations of cadmium and zinc in ambient air and precipitation and hence deposition fluxes are significantly underpredicted by the GKSS-model. This is in accordance with results from other modelers, who have used the same emission data [2], [30], [34], [74] indicating that the European emission inventories for these two substances are still incomplete. The model performance for arsenic is about the same as for lead but measurement data for arsenic are too scarce to draw any firm conclusions. As a follow-up activity, the GKSS-model for the long-range transport of particle-associated and inert heavy metals has been extended to mercury and mercury species [56], [60]. The chemical equation set used in this model is a simplified description of a complex series of chemical reactions and mass transfer processes of atmospheric mercury species. An update to earlier emission inventories including three most relevant species for atmospheric long-range transport, i. e. elemental gaseous mercury, particulate mercury and divalent inorganic mercury compounds [6] is used with the model. Model runs are focused on estimates of atmospheric mercury input into the North Sea [57] and the Baltic Sea [58] and on scenario runs concerning inorganic mercury compounds and the role of soot particles for the mercury load in Europe's atmosphere. Since the model has been successfully validated through comparison with observations from the Nordic Network [29], the calculated range of 5-12 tones for the North Sea and 6-13 tones for the Baltic are considered to be the most reliable data currently available.

Five years ago a study was conducted at the University of Hamburg investigating the ability of the three models TREND, HHLRT and GKSS-EMEP to simulate atmospheric long-range transport of lead to the North Sea [35]. The models have been applied for a defined time period using identical emission databases. Modeling results are compared against each other and with measured data from a network of seven monitors around the North Sea. In conclusion it was found that the models perform well in predicting total annual input but that mainly due to errors associated with precipitation fields over sea areas substantial differences can occur as far as spatial and temporal variability of predicted monthly wet deposition fluxes over the North Sea is concerned.

The International Institute for Applied System Analysis (IIASA) at Laxenburg, Austria, has developed a set of two numerical models to quantify the long-range transport of particle-associated and inert heavy metals in Europe [4], [5]. The metals selected are those which have been recommended by PARCOM and HELCOM for monitoring, i. e. lead, cadmium, zinc and arsenic. A modified version of the EMEP-model is used for generating time- and space-average inputs for a climatologic model called TRACE (TRace Metal Atmospheric Concentrations in Europe). The advantage of this combined approach is that the climatologic model can be conveniently run for all Europe and several years, and can thus make use of the very scattered measurements available for metals in Europe. Most TRACE model calculations for lead and arsenic agree within a factor of two of observed annual air concentrations and deposition. Zinc and cadmium computations clearly underestimate observations.

In recent years an Eulerian high-resolution non-hydrostatic mesoscale model called GESIMA (GEesthacht SImulation Model of the Atmosphere) has been developed at the GKSS Research Centre Geesthacht within an Umweltbundesamt research contract. The model is applicable for a variety of problems comprising large-eddy simulations, short-term weather prediction, or environmental problem studies such as pollutant transport. As a first application for atmospheric heavy metal transport the model has been used to simulate two episodes of atmospheric lead transport and deposition over the German North Sea coast [21].

One of the regional-scale Eulerian models currently in use for simulating heavy metal transport over Europe is the HMET (Heavy Metals Eulerian Transport) model, which has been developed to study the long-range and long-term transport, deposition and overall budget of As, Cd, Pb and Zn in Europe [10], [11], [12]. The main purpose of developing the model was the computation of atmospheric transboundary fluxes of heavy metals. The emission data used with the study are the above-mentioned recent updates to databases for arsenic, cadmium, lead and zinc in the EMEP-grid [7]. The model can be used both for long-term (one year or longer) as well as shorter (one week) episodes. Up to now, model applications are restricted to long-term averages, i. e. calculation of annual concentrations and deposition fluxes for 1985. Due to the small number of observations available for that year it was difficult to validate the model. The results indicate that model calculations for Cd and Pb are in good agreement with observations, but only satisfactory for As. Concentrations and deposition fluxes of Zn calculated by HMET are significantly lower than the observed values, but are in agreement with results from other models [5], [59].

The EMEP Meteorological Synthesizing Centre East (MSC-E) has developed an Eulerian model called MSC-E capable to calculate concentrations and deposition fluxes of heavy metals on both regional and hemispheric geographical scales [22]. Emission databases for lead, cadmium, zinc and arsenic [7] and mercury [6] are used with the models. In the mercury version, natural emissions and air-sea exchange processes of both organic and inorganic mercury species are treated but the chemical equation set used in this model is in the early stages of evaluation at the moment [23]. Recently, the three-dimensional Eulerian Large Particle MODel (LPMOD) and ASIMD have been developed and applied at the EMEP MSC-E to calculate Pb and Cd transport and deposition over Europe [53] using emission data bases for 1990 from the ESQUAD project [13]. These two model approaches employ different advanced numerical advection schemes but incorporate the same detailed description of the atmospheric boundary layer with respect to dry and wet deposition of particle-associated heavy metals. Observed Pb concentrations from the PARCOM air and precipitation monitoring network are quite well reproduced by both models. For Cd, the models overestimate concentrations in air and underpredict concentrations in precipitation. This is most probably due to both incomplete European emission inventories for Cd and insufficient precipitation fields at coastal sites obtained from numerical weather prediction models.

Recent progress in understanding the atmospheric mercury cycle has emphasized the need for direct modeling of the complex nonlinear mercury chemistry of the atmosphere by comprehensive Eulerian models. As a first step in this direction the cloud mixing, scavenging, chemistry and wet deposition components of the comprehensive three-dimensional Eulerian Acid Deposition and Oxidants Model (ADOM), originally designed for regional-scale acid precipitation and photochemical oxidants studies in North America and Europe [73], [70] are restructured to accommodate the most recent developments in atmospheric mercury chemistry [14],

[61]. The chemistry scheme of this ADOM mercury version was developed by systematic simplification of the detailed Chemistry of Atmospheric Mercury (CAM) process model, which was tested and evaluated in a sensitivity analysis aimed at determining the most important parameters and identifying the critical processes in the atmospheric cycling of mercury [63], [64].

A stand-alone version of the ADOM mercury model referred to as the Tropospheric Chemistry Module (TCM) has been developed and tested under different environmental conditions allowing changes to the chemistry and meteorology alone to be evaluated at different vertical TCM levels [61], [62]. Overall, the present development level of the TCM is such that its implementation in full ADOM mercury versions for Europe and North America is justified. These models with the TCM as their major component will be developed further according to advancements in the knowledge of atmospheric mercury processes to ensure that they maintain their capability to address effectively the scientific and policy questions that may arise over the next decade.

3. Conclusions

This review has produced strong evidence that long-range transport models with relatively simple formulations, extensively used in simulating European transboundary transport of acidifying pollutants and modified for heavy metals associated with particulate matter, are now in an operational stage. Results from a model intercomparison study for lead demonstrate their capability to predict annual and monthly means of concentrations in ambient air and in precipitation as well as dry and wet deposition fluxes within a factor of two of observations over spatial scales from about 100 km to continental [69]. Recent progress in understanding physico-chemical processes of atmospheric mercury and the availability of European emission data bases for different mercury species has permitted to investigate mercury transport and deposition in Europe's atmosphere by means of EMEP-type models and by comprehensive Eulerian modeling frameworks. Currently, efforts are made to implement atmospheric transport of heavy metals in complex mesoscale models, but pollutant transport in these models is in the early stages of evaluation at the moment. Although there have been significant advancements in the knowledge of heavy metal properties and in the understanding of the processes which determine control and fate of heavy metals in Europe's atmosphere, our knowledge is far from complete. Listed below are the main areas of uncertainties and required activities toward further development of transport models for heavy metals:

1) In general, the accuracy of calculated deposition fluxes could be improved simply by improving the model inputs such as emissions and meteorological data. There is evidence from model testing that the European emission inventories are still incomplete for some substances and that precipitation fields obtained from numerical weather prediction models are insufficient to calculate wet deposition patterns over sea areas in some cases.

2) Because there is no coordinated atmospheric monitoring program for heavy metals in Europe it was difficult for all model applications to assemble

measurements suitable for comparing to computed values. Hence, a reliable identification of the main sources of model uncertainties and a quantitative analysis of uncertainties in model results cannot be performed at present. Air and precipitation measurements of heavy metals are few and highly uncertain in some cases. To verify model predictions, high quality monitoring networks especially over sea areas and at coastal zones are required.

3) To improve model-derived estimates of heavy metal deposition to receptor areas such as the North Sea and the Baltic Sea it is essential that suitable emission inventories for countries surrounding these two sea areas are compiled. At the moment the quality of the emission data for some substances such as zinc and cadmium limits the confidence with which the results of different control options can be predicted.

4) Sensitivity analysis performed for two long-range and long-term models showed that deposition parameters and assumptions are no major sources of model uncertainties, as these models are relatively robust with respect to large variabilities in dry and wet deposition parameters [25], [5].

5) Model calculations underestimate Cd observations, but are correlated and within a factor of two of observations. Zn calculations greatly underestimate observations [11], [72], although calculations are correlated to measurements [5], [59], thus indicating that current Cd and Zn emissions in Europe are underestimated.

6) In case of mercury sufficient information on speciation and mechanisms and kinetics of chemicals reactions exists to model the main atmospheric pathways of mercury. It is well established now that redox reactions of elemental mercury [40], [41] and adsorption of oxidized mercury on soot particles in the aqueous phase [60] and subsequent wet deposition together with dry and wet deposition of particulate phase mercury emitted from industrial sources are the most important processes in polluted air masses over Europe. Values predicted by a model which contains simplified representations of these processes agree with observations from the Scandinavian network within a factor of two [60]. Additional laboratory and field measurements close to the main European industrial sources would be necessary to get a more complete picture of existing atmospheric mercury species, e. g. organic and inorganic gaseous mercury compounds. These species are hard to identify with currently available analytical techniques but they could have a significant influence on the deposition pattern close to sources.

7) Mesoscale modeling capabilities are necessary to quantify the subgrid scale variability of heavy metal concentrations simulated by regional scale models. Mesoscale models should be interfaced with large scale Eulerian models to obtain a 'nested' modeling system capable to simulate long-range transport induced, local effects with a spatial resolution of 20 km or less.

4. References

1. van Aalst, R. M., van Ardenne, R. A. M., de Kruk, F. J., and Lems, T. (1983a) Pollution of the NorthSea from the Atmosphere. TNO Report CL 182/152, Delft, the Netherlands.

2. van Aalst, R. M., Duyzer, J. H., and Veldt C. (1983b) Atmospheric deposition of lead and cadmium in the southern part of the North Sea. I. Emissions and preliminary model calculations. TNO Report R 83/222, Delft, the Netherlands.
3. van Aalst, R. M. (1988) Input from the Atmosphere, in W. Salomons, B. L. Bayne, E. K. Duursma and U. Förstner (eds.), *Pollution of the North Sea - AnAssessment*, Springer Verlag Berlin, pp. 275-283.
4. Alcamo, J., Bartnicki, J., and Olendrzynski, K. (1991) Modeling Heavy Metals in Europe's Atmosphere:A Combined Trajectory - Climatologic Approach, in H. van Dop and D. G. Steyn (eds.), *Air Pollution Modeling and its Application VIII*, Plenum Press, New York, pp. 389-398. .
5. Alcamo, J., Bartnicki, J., Olendrzynski, K. and Pacyna, J. (1992) Computing heavy metals in Europe's atmosphere - I. Model development and testing, *Atmospheric Environment* 26A, 3355-3370.
6. Axenfeld, F., Münch, J., and Pacyna, J. M. (1991) Europäische Test-Emissionsdatenbasis von Quecksilberverbindungen für Modellrechnungen. In: G. Petersen (1992). Belastung von Nord- und Ostsee durch ökologisch gefährliche Stoffe am Beispiel atmosphärischer Quecksilberverbindungen. Abschlußbericht des Forschungsvorhabens 104 02 726 des Umweltforschungsplanes des Bundesministers für Umwelt, Naturschutz und Reaktorsicherheit. Im Auftrag des Umweltbundesamtes, Berlin.
7. Axenfeld, F., Münch, J., Pacyna, J. M., Duiser, J. A. and Veldt, C. (1992) Test-Emissionsdatenbasis der Spurenelemente As, Cd, Hg, Pb, Zn und der speziellen organischen Verbindungen Lindan, HCB, PCB und PAK für Modellrechnungen in Europa. Abschlußbericht des Forschungsvorhabens 104 02 588 des Umweltforschungsplanes des Bundesministers für Umwelt, Naturschutz und Reaktorsicherheit. Im Auftrag des Umweltbundesamtes, Berlin.
8. Baart, A. C. and Diederen, H. S. M. A. (1991) Calculation of the Atmospheric Deposition of 29 Contaminants to the Rhine Catchment Area. TNO-rapport R 95/138. Instituut voor Milieuwetenschappen. TNO (IMW), 2600 JA Delft, The Netherlands.
9. Baart, A. C., Berdowski, J. J. M., and van Jaarsveld, J. M. (1995) Calculation of Atmospheric Deposition of Contaminants to the North Sea. TNO-rapport R 91/219. Instituut voor Milieuwetenschappen. TNO (IMW), 2600 JA Delft, The Netherlands.
10. Bartnicki, J., Modzelewski, H., Szewczuk-Bartnicka, H., Saltbones, J., Berge, E., and Bott, A. (1993) An Eulerian model for atmospheric transport of heavy metals over Europe: Model description and testing. DNMI Technical report no. 117. Norwegian Meteorological Institute, Oslo, Norway.
11. Bartnicki, J. (1994) An Eulerian model for atmospheric transport of heavy metals over Europe:Model description and preliminary Results, *Water, Air and Soil Pollution* 75, 227-263.
12. Bartnicki, J. (1996) Computing atmospheric transport and deposition of heavy metals over Europe: Country budgets for Europe. *Water, Air and Soil Pollution* 92, 343-374.
13. Berdowski, J. J. M., Pacyna, J. M., and Veldt, C. (1994) Emissions, in K. D. van den Hout (ed.) *The Impact of Atmospheric Deposition of Non-Acidifying*

Compounds on the Quality of European Forest Soils and the North Sea, TNO-rapport R 93/329, Instituut voor Milieuwetenschappen. TNO (IMW), 2600 JA Delft, the Netherlands.

14. Bloxam, R., Schroeder, W. H., and Petersen, G. (1996) An Eulerian model of mercury atmospheric transport, chemistry and deposition, in R. Ebinghaus, G. Petersen and U. von Tümpling (eds.), *Fourth International Conference on Mercury as a Global Pollutant, Book of Abstracts*, GKSS Forschungszentrum, Max-Planck-Strasse, D-21502 Geesthacht, Germany, p. 40
15. Brügmann, L. (1986) Contamination of the Baltic Sea Atmosphere by Metals - Concentration Levels and Inputs. - Proceedings of the Seminar on Diffusitivity, Transport and Deposition of Atmosperic Pollutants to the Baltic Sea, pp. 81-98. Neubrandenburg, GDR, May 7-8, 1986.
16. Bullock, O. R., Benjey W. G., and Keating M. H. (1996) The Modeling of Regional-scale Atmospheric Mercury Concentration and Deposition Patterns using RELMAP, *Environmental Toxicology and Chemistry*. (in press)
17. Cambray, R. S., Jefferies, D. F., and Topping G. (1975) An Estimate of the Input of Atmospheric Trace Elements into the North Sea and the Clyde Sea. AERE-Report R 7733, AERE, Harwell.
18. Cambray, R. S., Jefferies, D. F., and Topping G. (1979) The atmospheric input of trace elements to the North Sea, *Marine Science Communications* 5, 175-194.
19. Ebinghaus, R., Kock, H. H., Jennings, S. G., McCartin, P., and Orren, M. J. (1995) Measurements ofatmospheric mercury concentrations in northwestern and central Europe - Comparison of experimental data and model results, *Atmospheric Environment* 29, 3333-3344.
20. Eliassen, A. and Saltbones, J. (1983) Modelling the long-range transport of sulphur over Europe: a two year model run and some model experiments, *Atmospheric Environment* 17, 1457-1473.
21. Eppel, D. P., Kapitza, H., Claußen, M., Jacob, D., Koch, W., Levkov, L., Mengelkamp, H. -T., Resch, E., and Werrmann N. (1992) Untersuchung und Bewertung des Schadstoffeintrags über die Atmosphäre im Rahmen von PARCOM (Nordsee) und HELCOM (Ostsee). Teilbericht: Simulation des Transports und der Deposition atmosphärischen Bleis in der norddeutschen Küstenregion. Abschlußbericht des Forschungsvorhabens 104 02 583 des Bundsministers für Umwelt, Naturschutz und Reaktorsicherheit. Im Auftrag des Umweltbundesamtes, Berlin.
22. Galperin, M., Sofiev, M., Gusev, A., and Afinogenova, O. (1995) The Approaches to Modeling of Heavy Metals Transboundary and Long-Range Airborne Transport and Deposition in Europe, EMEP/MSC-E Technical Report 7/95, EMEP Meteorological Synthesizing Center-East, Kedrova str. 8-1, Moscow, Russia.
23. Galperin, M., Sofiev, M., and Mantseva, E. (1996) A model of the chemical transformation of mercury and its long-range atmospheric transport, in W. Baeyens, R. Ebinghaus, and O. F. Vassiliev (eds.), *Regional and Global Mercury Cycles, Fluxes and Mass Balances*, NATO-ASI Series, Series C: Mathematical and Physical Sciences, Kluwer Academic Publishers, Dordrecht, the Netherlands, pp. 219-227.

24. GESAMP (1989) The Atmospheric Input of Trace Species to the World Ocean. IMO/FAD/UNESCO/WMO/IAEA/UN/UNEP. WMO Reports and Studies 38.
25. Graßl, H., Eppel, D. P., Petersen, G., Schneider, B., Weber, H., Gandraß, J., Reinhardt, K. H., Wodarg, D., and Fließ J. (1989) Stoffeintrag in Nord- und Ostsee über die Atmosphäre. Abschlußbericht des Forschungsvorhabens 104 02 572 des Umweltforschungsplanes des Bundesministers für Umwelt, Naturschutz und Reaktorsicherheit. Im Auftrag des Umweltbundesamtes, Berlin.
26. HELCOM (1989) Deposition of Airborne Pollutants to the Baltic Sea Area 1983 - 1985 and 1986. Baltic Sea Environment Proceedings No. 32. Baltic Marine Environment Protection Commission - Helsinki Commission -.
27. HELCOM (1991) Airborne Pollution Load to the Baltic Sea 1986-1990. Baltic Sea. Environment Proceedings No. 39. Baltic Marine Environment Protection Commission - Helsinki Commission -.
28. van den Hout K. D. (ed.) (1994) The Impact of Atmospheric Deposition of Non-Acidifying Compounds on the Quality of European Forest Soils and the North Sea. TNO-rapport R 93/329. Instituut voor Milieuwetenschappen. TNO (IMW), 2600 JA Delft, The Netherlands.
29. Iverfeldt, Å. (1991) Occurrence and Turnover of Atmospheric Mercury over the Nordic Countries, *Water, Air and Soil Pollution* 56, 251-165.
30. van Jaarsveld, J. A., van Aalst, R. M., and Onderdelinden, D. (1986) Deposition of metals from the Atmosphere into the North Sea: model calculations. Report No. 84 20 15 002, National Institute of Public Health and Environmental Protection (RIVM), Bilthoven, The Netherlands.
31. van Jaarsveld, J. A. and Onderdelinden, D. (1989) Dispersion and deposition of dioxins, dibenzofurans and heavy metals emitted by a municipal waste incinerator (in Dutch). Report No. 73 84 73 007, National Institute of Public Health and Environmental Protection (RIVM), Bilthoven, the Netherlands.
32. van Jaarsveld, J. A. (1990) An operational atmospheric transport model for priority substances; specification and instruction for use. Report No. 222 501 002, National Institute of Public Health and Environmental Protection (RIVM), Bilthoven, the Netherlands.
33. Kapitza, H. and Eppel, D. P. (1992) The Non-Hydrostatic Mesoscale Model GESIMA. Part I: Dynamical Equations and Tests, *Beiträge zur Physik der Atmosphäre* 65, 129-146.
34. Krell, U. and Roeckner, E. (1988) Model simulation of the atmospheric input of lead and cadmium into the North Sea, *Atmospheric Environment* 22, 375 - 381.
35. Krell, U. (1991) Vergleich von Modellen für den Eintrag von Spurenstoffen aus der Atmosphäre in die Nordsee - Beispiel Blei. GKSS Forschungszentrum Geesthacht, externer Bericht GKSS 91/E/79.
36. Krüger O. (1996) Atmospheric deposition of heavy metals to North European marginal seas: Scenarios and trend for lead, *GeoJournal* 39. 2, 117-131.
37. Lehmhaus, J., Roeckner, E., Bernhardt, I., and Pankrath, J. (1984) A Monte Carlo model for the simulation of long-range transport of air pollutants, in C. de Wispelaere (ed.), *Air Pollution Modeling and its Application III*, Plenum Press, New York, pp. 149-163.

38. Lindqvist, O., Johansson, K., Aastrup, M., Andersson, A., Bringmark, L., Hovsenius, G., Hakanson, L., Iverfeldt, A., Meili, M., and Timm, B. (1991) Mercury in the Swedish Environment-Recent Research on Causes, Consequences and Corrective Methods, *Water, Air and Soil Pollution* 55.
39. Michaelis, W., Schönburg, M., and Stößel, R. -P. (1989) Deposition of Atmospheric Pollutants into a North German Forest Ecosystem, in H. -W. Georgii (ed.), *Mechanisms and Effects of Pollutant-Transfer into Forests,* Kluwer Academic Publishers, pp 3-12.
40. Munthe, J., Xiao, Z. F., and Lindqvist, O. (1991) The Aqueous Reduction of Divalent Mercury by Sulfite, *Water, Air and Soil Pollution* 56, 621-630.
41. Munthe, J. (1992) The Aqueous Oxidation of Elemental Mercury by Ozone, *Atmospheric Environment* 26A, 1461 - 1468.
42. OECD (1979) The OECD Programme on Long-Range Transport of Air Pollutants. Measurements and Findings, 2nd edn. Organization for Economic Co-operation and Development, Paris.
43. Ottar, B. and Pacyna, J. M. (1984) Sources of Ni, Pb and Zn during the Arctic Episode in March 1983, *Geophysical Research Letters* 11, 441-444.
44. Ottar, B., Pacyna, J. M. and Berg, T. C. (1985) Aircraft measurements of air pollution in the Norwegian Arctic, *Atmospheric Environment* 20, 87-100.
45. Pacyna, J. M., Semb, A., and Hansen, J. E. (1984) Emission and long-range transport of trace elements in Europe, *Tellus* 36B, 163-174.
46. Pacyna, J. M. (1984) Estimation of the atmospheric emissions of trace elements from anthropogenic sources in Europe, *Atmospheric Environment* 18, 41-50.
47. Pacyna, J. M., Ottar, B., Tomza, U., and Maenhaut, W. (1985) Long-Range Transport of Trace Elements to Ny Ålesund, Spitsbergen, *Atmospheric Environment* 19, 857-865.
48. Pacyna, J. M., Bartonova, A., Cornville, P., and Maenhaut, W. (1988) Modelling of Long-Range Transport of Trace Elements. A Case Study, *Atmospheric Environment* 23, 107-114.
49. Pacyna, J. M. (1988) Atmospheric Lead Emissions in Europe 1982. NILU-Report 18/88. Norwegian Institute for Air Research, Lillestrom, Norway.
50. Pacyna, J. M. and Münch, J. (1988) Atmospheric Emissions of Arsenic, Cadmium, Mercury and Zinc in Europe 1982. NILU-Report 17/88. Norwegian Institute for Air Research, Lillestrom, Norway.
51. Pacyna, J. M., and Ottar, B. (eds.) (1989) *Control and Fate of Atmospheric Trace Metals*, NATO ASI Series. Series C: Mathematiacal and Physical Sciences - Vol. 268, Kluwer Academic Publishers.
52. PARCOM (1988) Interim Report on Atmospheric Input to the North Sea. Convention for Prevention of Marine Pollution from land-based Sources. Sixth Meeting of the Working Group on the Atmospheric Input of Pollutants to Convention Waters, Paris, November 15-17, 1988, (ATMOS 5/6/4-E).
53. Pekar, M. (1996) Regional Models LPMOD and ASIMD. Algorithms, Parameterization and Results of Application to Pb and Cd in Europe Scale for 1990, EMEP / MSC-E Report 9/96, EMEP Meteorological Synthesizing Center-East, Kedrova str. 8-1, Moscow, Russia.

54. Peters, L. K., Berkowitz, C. M., Carmichael, G. R., Easter, R. C., Fairweather, G., Ghan, S. J., Hales, J. M., Leung, R. L., Pennell, W. R., Potra, F. A., Saylor, R. D., and Tsang, T. T. (1995) The current state and future direction of Eulerian models in simulating the tropospheric chemistry and and transport of trace species: A review, *Atmospheric Environment* 29, 189 - 222.
55. Petersen, G., Weber, H., and Graßl, H. (1989) Modelling the Atmospheric Transport of Trace Metals from Europe to the North Sea and the Baltic Sea, in J. M. Pacyna and B. W. Ottar (eds.), *Control and Fate of Atmospheric Trace Metals*, NATO-ASI Series, Series C: Mathematical and Physical Sciences - Vol. 268, Kluwer Academic Publishers, Dordrecht, the Netherlands, pp. 57-83.
56. Petersen, G. (1992a) Belastung von Nord- und Ostsee durch ökologisch gefährliche Stoffe am Beispiel atmosphärischer Quecksilberverbindungen. Abschlußbericht des Forschungsvorhabens 104 02 726 des Bundesministers für Umwelt, Naturschutz und Reaktorsicherheit. Im Auftrag des Umweltbundesamtes, Berlin. Externer Bericht GKSS 92/E/111.
57. Petersen, G. (1992b) Atmospheric Input of Mercury to the North Sea. Paris Convention for the Prevention of Marine Pollution. Ninth Meeting of the Working Group on the Atmospheric Input of Pollutants to Convention Waters. London: 5-8 November 1991 (PARCOM-ATMOS 9/13/1).
58. Petersen, G. (1992c) Atmospheric Input of Mercury to the Baltic Sea. Baltic Marine Environment Protection Commission - Helsinki Commission -. Ninth Meeting of the Group of Experts on Airborne Pollution of the Baltic Sea Area (EGAP). Solna, Sweden: 19-22 Mai 1992 (HELCOM-EGAP 9/2/5).
59. Petersen, G. and Krüger, O. (1993) Untersuchung und Bewertung des Schadstoffeintrags über die Atmosphäre im Rahmen von PARCOM (Nordsee) und HELCOM (Ostsee) Teilvorhaben: Modellierung des großräumigen Transports von Spurenmetallen. Abschlußbericht des Forschungsvorhabens 104 02 583 des Bundesministers für Umwelt, Naturschutz und Reaktorsicherheit. Im Auftrag des Umweltbundesamtes, Berlin. Externer Bericht GKSS 93/E/28.
60. Petersen, G., Iverfeldt, A. and Munthe, J. (1995) Atmospheric Mercury Species over Central and Northern Europe. Model Calculations and Comparison with Observations from the Nordic Air and Precipitation Network for 1987 and 1988, *Atmospheric Environment* 29, 47-67.
61. Petersen, G., Munthe, J., and Bloxam, R. (1996) Numerical modeling of regional transport, chemical transformations and deposition fluxes of airborne mercury species, in W. Baeyens, R. Ebinghaus, and O. F. Vassiliev (eds.), *Regional and Global Mercury Cycles, Fluxes and Mass Balances*, NATO-ASI Series, Series C: Mathematical and Physical Sciences, Kluwer Academic Publishers, Dordrecht, the Netherlands, pp. 191-217.
62. Petersen, G., Munthe, J., Pleijel, K., Bloxam, R., and Vinod Kumar, A. (in press) A Comprehensive Eulerian Modeling Framework for Airborne Mercury Species : Development and Application of a Tropospheric Chemistry Module, *Atmospheric Environment* 31.
63. Pleijel, K. and Munthe, J. (1995a) Modelling the Atmospheric Mercury Cycle - Chemistry inFogDroplets, *Atmospheric Environment* 29, 1441-1457.

64. Pleijel, K. and Munthe, J. (1995b) Modeling the Atmospheric Chemistry of Mercury - The Importance of a Detailed Description of the Chemistry of Cloud Water, *Water, Air and Soil Pollution* 80, 317-324.
65. Remoudaki, E., Bergametti, O., and Buat-Ménard, P. (1991) Temporal Variability of Atmospheric Lead Concentrations and Fluxes over the Northwestern Mediterranean Sea, *Journal of Geophysical Research* 96, No. D1, 1043-1055.
66. Rühling, A., Rasmussen, L., Pilegaard, K., Mäkinen, A. and Steinness, E. (1987) Survey of Atmospheric Heavy Metal Deposition in the Nordic Countries in 1985. Report NORD 1987: 21, prepared for 'The Steering Body for Environmental Monitoring', The Nordic Council of Ministers.
67. Schneider, B. (1992) Untersuchung und Bewertung des Schadstoffeintrags über die Atmosphäre im Rahmen von PARCOM (Nordsee) und HELCOM (Ostsee) - Teilvorhaben: Messungen von Spurenmetallen - Abschlußbericht des Forschungsvorhabens 104 02 583 des Umweltforschungsplanes des Bundesministers für Umwelt, Naturschutz und Reaktorsicherheit. Im Auftrag des Umweltbundesamtes, Berlin.
68. Shannon, J. D. and Voldner, E. C. (1995) Modeling Atmospheric Concentrations of Mercury and Deposition to the Great Lakes, *Atmospheric Environment* 29, 1649-1662.
69. Sofiev, M., Maslyaev, A., and Gusev, A. (1996) Heavy metal model intercomparison. Methodology and results for Pb in 1990. EMEP/MSC-E Technical Report 2/96, EMEP Meteorological Synthesizing Center-East, Kedrova str. 8-1, Moscow, Russia.
70. Stern, R., Scherer, B., and Schubert, S. (1990) Modeling of Photochemical Oxidant Formation in North-Western Europe with special Emphasis on the non-linearity issue in the SOX / NOX / VOC System. Presented at the EMEP Workshop on the Progress of Transport Modelling of Nitrogen Compounds, Potsdam, Germany, October 1990.
71. Stössel, R. P. (1987) Untersuchungen zur Naß- und Trockendeposition von Schwermetallen auf der Insel Pellworm. GKSS Forschungszentrum Geesthacht, externer Bericht GKSS 87/E/34.
72. Swietlicki, E., Bartnicki, J., Jalkanen, L., Kemp, K., Wahlin, P., and Krejci, R. (1996) Heavy Metals in the Atmosphere. Comparison of source-receptor relationship estimated by means of receptor and meteorological dispersion modeling, Lund University Report No. LUTFD2 / (TFKF-3079) / 1-29 / (1996). Lund University, Division of Nuclear Physics, Sölvegatan 14, Lund, Sweden.
73. Venkatram, A.,, Karamchandani, P. K., and Misra P. K. (1988) Testing a Comprehensive Acid Deposition Model, *Atmospheric Environment* 22, 737-747.
74. Warmenhoven, J. P., Duiser, J. A., de Leeuw, L. T., and Veldt, C. (1989) The Contribution of the Input from the Atmosphere to the Contamination of the North Sea and the Dutch Wadden Sea. TNO Report R 89/349A, Delft, The Netherlands.
75. Xiao, Z. F., Munthe, J., Schroeder, W. H, and Lindqvist, O. (1991) Vertical fluxes of volatile mercury over forest soil and lake surfaces in Sweden, *Tellus* 43B, 267-279.

MEASUREMENTS AND MODELS FOR TRACE ELEMENT INPUT AND TIME DISTRIBUTION RECORDED IN FOREST SOIL AND BOG ECOSYSTEMS

W. R. SCHELL
Dept. of Ciencies Ambientals, Universitat de Girona, SPAIN[1]

I. LINKOV
Department of Physics, Harvard University, Cambridge, MA 02138, USA

M.J.V. NOVAK
Czech Geological Survey, Klarov 3, 118 21 Prague 1, CZECH REPUBLIC

1. Abstract

Atmospheric fallout of pollutant effluent from chemical industries, metal refining plants, coal and oil combustion sources and the nuclear industry are widely distributed in forests and wetland environments. These ecosystems are repositories which collect and store the waste products of man's industrial activities and provide a source of samples which may record the regional history of pollutant input to the ecosystem and to man. Measurements of such toxic elements must be accompanied by detailed quality control procedures for sampling and analysis in order to enable a critical evaluation of the results. In plant nutrition, the local ecosystem is in general equilibrium with precipitation, nutrient and trace metal uptake from soil, and plant litter decomposition on an annual cycle. Distribution coefficients and soil-to-plant transfer factor measurements of elements provide equilibrium values but do not represent conditions where dynamic, long-term processes dominate. The fallout from nuclear weapons tests and the Chernobyl NPP reactor accident, however, provided tracers for identifying the kinetic processes which make up the total equilibrium of plant growth. In forests, the complications of long term-growth and nutrient availability can be separated into compartments representing deposition onto leaves, wash-off by precipitation, stem flow, understory and litter accumulation and, finally, infiltration into the root zone for plant uptake. Studies of cores collected from ombrotrophic bogs, forest floor and lake sediments have shown the rate of atmospheric deposition and transport of elements. The bog and wetland environments provide a highly organic medium where deposition and migration of elements can be studied without the complications of the forest roots. Models which provide insight into transport processes of elements have been made in order to evaluate and predict long term

1 Present address: Department of Environmental and Occupational Health, University of Pittsburgh, Pittsburgh, PA 15261, USA

I. Linkov and R. Wilson (eds.), Air Pollution in the Ural Mountains, 99–114.

effects on ecosystems and to reconstruct the element dose to man from breathing the toxic metallic fumes from industrial sources.

2. Introduction

The rate of migration of trace elements and radionuclides through the upper soil layers may be variable and their availability to plants may be a problem in the future. To estimate element deposition and migration rates over time in bog and forest ecosystems, radioactive tracers that have specific known input functions may be utilized. Input times for the tracers are from the atmospheric nuclear weapons testing period and the instantaneous input from the Chernobyl NPP and Kyshtym accidents in Ukraine and Russia, respectively that contaminated much of the Northern Hemisphere. The development of a chronology for regional deposition of chemicals introduced into the environment from anthropogenic and natural sources provides information on quantitative measurement of pollutants in ecosystems. Dating of the regional history of metals pollution provides data which can be used to determine the contamination effects on man and the ecosystem over the past 200 years. Reconstruction of the exposure dose to man from living in an environment contaminated by trace metals and radionuclides may now be possible.

Such chronological data of trace elements and radionuclides can be obtained in a given region using atmospheric tracers such as the natural radionuclide, ^{210}Pb, and man-produced radionuclides such as ^{137}Cs, ^{90}Sr, and $^{239,240}Pu$. These radionuclides move with other particulate matter in the atmosphere and are used to determine the deposition rate of pollutant material in given regions. The deposition of fallout radionuclides has been well documented for the northeastern USA by the Department of Energy - Environmental Measurements Laboratory in New York City, and in Europe by the Swedish Defense Ministry and the UK Atomic Energy Authority, Harwell [1]. These fallout tracers provide time and flux data that can be used to estimate the trace metals deposition, provided a suitable geochemical repository is found. Such repositories for certain elements may be lakes as well as marshlands, bogs, forests and other organic rich ecosystems. Biological activity in, for example, a bog decomposes plant matter that becomes humified at increasing depths. When organic matter decomposes, inorganic sediments of clay-like material may remain to further retard migration [2]. Little or no mixing of elements occurs below the active root zone where anoxic conditions exist, except by diffusion and advection due to water transport by precipitation. Trace elements are strongly absorbed in peat, which has a cation exchange capacity of 100-1000 times greater than that of clay soil. Humification of certain peats produces much greater cation exchange capacity, which may further limit the migration of elements.

Recent reports of Eastern Europe have illustrated that heavy metals emitted without environmental controls have had severe detrimental effects on the health of the regional ecosystems and population [3]. The effects include respiratory problems, neurological damage, heart and kidney diseases, birth defects, and other such diseases. Similar problems existed in the early days of Pittsburgh, Philadelphia and other heavy industrial population centers of the eastern USA and New York before pollution controls on emissions were implemented. The lead, cadmium, chromium and other trace elements measured in the bog core samples are indicators of industrial effluents from nearby coal combustion, mining and steel production, gasoline consumption and waste disposal products. Fly ash from coal burning is shown by the increased deposition of aluminum, barium, magnesium, calcium, vanadium, and uranium [4]. The dating methods developed provide the only available means to reconstruct previous chemical input levels and to assess the effectiveness of present day clean-up attempts.

3. Materials and Methods

3.1. SAMPLE SITES

The sampling sites selected are near sources of pollutants at a mountain top (ombrotrophic/minerotrophic) bog near Pittsburgh, PA, at a marshland on the Hudson River, New York, at blanket bogs in Ireland near both the west coast and Dublin, and in the Czech Republic at a relatively pristine site in southern Bohemia, Jezerni slat, and at one of the most polluted regions of Europe, Bozi Dar in northern Bohemia. These sites were selected because they record the pollutant input primarily from atmospheric deposition processes. The Czech Republic site Jezerni slat, is a bog with 1 m thick floating mat on top of a lake [5], [6], located in the relatively pristine Sumava Mountains of southern Bohemia at an elevation of 1070 m with 760 mm/y precipitation. Bozi Dar Bog, 50°24' N, 12°54' E, is located in the Krusne hory Mountains of northern Bohemia near the center of the heavy industries of central Europe and has precipitation of 850 mm/y. The peat deposit reaches 5.5 m depth and consists of *Sphagnum*-derived material that has had little growth in recent years [7].

3.2. SAMPLE COLLECTION AND ANALYSIS

Sediment cores were taken by driving a 10-cm diameter sharpened polyvinyl chloride tube vertically into a site away from tree roots and hummocks. The cores were sectioned into 1 or 2 cm thick layers, often in the field, weighed, dried at 70°C or freeze dried, and homogenized upon arrival in the laboratory. Particular care was taken to avoid contamination by using double plastic bags in field collections and by minimizing sample exposure to the laboratory environment. The first work on the cores was to determine if the ^{210}Pb decay profile was continuous. Extensive radiochemical and element analyses were then initiated on suitable cores.

The distribution of the natural radionuclide, ^{210}Pb, with a half-life of 22.3 years, in a sediment or bog core provides the time scale for atmospheric chemical input over the past 200 years. After dissolution with aqua regia, each layer was spiked with ^{208}Po, as a chemical yield tracer, and the solution electroplated onto silver discs for the analysis of the ^{210}Pb by measuring the ^{208}Po and ^{210}Po content. Measurements of the ^{210}Pb profile in each 1-2 cm section of a core provide the first work in validating the method to see if the ^{210}Pb decay profile was continuous. Aliquots were then taken for the various radiochemical and elemental analyses. Standard radiochemical techniques were used to analyze for the radionuclides ^{210}Pb, ^{137}Cs, ^{90}Sr, and ^{239,240}Pu [8], [9]. These radionuclides are from global fallout and have been deposited at well defined periods during nuclear weapons testing [10] and following the

Chernobyl NPP accident in 1986. Data for these fallout radionuclides are used to compare and evaluate the mobility of certain elements which may not be tightly bound to the matrix. Additional aliquots of each layer were taken for ICP-MS and ICP-AES analysis where up to 60 elements can be analyzed simultaneously. The ICP-MS data from the Czech Republic have been re-calculated and values for several metals can now be compared with independent Pb measurements by Vile *et al.* [7] on nearby cores. Duplicate samples were taken, where possible, and quality control procedures used to validate the analytical results. The ^{210}Pb data, used to construct the chronology for the sediment layers, applied the constant rate of supply (CRS) interpretation [11] with the minimum variance treatment, the CRS/MV model [12], [13]. The appropriate errors were propagated giving a time error for each section. Where different slopes of the data lines were found, piecemeal regression analysis was made. This method provides a confidence interval over the entire data set giving an uncertainty of 2-10 years for the core. Distribution coefficients, K_d, values for adsorption and desorption were determined using radioactive isotopes of the metals with the peat/water mixture at pH 4 by using the constant shaking method at a concentration of 200 mg/l [14].

3.3. ADVECTION AND DIFFUSION OF ELEMENTS IN CORES

The six major mechanisms for vertical movement of trace metals and radionuclides are: precipitation, infiltration, evapotranspiration, mineral solubility, gravity and capillary effects. The hydrologic cycle, particularly in the unsaturated zone, is difficult to understand as a result of the complex dynamic interactions of precipitation, surface water, soil moisture, vegetation, soil properties, and weather [15]. However, many difficulties in the availability of these data can be overcome through approximations. For the purpose of the model used here, the physical system is described as one-dimensional with two average forces of movement occurring, i.e., advection and dispersion; these forces are dependent upon the concentration of radionuclides in the soil and the dynamic forces of transport which are acting on the radionuclides or trace metals [2].

The advective/dispersive model is used to predict changes in radionuclide concentration in the soil layers over time. Advection refers to the components of motion due to kinetic movement, such as precipitation, infiltration and evapotranspiration. Dispersion refers to the components of motion due to molecular diffusion, turbulent diffusion, shear dispersion, and gravitational and capillary forces [16], [17]. Molecular diffusion is random molecular motion which, in water, has a value of about 10^{-5} cm^2/sec. Turbulent diffusion is random mixing due to turbulent fluid motion and has values between 10^{-2}-10^{-6} cm^2/sec. Shear dispersion is mixing due to velocity gradients [18]. Since the bogs under investigation are ombrotrophic, turbulent fluid motion and shear dispersion are essentially absent. The advective/dispersive model also takes into account the processes of adsorption, absorption, desorption and radioactive decay.

The one-dimensional transport equation describing the rate migration of radionuclides in a porous solid matrix is [19]:

$$\frac{\partial C}{\partial t} = \frac{D\partial^2 C}{\partial x^2} - \frac{U\partial C}{\partial x} - \frac{\rho\partial S}{\theta\partial\tau} - \lambda C - \frac{\lambda\rho S}{\theta} \qquad (1)$$

The partitioning of a radionuclide or metal between liquid and solid phase is defined by the distribution coefficent, K_d as:

$$K_d = \frac{radioactivity\_adsorbed}{radioactivity\_dissolved} = \frac{S}{C} \tag{2}$$

The retardation factor, which accounts for the slower movement, due to matrix sorption, of the radionuclide or element than the water (precipitation) which flows unimpeded, is:

$$R = 1 + \frac{\rho K_d}{\theta} \tag{3}$$

The one dimensional solute transport equation can then be written by:

$$\frac{\partial C}{\partial t} = \frac{D\partial^2 C}{R\partial x^2} - \frac{U\partial C}{R\partial x} - \lambda C \tag{4}$$

where λ = decay constant (sec $^{-1}$)
C = concentration in solution (pCi/ml)
S = concentration sorbed on solid matrix (pCi/g)
ρ = density in gm/cm^3 (dry) of the bulk soil
θ = moisture content in ml water per cm^3 (wet) of the bulk soil
U = advection velocity (cm/sec)
D = diffusivity (cm^2/sec)

Therefore, the rate of change with time of the concentration in solution is equal to the rate of change due to dispersion minus the rates of change due to advection, geochemical reactions (such as adsorption desorption and precipitation) and radioactive decay. The total adsorption (S) is the amount of a radionuclide sorbed or precipitated per unit mass of the porous medium. The partitioning of a radionuclide between the liquid and solid phase is experimentally determined and is defined as the distribution coefficient, K_d. This equilibrium sorption isotherm is a mathematical relationship which describes the radionuclide distribution between the liquid and solid phase. The effect of the K_d value on radionuclide migration is shown by Equation 2 [20].

It is apparent that as K_d increases, the radionuclide migration rate decreases. Thus, the migration of the radionuclides is retarded with respect to the migration of the water. The effect of K_d is taken into account by the retardation factor, R (Equation 3). Equation 4 shows that the effective diffusivity and effective advection velocity are reduced by the factor I/R.

4. Results and Discussion

4.1. TESTS OF DATING CORE DEPTHS IN LAKE AND BOG SAMPLES

To test the dating method for lakes and bogs, the complex recent history of Lake Washington as recorded in the sediment in the deep basin was used [9]. The construction of navigational locks from the Puget Sound to the lake was completed in 1916, when the lake level was reduced three meters and the Green River was channeled into the south of the lake. This caused significant redistribution of sediment which deposited throughout the lake at the specific time. The CRS model predicts the sediment deposition at 1923 whereas the minimum variance addition to the model predicts the sediment deposition at 1912 $\pm$ 4 years, in excellent agreement with the actual construction time.

Another anthropogenic test of the date is by use of records from the ASARCO smelter which started operations to produce Pb in the early 1890's with no appreciable control of emissions. Lake Washington is located about 35 km north of the smelter and the prevailing winds are often directed from the smelter to the lake. Two major time periods are related to the Pb concentration from coal burning (about 50 mg g^{-1}) and to the increase in production and emissions from the smelter reaching a maximum at about 1913 and then decreasing when the production changed from lead to copper [21], [22]; the second time period is a record of the use of internal combustion engines with the tetraethyl lead additive which began in the 1920's and the effect of emission controls starting in the early 1970's.

The results of these studies on dating bog cores establish methods to determine the inputs of heavy metals and their consequent exposure to man and ecosystems at specific regions. The cores from bogs and wetland ecosystems serve as undisturbed repositories where the deposition history of selected toxic metals are recorded. After dating the cores, the fluxes that have been deposited are validated using historical data and the analyses of nuclear weapons test fallout radionuclides ^{137}Cs and ^{239,240}Pu on the same cores. The concentrations of Pb, Cd, Cr and other trace metals as well as C, H, N, and S are determined on the dated cores and the history of their deposition is evaluated [23].

4.2. COMPARISONS OF RADIONUCLIDE AND TRACE METAL MIGRATION IN MARSHLAND

To compare one element's behavior with another in bog systems, it is necessary to utilize several indicators in model building. Core samples taken at the different bog types, sectioned and analyzed for trace metals and radionuclides, provide comparisons of mobility for the different elements.

The effect of the different amounts of organic matter is shown by the relative transport through the core length. Table 1 shows profiles of Cs, Pu and Sr in organic-rich ombrotrophic bogs with the highest levels at the surface. The expected peak activity of ^{137}Cs in 1963 is not found. It has diffused both upward to the surface maxima and downward to levels representing ages of 1850. It is evident that the Pu and Cs follow closely the same activity profile with depth giving a standard error of the ratio of only 12%.

In contrast, the organic-poor bogs, Table 2, show the maximum in ^{137}Cs concentrations in the mid-1960's, a low deposition at the surface but with Cs diffusing down to depth of the 1915 layer giving a standard error for the Pu/Cs ratio of 67%. The ^{90}Sr data in the organic-rich bog measured also shows an apparent constant ratio of ^{90}Sr/^{239,290}Pu to the measured depth of 10 cm giving a standard error of 18%. The similarity of the Pu, Cs and ^{90}Sr profiles in bog cores would not be expected since the elements have such different chemical properties. One possible explanation for the measurement is the formation of colloidal-sized particles or organic complexes which move as discrete micelles in the organic-rich bog environment and may not be present in minerotrophic bogs.

Figure 1 shows the deposition history of alkaline earth metals at two organic-rich (ombrotrophic) bog sites in the Czech Republic where levels of these metals can be compared

TABLE 1. Deposition and migration of radionuclides at organic rich site, Black Rock Forest Bog, NY

Depth (cm)	^{210}PbYr (AD)	C (%)	^{137}Cs Bqm^{-2}	^{90}Sr Bqm^{-2}	^{239,240}Pu Bqm^{-2} (E-2)	Pu/Cs (E-2)	Pu/Sr (E-2)	Cs/Sr
0-2	1978.3	25.6	223±49	65±5	2.3±0.2	1.03±0.	3.6±0.8	3.4±0.8
2-3	1966.5	25.1	217±46	50±2	2.5±0.3	1.15±0.3	5.0±1.0	4.3±0.9
3-4	1957.2	25.3	232±49	64±4	2.8±0.3	1.22±0.3	4.4±0.9	3.6±0.8
4-5	1948.8	22.2	177±38	--	1.7±0.2	0.96±0.2	--	--
5-6	1940.4	23.5	274±53	72±6	2.5±0.3	0.91±0.2	3.5±0.7	3.8±0.8
6-7	1931.9	20.7	171±42	--	2.1±0.2	1.22±0.3	--	--
7-8	1923.5	36.9	174±54	--	--	--	--	--
8-9	1915.1	24.5	122±44	46±4	1.5±0.2	1.23±0.5	3.2±1.0	2.6±1.0
9-10	1906.7	22.5	154±50	52±6	1.8±0.2	1.16±0.4	3.5±1.0	2.9±1.0
10-11	1898.2	22.4	78±43	--	--	--	--	--
11-12	1889.8	21.5	68±21	--	--	--	--	--
12-13	1881.4	21.3	58±19	--	--	--	--	--
13-14	1873.0	21.4	33±24	--	--	--	--	--
mean±SD		--	--	--	--	1.12±0.13	3.86±0.68	3.43±0.62

TABLE 2. Deposition and migration of radionuclides at organic poor site Manns Bog, VA

Depth (cm)	^{210}Pb Year (AD)	Carbon (%)	^{137}Cs Bqm^{-2}	^{239,240}Pu Bqm^{-2} (E-2)	Pu/Cs (E-2)
0-2	1981.6	11.6	29±7	0.4±0.1	1.38±0.78
2-4	1978.8	6.61	69±13	1.9±.04	2.75±0.52
4-6	1974.4	5.13	78±17	3.5±0.2	4.48±1.01
6-8	1969.1	3.41	288±26	6.4±0.2	2.22±0.21
8-9	1962.6	3.76	309±19	6.3±0.3	2.00±0.16
9-10	1951.9	2.58	530±30	6.9±0.3	1.30±0.93
10-11	1940.2	1.53	297±36	3.1±0.2	1.04±0.14
11-12	1933.9	4.06	188±18	1.4±0.1	0.74±0.88
12-13	1921.5	1.58	166±34	2.1±0.4	1.26±0.35
13-14	1909.5	1.24	38±32	1.6±0.3	4.21±3.63
14-15	1892.9	1.48	93±28	0.5±0.3	0.54±0.28
15-16	1855.6	.49	29±69	NA	NA
16-17	1810	.47	30±55	NA	NA
mean ±SD					1.99±1.33

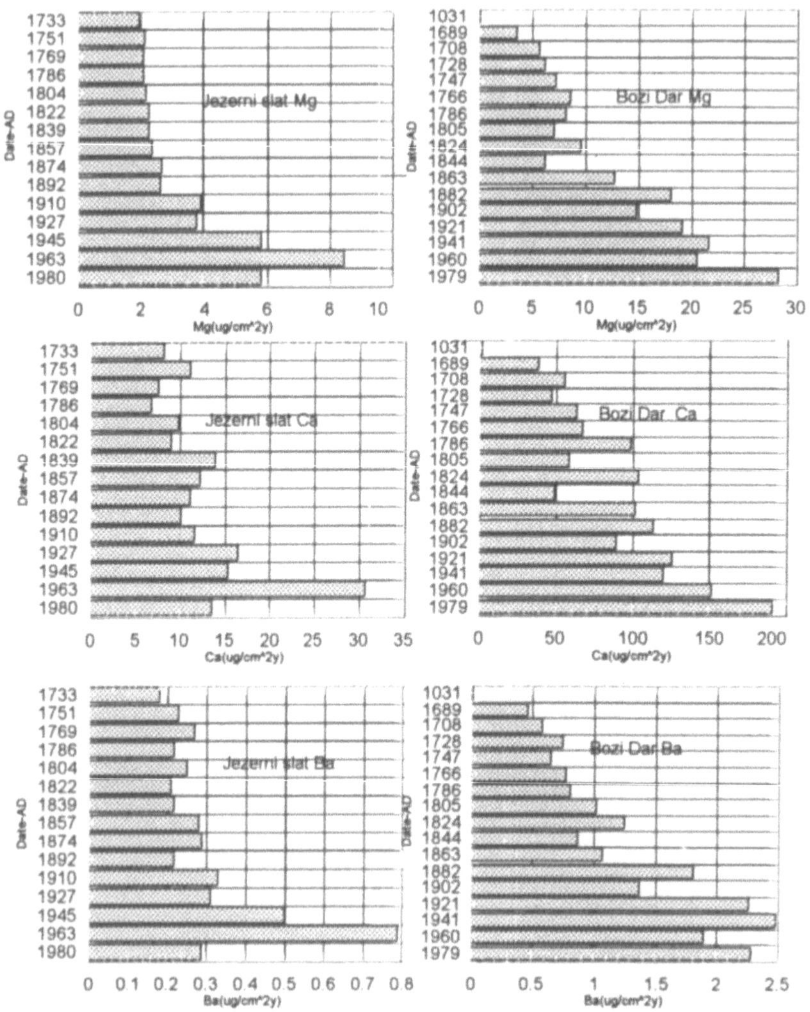

FIGURE 1. Deposition history of alkaline earth metals at the "pristine" Jezerni slat and contaminated Bozi Dar peat bogs in the Czech Republic

with the Pb deposition at USA and Ireland [24]. The "pristine" site near the Austrian border, Jezerni slat, has highest deposition near the top of the core with the deposition rate 4-10 times less than that at the highly contaminated site, Bozi Dar. It appears that the alkaline earth elements reflect the industrial activities resulting in atmospheric pollution deposition since 1700 when compared with the Pb deposition history, Figure 2. The alkaline earth family of elements was chosen to test correlation between strontium using the tracer radionuclide, ^{90}Sr which has been measured at other locations. The bi-valent cations appear to be strongly complexed with the peat material and little loss over a time period of 300 years has occurred due to solubilization and vertical transport by precipitation. Other trace metals, Cu, Cd, Zn and Ni are retained and show pollution from heavy industries over the past 300 years (Figure 3).

The mono-valent cation Na, appears to be dispersed throughout the bog core depths and cannot be used in pollution history. Schell [4] measured the Cs, K, Rb, Na at a minerotrophic bog site in Pennsylvania and found strong correlation with depth within the family of mono-valent elements. The ^{137}Cs tracer was found to be mobile in the upper layers of organic-rich bog cores measured in the USA and Ireland. At ombrotrophic/minerotrophic bogs with decreasing amounts of organic matter with depth, Cs has less mobility than in the organic-rich bogs measured.

It can be concluded that mono-valent cations are more mobile than bi-valent cations and that organic-rich areas contaminated by ^{90}Sr would be effected much longer than areas effected by ^{137}Cs. In soils below the organic layer, which contain certain clay minerals, the precipitation-induced migration of ^{137}Cs would be very low and Cs would be retained.

As Kadlec and Keoleian [25] show, peat acts both as an ion exchanger and a complex ligand matrix for heavy metal chelation. The principal functional groups are the carboxylic and phenolic hydroxyl groups [26]. The values measured for the cation exchange capacity for sedge peat were 1.2 Meq/g and 1.6 Meq/g. This high exchange capacity can be compared to the capacity of commercial ion exchange resins such as Dowex 50 of 2 Meq/g wet. In comparing the cation exchange of mixtures of element ions, Wolf et al. [27] showed that H^+ was tightly bound to the peat compared to Ca^{+2}. Heavy metal ions can displace Ca^{+2} more easily than H^+. The peat has a higher exchange capacity for heavy metals than for Ca^{+2}. For bivalent ions, the stability complexes follow the selectivity order [28], [20]:

$$Pb^{+2} > Cu^{+2} > Ni^{+2} > Co^{+2} > Zn^{+2} > Cd^{+2} > Fe^{+2} > Mn^{+2} > Mg^{+2}$$

Mono-valent cations are less strongly exchanged than bivalent cations on peat execpt for Ag^+, Tl^+, Hg^+, Cu^+.

The distribution coefficient (K_d) for heavy metals on peat is strongly sensitive to pH and to the type of peat. At pH 2, the K_d for Cu^{+2} ion on sedge peat is 50 and at pH 5.5 it is 300. Deionized water has virtually no capability for exchanging Cu^{+2} due to the absence of counter-ions. The K_d for Ni is 200 and for Cd is 900 at pH 5.5, giving a selectivity order of Cd > Cu > Ni. Thus, the pH-dependent K_d becomes an important variable in metals transport modeling.

In summary, ion exchange and complex organic chemical reactions dominate the transport of radionuclides through the organic-rich unsaturated zone of bogs and forests. Initial uptake by the foliage dominates the atmospheric transfer of radionuclides and trace

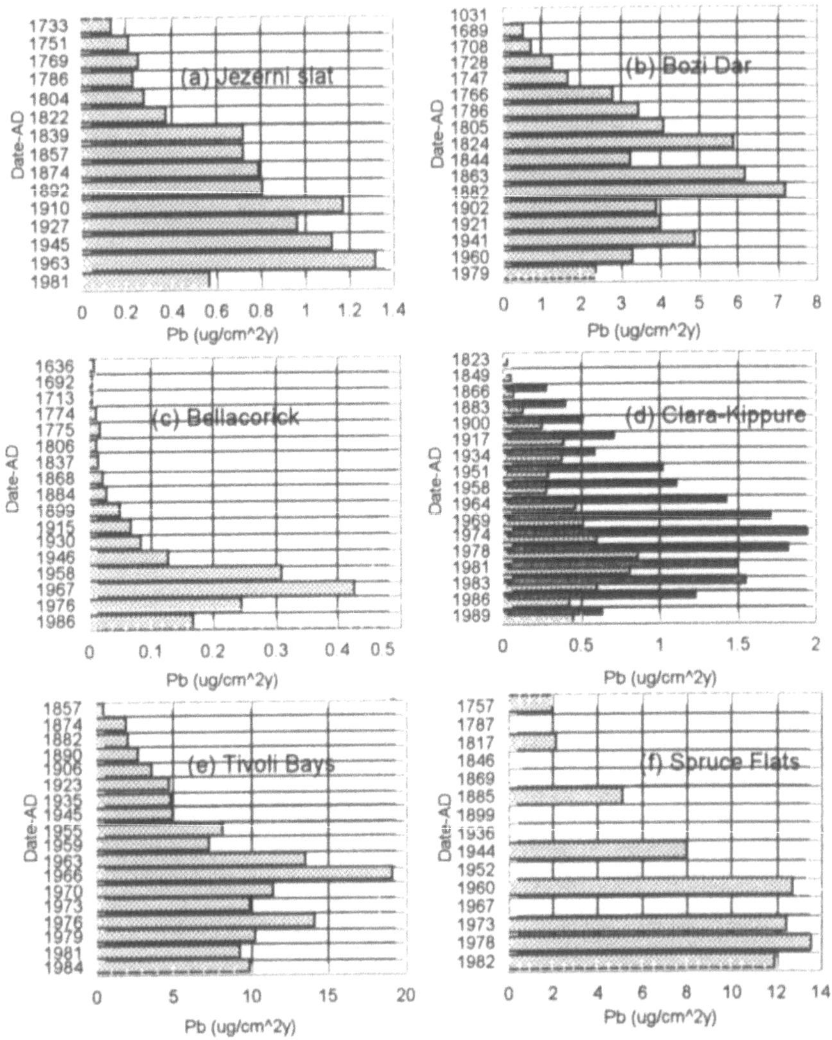

FIGURE 2. Deposition history of lead at peat bogs in the (a) Czech Republic, (b) Ireland and (c) USA .

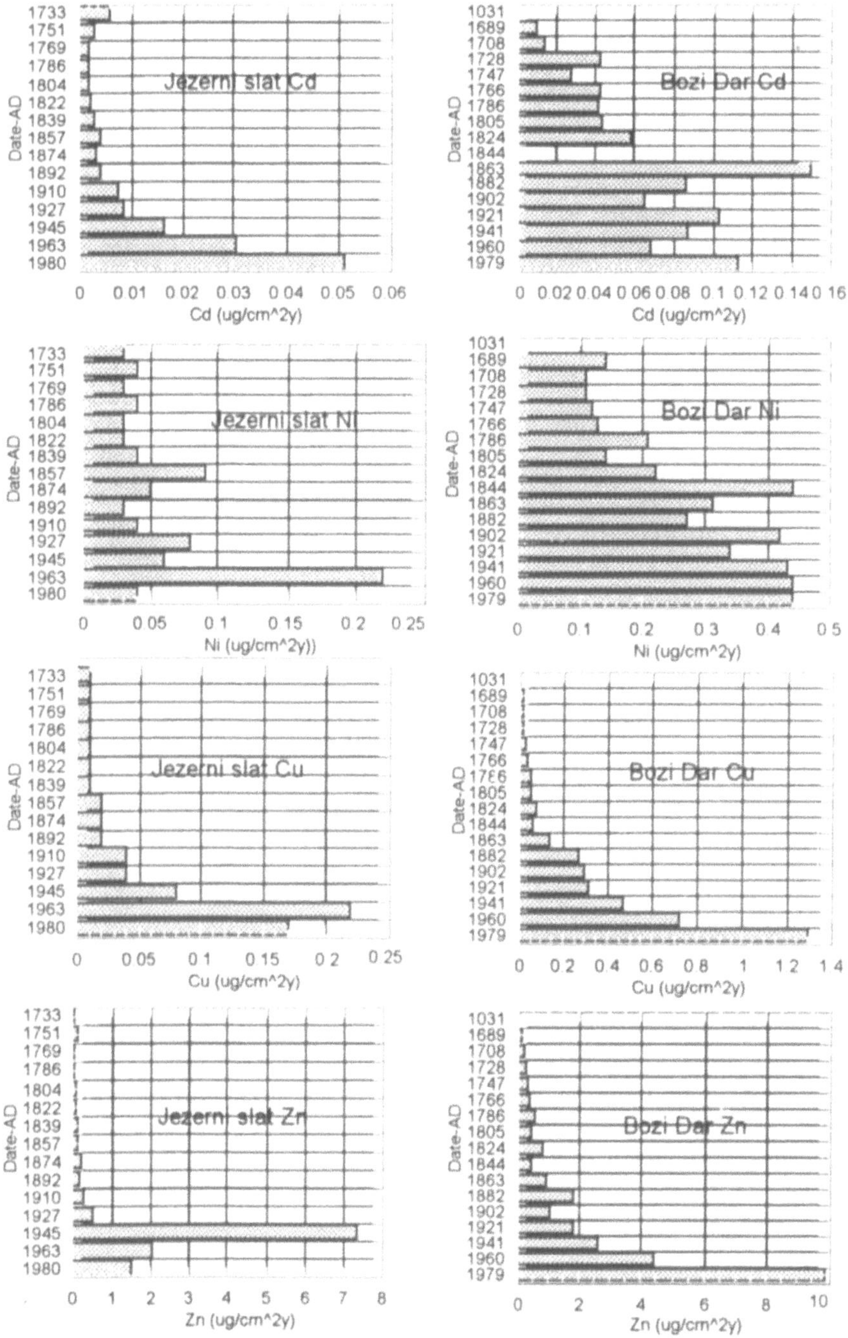

FIGURE 3. Deposition history of the coinage metals Cu, Cd, Zn and Ni in peat bogs at the "pristine" Jezerni slat and contaminated Bozi Dar in the Czech Republic

elements to the trees. These nutrients would be recycled due to their specific requirements in plant growth and the radionuclides would remain in the active zone until removed by radioactive decay. During the 30 years since nuclear weapons testing, ^{137}Cs has now been transported in ombrotrophic peat bogs to about 16 cm. In woodland soils in the UK, Japan, and in Pennsylvania, USA, greater than 90% of the ^{137}Cs is found above 20 cm [30].

In assessing the interaction of heavy metals with sorption material in aqueous solution, the basic analytic tool is the distribution coefficient, K_d. This experimental coefficient is determined by measuring the concentration in the sediment matrix divided by the concentration in aqueous media, e.g., (μg/g)/(μg/ml), in units of ml/g. This coefficient is only an indicator of a geochemical system since it is expected that equilibrium is achieved in the laboratory measurements. The distribution coefficients for the high organic content bogs for ^{137}Cs have some variation but are much smaller than for the low organic content (minerotrophic) bogs from the surface to 22 cm depth. It is not the mobility of Cs which is the interesting process but it is the upward migration as well as the downward transport which provides insight into the transport of metals by water through infiltration as well as the upward transport by evaporation. This process can be modeled relative to the migration of the several elements. The values of ρ and θ and K_d for ^{137}Cs in the typical organic-rich and organic-poor bogs were determined in the laboratory to be [32]:

Organic rich (ombrotrophic)	Organic poor (minerotrophic)
$\rho = 0.6$ g/cm^3	$\rho = 0.6$ g/ cm^3 at surface
	$\rho = 1.6$ g/ cm^3 at 20 cm.
$\theta = 0.89$ ml/ cm^3	$\theta = 0.8$ ml/ cm^3 at surface
	$\theta = 0.18$ ml/ cm^3 at 30 cm.
$K_d = 500$ at 6-8 cm	$K_d = 29000$ at surface
$K_d = 200$ at 20-21 cm	$K_d = 10000$ at 20-22 cm
Advection = -1.5 E-6 cm sec^{-1}	Advection = -1.5 E-6 cm sec^{-1}
Diffusivity = 8.0 E -6 cm^2 sec^{-1}	Diffusivity = 2.0 E -5 cm^2 sec^{-1}

Equation (4) assumes that a linear equilibrium sorption isotherm exists in the terrestrial environment as opposed to a more complex kinetic/equilibrium type isotherm. The justification for this assumption is as follows. First, the use of the empirical linear sorption isotherm implies that the concentration of solute in the solution is low and that equilibrium is attained rapidly [19]. The concentrations of radioactive solutes in the bogs are low (0 to 1 x 10^{-14} gm/ml) which satisfies the first condition. Secondly, laboratory experiments show that equilibrium is attained in 15-30 days for the peat under investigation [14]. Since the times under consideration are in years, the second condition is met. In addition, many studies indicate that ^{137}Cs does follow the linear sorption isotherm in organic/inorganic media [31]. The greater K_d values for minerotrophic bogs indicate that ^{137}Cs is more strongly sorbed in inorgani- rich (clay) bogs than in an organic-rich (humic) bog. This is not surprising, as Sanchez et al. [32] found that clays exhibit a higher ion exchange capacity for alkali metals than humic material. The values calculated by successive approximations of the ^{137}Cs deposition profile provide an estimate for the "global" advection term of -1.5 x 10-6 cm^{2}sec^{-1} in the organic-rich bog; the diffusivity term is calculated to be 8.0 x 10^{-6}cm sec^{-1}. The negative sign for advection indicates a net migration upward. In a natural system, the advection depends on evapotranspiration, temperature, precipitation, infiltration, water content and other processes. These interactive processes are averaged over relatively long spacial distances and temporal periods and a single advective term can be used to accurately represent these processes. This simplification greatly reduces the complexity of the model. In comparing two quite different bogs with high and low mineral content using a constant global advection term, the diffusivity was calculated to be 2.0 x 10^{-5} cm^2 sec^{-1} for the organic-poor bog.

The test of model structure was made on a ^{137}Cs profile comparing measured and predicted values for each level. The model predicted results were within 20% error on (predicted /measured) values [17].

4.3. RECONSTRUCTION OF REGIONAL EXPOSURE HISTORY TO TOXIC METALS

This dating information can be used to estimate the amount of pollutants inhaled by man at different time periods. The medical experts are well aware of the neurological problems which have been identified in people exposed to high levels of lead and mercury, for example. Numerous public health problems and reduction in intelligence quotient, IQ, in children have been found from exposure to urban and industrial effluents and from other sources prior to the imposition of environmental controls [33].

The following is a preliminary method for calculating the respiratory Pb dose to man exposed by breathing air contaminated with Pb. The use of a repository, such as a lake core, marshland or bog core which is dated by ^{210}Pb methods can establish the time history provided that the amount is determined in deposition units, i.e., amount cm^{-2} yr^{-1}. In urban air, Pb is found predominantly in the 0.25 mm particle sizes which has an approximate deposition velocity of about 0.3 cm sec^{-1} [34]. If one evaluates an exposed population, breathing is the exposure route and the respiratory tract where lead accumulates becomes important. If the deposition amount measured in cores from a lake or bog is 20 mg cm^{-2} yr^{-1} and the average deposition velocity for the particle sizes of lead present in the atmosphere is about 0.3 cm/sec it is estimated that the lead concentration in air is 2.12 μg m^{-3}.

If Standard Man inhales about 8,395 m^3 yr^{-1} air and the human respiratory system retains about 20% of the deposited particles of this size range, it is possible to estimate how much lead is retained by the human system by inhalation from the atmosphere per year [35]. The other sources of lead exposure are by food, dust and paint fragments entering the digestive system.

For example, if it is assumed that the average measured Pb concentration in air particulates is 1 μg m^{-3}, then the amount per year accumulated in the body would be 0.001679 g-yr^{-1}. Many cities have at least 20 times this level of Pb in the air or, in terms of exposure dose to man, 0.0336 g yr^{-1}. Over a 70-year lifetime, an adult would have accumulated from breathing 2.35 g Pb distributed in vital organs and bones of about 70 kg, giving an average concentration of 3.3 10^{-5} g kg^{-1}. Most of this Pb will accumulate in bone. A child inhales about 5475 m^3 yr^{-1} of air and, at an atmospheric concentration of 20 μg m^{-3}, would accumulate 0.0219 g yr^{-1}. If this is accumulated over the first 10 years of life and in a 20 kg body, an average concentration of 1.1 10^{-5} g kg^{-1} is calculated. Pb tissue deposition may vary due to metabolic processes as the weight increases and the minerals needed for growth and development are redistributed. The rapid growth of a child would distribute Pb in the rapidly growing organs and tissue as well as in the bone. Thus, the effects of such concentrations on the brain of a child may be quite different from that of a mature human. It is not yet known if this additional exposure (oral ingestion is the primary exposure pathway for lead) poses a significant health problem. Further studies are needed to evaluate the pharmacokinetics of lead in children and possible psychometric outcomes. This research may ultimately explain why children in many urban areas have a mean IQ 5 points lower than control children [33].

5. Conclusions

Results of atmospheric pollutant deposition history may be used to validate models for predicting transport and dispersion of metals and to suggest remediation methods for contaminated lands. The historical information presented here can help in setting local metals emission limits, give credence to the regulations already written, and document past pollution practices. Long-range transport of radioactive fallout and other pollutants has been demonstrated from nuclear weapons tests, the Chernobyl NPP accident, and trace metal deposition. The chronological deposition provides a means to estimate the exposure history of the pollutant dose to man since 1700.

6. Acknowledgements

The first author wishes to express his thanks for partial financial support from the Ministerio de Education y Ciencia of the Spanish government, program of sabbatical leave, SAB95-0228, contract CLI 95-1867 at the Universitat de Girona, and the University of Pittsburgh.

7. References

1. Krey, P.W. and Krajewski, B. (1970) Tropospheric scavenging of ^{90}Sr and ^{3}H, in R.J. Engleman and W.G.N. Slinn (eds.), Precipitation Scavenging, USAEC Conf. 700601.
2. Schell, W.R., Tobin, M.J. and Massey, C.D. (1989) Evaluation of trace metal deposition history and potential element mobility in selected cores from peat and wetland ecosystems, Sci. Total Environ. 87/88, 19-42.
3. Meyer, R. (1983) Interaction of forest canopies with atmospheric constituents: aluminium and heavy metals, in P. Ulrich and J. Pankrath (eds.), Effects of Accumulation of Air Pollutants in Forest Ecosystems, Reidel, Dordrecht, pp. 47-55.
4. Schell, W.R. (1987) A historical perspective of atmospheric chemicals deposited on a mountain top bog in Pennsylvania, Int. J. Coal Geol. 8, 147-173.
5. Novak, M.J.V. (1990) Speciation of sulfur and vertical $\delta^{34}S$ profiles in *Sphagnum*-dominated wetlands: indirect evidence for openness of the system toward buried sulfur, M.S. Thesis, Villanova University, Villanova, PA.

6. Novak, M., Wieder, R.K. and Schell, W.R. (1994) Sulfur during early diagenesis in *Sphagnum* peat: insights from $\delta^{34}S$ ratio profiles in ^{210}Pb-dated peat cores, Limn. and Oceanog. 39 (5), 1172-1185.
7. Vile, M.A., Novak, M.J.V., Brizova, E., Wieder, R.K. and Schell, W.R. (1995) Historical rates of atmospheric Pb deposition using ^{210}Pb dated peat cores: corroboration, computation, and interpretation, Water, Air and Soil Pollut. 79, 89-106.
8. Volchok, H. and de Planque, G. (eds.) (1983) EML Procedures Manual HASL-3000, Environmental Measurements Laboratory, U.S. Department of Energy, New York, New York.
9. Schell, W.R. and Barnes, R.S. (1986) Environmental isotope and anthropogenic tracers of recent lake sedimentation, in J.C. Fontes and P. Fritz (eds.), Handbook of Environmental Isotope Geochemistry, The Terrestrial Environment, Elsevier Scientific Pub. Co., Amsterdam, 2B, 169-206.
10. Hardy, E.P., Meger, M.W., Allen, J.S. and Alexander, L.T. (1968) Strontium-90 on the Earth's surface, Nature 219, 584-587.
11. Appleby, P.G. and Oldfield, F. (1978) The calculation of ^{210}Pb dates assuming a constant rate of supply of unsupported ^{210}Pb to the sediment, Catena 5, 1-8.
12. Tobin, M.J. and Schell, W.R. (1988) Recent developments in sedimentation modeling and the statistical reliability of the ^{210}Pb method, Paper presented at the Trace Metals in Lakes Conference, McMaster University, Ontario, Canada.
13. Schell, W.R. and Tobin, M.J. (1994) ^{210}Pb dating using the CRS-MV model with historical data to test and evaluate accuracy, in M. Garcia-Leon and R. Garcia-Tenorio (eds.), Low-Level Measurements of Radioactivity in the Environment, World Scientific Pub. Co., Singapore, pp. 355-36.
14. Schell, W.R., Sanchez, A.L., Underhill, D.D. and Thomas, E.D. (1986) Natural geochemical repositories for wastes: field and laboratory measurements of diffusion and distribution coefficients in age-dated peat bogs, in T.H.E. Sibley and C. Myttenaere (eds.), Application of Distribution Coefficients to Radiological Assessment Models, Elsevier Applied Sci. Pub., London, pp. 207-223.
15. Eagleson, D.E. (1978) Climate, soil and vegetation, Water Resources Research 14, 705-776.
16. Massey, C.D. (1987) A mathematical model using radioactive fallout for predicting radionuclide waste migration, Master of Science Thesis, University of Pittsburgh.
17. Schell, W.R. and Massey, C.D. (1988) Environmental radionuclide tracers of transport and diffusion in organic-rich soil of the unsaturated zone, in Proceedings of the Isotope Techniques in Water Resources Development Symposium, 30 March - 3 April 1987, IAEA-SM-299/124P, 781-783.
18. Small, M. (1986) Notes from course entitled Mathematical Modeling of Environmental Systems offered at Carnegie Mellon University, PA.
19. Jackson, R.E. and Inch, K.J. (1980) Hydrogeochemical processes affecting the migration of radionuclides in a fluvial sand aquifer at the Chalk River nuclear laboratories, National Hydrology Research Institute, NHRI, Paper No. 7, Ottawa, Canada.

20. Cameron, D.R. and Klute, A. (1977) Convective-dispersive solute transport with combined equilibrium and kinetic adsorption model, Water Resources Research 13, 183-188.
21. Barnes, R.S., Birch, P.B., Spyradakis, D.E. and Schell, W.R. (1979) Changes in the sedimentary history of lakes using lead-210 as a tracer of sinking particulate matter, Isotope Hydrology 1978, IAEA, Vienna, 975-898.
22. Crecelius, E.A. and Piper, D.Z. (1973) Particulate lead contamination records in sedimentary cores from Lake Washington, Seattle, WA, Environ. Sci. Technol. 7, 1053-1055.
23. Wieder, R.K., Novak, M.J.V., Schell, W.R. and Rhodes, W. (1994) Rates of peat accumulation over the past 200 years in five *Sphagnum*-dominated peatlands in the United States, J. Paleolimnol. 11, 1-13.
24. Schell, W.R., Tobin, M.J., Novak, M.J.V., Wieder, R.K., and Mitchell, P.I. (in press) Deposition history of trace metals and fallout radionuclides in wetland ecosystems using ^{210}Pb chronology, Water, Air and Soil Pollut.
25. Kadlec, R.H. and Keoleian, G.A. (1986) Metal ion exchange on peat, in C.H. Fuchsman (ed.), Peat and Water, Elsevier Applied Sci. Pub. Ltd., London, pp. 61-93.
26. Gamble, D.S., Schnitzer, M. and Kerndorf, H. (1983) Multiple ion exchange equilibria with humic acid, Geochemica et Cosmochemica Acta 43, 1311-1323.
27. Wolf, A., Bunzl, K., Dierl, F. and Schmidt, W.F. (1977) The effect of Ca^{+2} ions on the absorption of Pb^{+2}, Cu^{+2} and Zn^{+2} by humic substances, Chemosphere 5, 207-213.
28. Bunzel, K., Schmidt, W. and Sansoni, B. (1976) Kinetics of ion exchange in soil organic matter. IV Absorption and desorption of Pb^{+2}, Cu^{+2}, Cd^{+2}, Zn^{+2} and Ca^{+2} by peat, J. Soil Sci. 27, 32-41.
29. Irving, H. and Williams, R.J.P. (1948) Order of stability of metal complexes, Nature 162, 746-747.
30. Schell, W.R., Breslin, J.P., Tobin, M.J. and Schell, C.W. (1988) Recycling and removal of radionuclides in forest soil resulting from nuclear accidents, in IV International Symposium of Radioecology, The Impact of Nuclear Origin Accidents on the Environment, Volume 1, Cadarache, France. Commission of European Communities, Brussels-Luxembourg.
31. Pietrzak, R.F. and Dyal, R. (1982) Evaluation of isotope migration - land burial. NUREG/CR-2192, Vol. 1, Nos. 3-4, Nuclear Regulatory Commission, Washington D.C.
32. Sanchez, A.L., Schell, W.R., and Thomas, E.D. (1988) Interactions of ^{57}Co, ^{85}Sr and ^{137}Cs with peat under acidic precipitation conditions, Health Physics 54, 317-327.
33. Needleman, H., Greiger S. and Frank, R. (1985) Lead and IQ scores, Science 227, 701-704.
34. ICRP-30 (1979) International Commission on Radiological Protection, Report of Committee II, Limits for intakes of radionuclides by workers.
35. Casarett, L.J. and Doull, J. (1975) Toxiciology, Macmillan Publishing Co., Inc., New York.

THE USE OF PORTABLE INSTRUMENTATION FOR ASSESSING ENVIRONMENTAL CONTAMINATION IN THE URALS REGION OF RUSSIA

T. M. SPITTLER
U. S. Environmental Protection Agency
60 Westview St., Lexington, MA 02173, USA

Abstract

Two of the principal environmental contamination problems facing the mining regions of the Urals are pollution of air and water by volatile organic chemicals and pollution of air, water and soil by heavy metal wastes from mining and heavy industry such as steel production. Because of the extent of these problems and the large areas affected in the Urals region the use of on-site analysis instruments would provide a powerful tool for making rapid assessments of air, water and soil contamination and for determining where to sample when samples are to be returned to a central laboratory for confirmatory analysis or for further intensive testing.

The first instrument is a very sensitive gas chromatograph using photoionization as the detection principle. This instrument has been widely used in America to perform rapid assessment of air, water and soil contamination from gasoline and other fuel leaks, hazardous waste disposal sites and workplace monitoring to protect workers from unnecessary exposure to volatile chemicals. The instrument has sensitivity to aromatic compounds and ethylene type compounds such as trichloroethylene (TCE) and vinyl chloride (VCM) down to the pictogram range. This results in a sensitivity to benzene and TCE of about 1-10 ppb in air.

Water and soil are analyzed using the simple technique of analyzing headspace in a closed vial for water or by placing soil in a vial with appropriate amounts of methanol or water and analyzing the headspace above the aqueous slurry or by dilution of the methanol extract in water and again using headspace analysis. Using the simple principle of Henry's Law, water can be analyzed for volatile organic down to the ppt range for aromatics and the low ppb range for alkanes (e.g. carbon tetrachloride, etc).

Detection of volatile contamination from spills or leaks of volatile organic chemicals can be done using another simple field technique. The technique is called "Soil Gas Analysis" and consists of making a small borehole in the area of a suspected contamination incident and sampling the soil gas, using the field gas chromatograph to produce real time data on the nature, spread and concentration of the spilled

I. Linkov and R. Wilson (eds.), Air Pollution in the Ural Mountains, 115–126.

material in the soil. This technique has found very widespread use in America where many buried fuel tanks have been found leaking with the consequent problem of widespread contamination of ground water supplies.

The second instrument is a portable X-Ray Fluorescent (XRF) Analyzer. This instrument uses a low level radioactive source (Cd 109) to excite heavy metal atoms in a sample which then fluoresce to produce characteristic (Raleigh) scattered peaks which can be used for both qualitative and quantitative analysis. With this instrument, the presence of lead and other toxic heavy metals can be readily detected in the field to determine where soil has been contaminated by mining and metallurgical operations, leaving the soil at levels that are dangerous especially to children playing in and around these operations.

1. Field Gas Chromatography

There are several instruments on the market, which can perform field analysis of volatile organic compounds. However, because of the need for special gases and large power requirements of several of these instruments, this discussion will be limited to the use of the Photovac 10S instrument. This instrument can be operated from an internal rechargeable power supply and can also be operated from an ordinary 12-volt automobile battery (using the cigarette lighter outlet). The only carrier gas required is clean nitrogen or clean ambient air. Because it is often difficult to obtain nitrogen of sufficient purity, we have operated the equipment in Russia using a simple device, which is inexpensive and practical. The gas supply contains a 12-volt d.c. compressor, a small holding tank and a canister containing clean activated charcoal. The tank is filled to about five atmospheres (100 psi) by using the cigarette lighter adapter of any automobile. This supply (in a two-liter tank) can operate an instrument for several hours and needs only to be reattached to the cigarette lighter adapter to recharge the tank when pressure drops to about two atmospheres.

1.1 AIR ANALYSIS

Because of the extraordinary sensitivity of the Photovac GC, ambient air can be analyzed by direct injection rather than by the more costly and time-consuming technique of preconcentration. A 1 cc sample of air contains about 1 mg by weight. Since the instrument can detect a few picograms of aromatic and ethylenic compounds, this represents an ambient concentration of 1-10 ppb of benzene, toluene, TCE etc. in the sample. The sensitivity to organic molecules which do not contain a double bond (e.g. carbon tetrachloride, methylene chloride, hexane etc.) is about 50 times poorer. Hence, these compounds can be detected in air samples in the range of 50-100 ppb. For most ambient air purposes, this is still adequate sensitivity to assist the field chemist in tracking plumes or monitoring volatile solvents in workplace situations.

Since this type of sampling is not time averaged, it is often necessary to take several samples over some predetermined time period to obtain time-weighted averages. Many of the Photovac instruments are equipped with a program that allows this kind of periodic automated sampling and sample averaging to obtain a time-weighted average level for air concentration. It is obvious that where higher concentrations are expected smaller samples can be injected into the instrument in order to keep the response within a calibration range.

1.2 WATER ANALYSIS

Analysis of water samples contaminated by volatile organic chemicals is very simple. The sample is collected in a volatile organic analysis (VOA) vial. This is simply a clean glass vial with a septum closure. The side of the septum facing the sample should be of teflon, and the cap should provide easy access to puncturing the septum and sampling without any need to open the bottle once the sample is collected. If the sample is to be analyzed within a few hours of collection, there is no need to place any kind of preservative in the sample. If the sample is to be held for an extended period of time (2 days or longer), it must be preserved against biodegradation. This is particularly important for aromatic hydrocarbon contaminants. Figure 1 shows a study of water samples contaminated with toluene and held at 4 degrees centigrade. Rapid biodegradation occurs for all aromatics except o-xylene in this study. Since bacterial populations differ from place to place, other patterns of degradation may occur. Even chlorinated solvents such as TCE and perchloroethylene (PCE) have been shown to undergo degradation to daughter products.

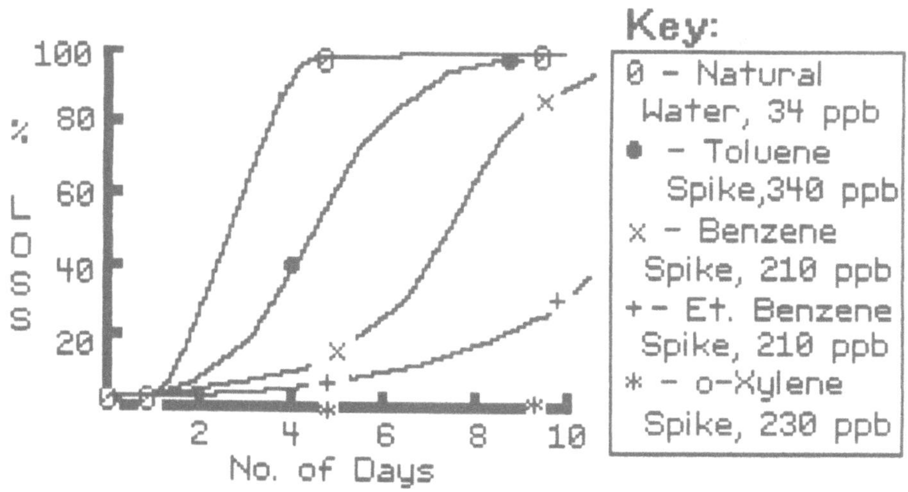

FIGURE 1. Biodegradation of aromatic compouonds in unpreserved warer held at 4 degrees centigrade.

One of the best preservatives of water samples is mercuric chloride. The water sample should be preserved by adding enough mercuric chloride to make the solution about 20-40 ppm. This is done in the field by using a saturated solution of mercuric chloride in water from which a small amount is added to each sample at the time of collection. Mercuric chloride is soluble in water to about 7 wt %. When 10 ul of this concentrate is added to 30 cc of water, the final sample will contain about 24 ppm of mercuric chloride. Samples can also be preserved by adding strong mineral acid to make the final sample less than pH 2. However, when samples of groundwater contain high levels of carbon dioxide, the addition of acid causes both evolution of carbon dioxide and the dissolved organic solvents which should be measured.

1.2.1 *Headspace Analysis*

Headspace analysis is the most convenient and simplest technique to analyze water samples for volatile organic contaminants. The basic headspace principle is shown in Figure 2. Many authors have reached agreement that when a sample vial contains 25% headspace, the headspace technique is optimized for sensitivity and ease of sampling. Hence, when sample vials (VOA vials) are submitted for analysis filled, one simply inserts a hypodermic needle to relieve the vacuum and then with a syringe, withdraws 25% of the water. The sample is shaken vigorously for about 60 second to achieve Henry's Law equilibrium and an appropriate sized aliquot of headspace vapor is withdrawn and injected into the GC. It is important to flush the syringe several times before any attempt is made to perform a quantitative transfer of vapor into a GC. This simple process allows for adsorption losses to the walls of the syringe to be minimized. With good technique one can reproducibly sample headspace with about 2-5% precision.

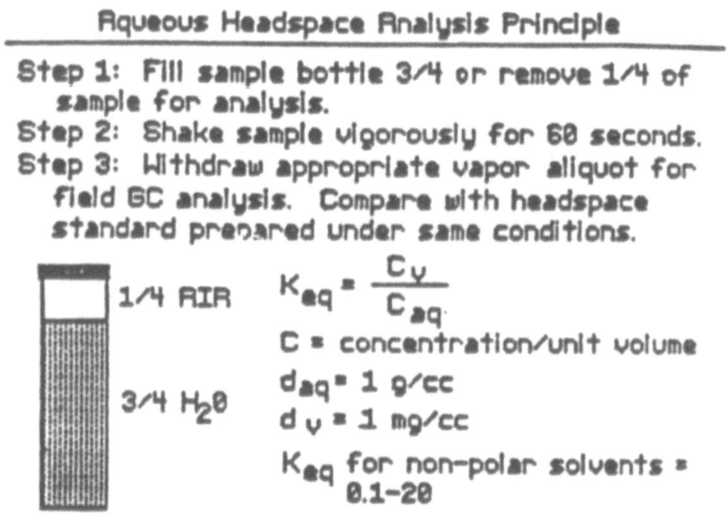

FIGURE 2. The Aqueous Headspace Analysis principle

Quantitation of dissolved solvents in water is achieved by simply analyzing a known standard and an unknown under identical conditions of temperature, sampling conditions and analysis conditions. For example; a standard containing 100 ppb of benzene is analyzed by injecting 100 ul of vapor and recording a response of 50% chart for the benzene peak. An unknown sample has a peak at exactly the same retention time but is 75% of chart and was achieved by injecting 200 ul of headspace vapor. After correcting for the use of twice the sample volume, one concludes that the unknown sample contains 75 ppb of benzene. The Photovac GC can be programmed using known standards and exactly reproduced sampling conditions and will yield quantitative data with excellent precision and accuracy. Both precision and accuracy are most affected by good sampling and injection technique.

1.2.2 *Two Column Chromatography*

Positive identification of peaks by retention time is often the only requirement when samples contain very few contaminants and there is supplementary information on the possible identification of contaminating organic chemicals. Where more certainty is required, there is a technique for positive identification, which relies on identifying the retention time of an unknown solvent peak by performing analysis of standard and unknown on two different chromatographic columns. Figure 3 shows chromatogram of a mixture of chlorinated organic volatiles, which have been run on two different chromatographic columns. This technique of two-column chromatography is widely used by chemists for routine positive identification of mixtures of organic chemicals. One of the important features of the Photovac GC is that it incorporates this technique into a field instrument.

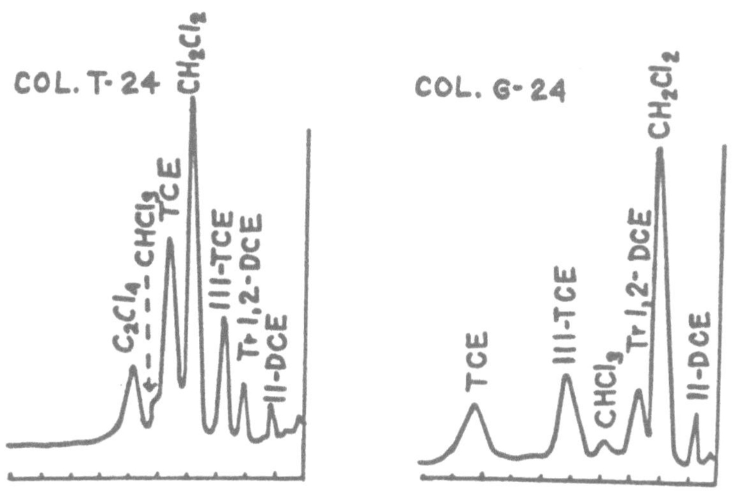

FIGURE 3. Chlorinated organic volatiles run on two different GC columns

A second important use of the two-column feature of the Photovac GC is that a long and short column of the same or similar packing material can be used to greatly enhance the efficiency of field sampling and analysis. Figure 4 shows this technique in practice. A mixture of four chlorinated compounds is shown analyzed on a short and a long column. The long column analysis provides very clean separation of the peaks, but the entire analysis time is about 17 minutes. The same sample, when analyzed on a short column achieves adequate separation in about one tenth of the time. This technique is often used in the field when many samples from a site must be analyzed, and the principal concern is to know the relative concentration of volatiles as one progresses around a site or analyzes several water samples from the same contaminated aquifer. This technique is also very useful for doing soil gas profiling (which is explained below).

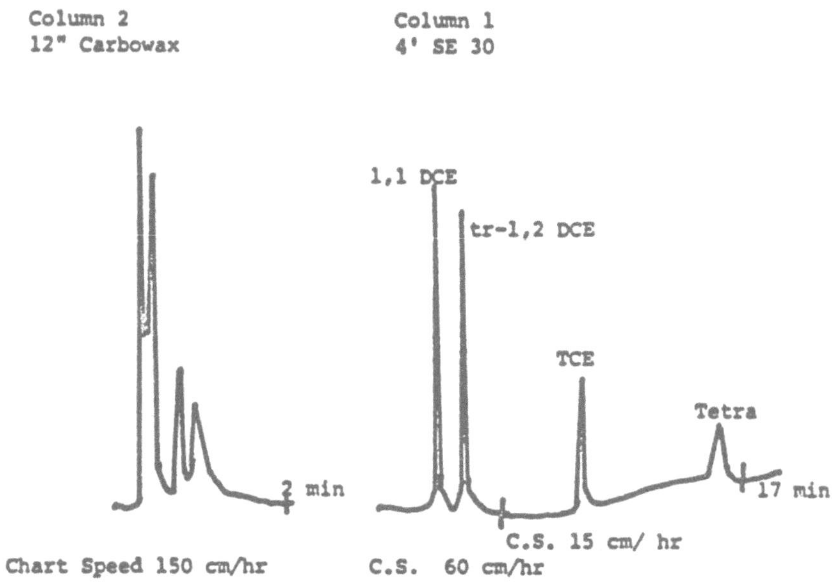

FIGURE 4. The use of a long and short column for greater efficiency in field GC work

1.3 SOIL ANALYSIS

Soil samples are analyzed in two ways. The first technique is a simple screening analysis. A suspect contaminated soil is placed in a plastic bag and sealed. The bag is shaken to break up the soil structure and allow volatiles to evaporate from the soil. A syringe needle is then used to puncture the plastic bag and withdraw an appropriate sample volume for injection into the GC. The sample size should be adjusted to the suspected volatile contamination level in the soil. If volatiles can be smelled, it indicates that contamination is severe and a very small sample of vapor should be injected into the GC. If no odor is detected from the soil, a correspondingly larger sample (up to 1 cc) may be injected. Once contamination has been detected, and the range of high or low level of contamination is determined by preliminary screening, a more precise procedure should be used.

1.3.1 *High Level Contamination*

High level contamination indicates that the sample will require significant dilution before a direct injection is made into the GC. This is done by placing about five grams of soil into a septum vial which already contains 10 cc of methanol. To prevent loss of volatiles, the soil sample should be collected and placed in the methanol as efficiently as possible. The procedure now used in America is to use a plastic 10 cc syringe which has had the tip of the syringe cut off with a sharp knife (Figure 5). This results in a clean calibrated barrel and plunger. The plunger is then drawn back to about the 5 cc mark and the syringe (sampler) is plunged into a fresh soil surface. A core of soil is thus obtained with a definite volume. The barrel of the syringe is immediately wiped clean with a paper towel or clean cloth and the sample is ejected into the sampling vial with no disruption of the soil structure. This technique yields the best data on the actual concentration of volatiles contaminating the soil. Once the sample is covered with methanol there is no possibility of loss of volatiles since all organics are highly soluble in methanol. The vial is capped with a septum cap and shaken vigorously to break up the soil structure and rapidly dissolve all organic into the methanol.

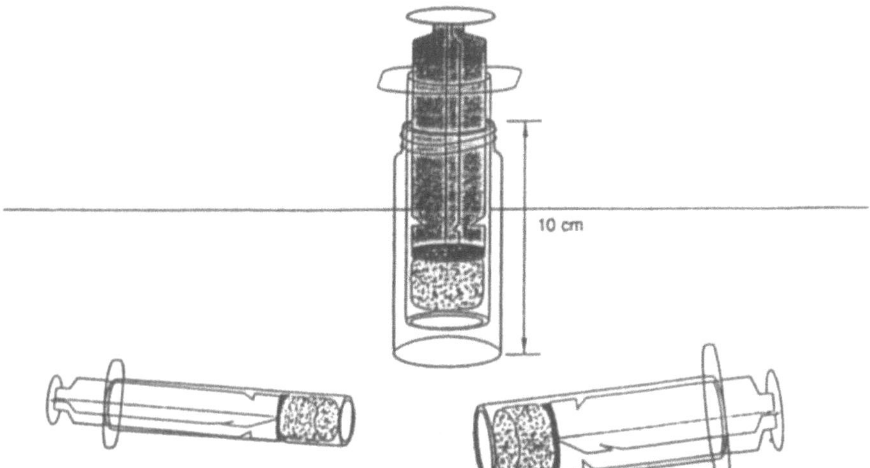

FIGURE 5. Using plastic syringes for soil sampling

This sample can now be held indefinitely for analysis without any loss of volatiles. If analysis is to be performed immediately, a small volume of the methanol is removed from the vial and injected into a measured amount of clean, volatile-organic free water. Any dilution factor may be used up to about 1% methanol extract in water. The diluted methanol sample is then shaken and used for headspace analysis as described above. It is only necessary to prepare a standard containing the volatiles of concern at a known concentration and under identical conditions of headspace/water ratio, temperature and instrument analysis conditions.

1.3.2 *Low Level Contamination*

When preliminary testing or knowledge of the contamination situation indicates that levels of volatile organic in soil are low (1 ppm or less), a slightly different procedure is used to analyze for volatile contamination. Using the same sampling device described above, a sample core of about 5 cc is injected into a septum bottle containing a measured amount of clean water and the bottle is immediately sealed. If the sample is to be held for any length of time beyond a few hours, it is necessary to preserve the sample against biodegradation as described above for water samples. Analysis is performed exactly as described above for headspace analysis of water samples. Using this technique it is possible to determine levels of organic contamination in soil below the ppb level. While such low levels are of no concern either as regards health or groundwater contamination, this remarkable sensitivity enables a field investigator to track contamination in soil for great distances, and provides excellent data for determining the direction of contaminant migration long before a spill of chemicals has had time to reach sensitive water supplies.

1.4 SOIL GAS ANALYSIS

In addition to the soil analysis techniques described above, there is another method which has been used extensively in America and elsewhere for about 13 years. This method involves the sampling of vapor in the vadose zone to determine the relative levels of contamination present in and near a suspect spill, leaking underground storage tank or other source of contamination of the soil by volatile organic chemicals. While recent research has been done to correlate soil analysis as described above with soil gas analysis, it is not necessary to establish this direct correlation in order to derive the principal benefits of this technique.

Soil gas analysis is performed by inserting into the soil a sampling tube which can be used to draw air (or soil gas) from the vadose zone for direct GC analysis in the field. There are countless variations on the exact technique for this kind of sampling, but all have in common a few key points. First, the sample must be withdrawn in a way that vapor is sampled at the bottom of the hole or sampling device rather than drawing air down the sides of the sampling tube because the hole is open to the atmosphere. This is readily accomplished with driven probes which are self-sealing as they penetrate to the desired sampling depth. When a hole is bored and a tube is inserted into the borehole, it is necessary to seal the hole around the sampling tube at the surface so that the sampling pump does not short-circuit the desired sampling region by drawing air down through the annular space between the hole and the sampling tube.

The second important factor in soil gas sampling is to assure that the vapor drawn up for sampling is representative of the vapor at the bottom of the sampling hole. To assure that equilibrium is achieved between where the sample is taken for analysis and the vapor at the bottom of the hole, it is advisable to first test the sampling procedure in the following manner. Using a rather low flow pump, air is removed from the sampling hole until about one or two volumes of the sampling device are pumped out. A sample is then injected into the GC and the principal peak is noted. This procedure is then repeated a few times until there is no further increase in the principal contaminant peak on the GC chromatogram. Figure 6 demonstrates this procedure. If this procedure is not used, it is possible to sample such a large volume of soil gas that the sampling depletes the vapor present around the sampling hole and the analysis results are much lower than the actual level of contamination present.

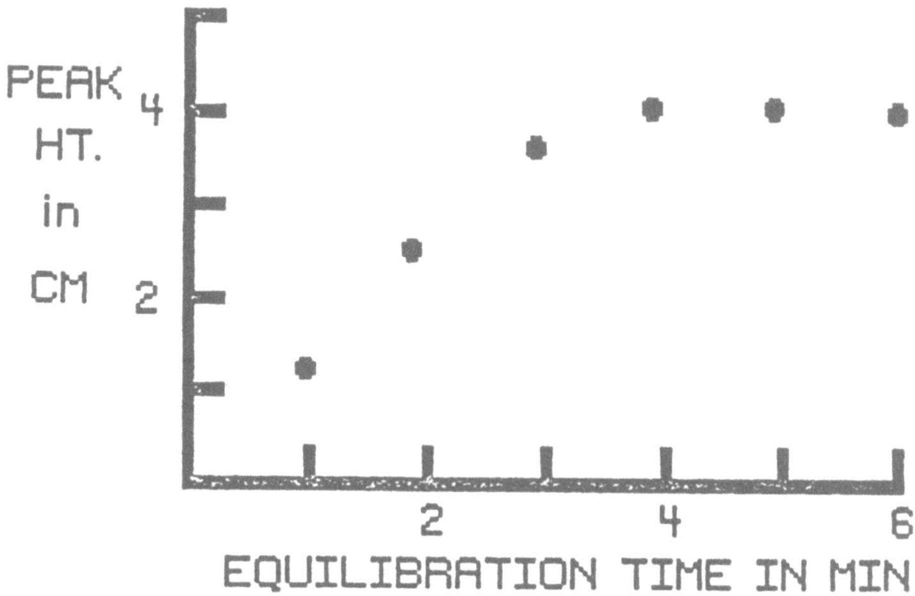

FIGURE 6. Establishing equilibration time for soil gas analysis

This soil gas technique does not give data, which are an exact measure of soil contamination. There are several factors, which influence the level of contamination in soil gas at any particular site. Soil moisture will affect the Henry's Law equilibrium between the soil, moisture and vapor space. Also, the presence of high levels of organic matter greatly influences the distribution of organic vapor in a soil-water-air matrix. However, when several soil gas measurements are made under the same conditions, the relative values obtained are very valuable date for the determination of the vertical and horizontal concentration gradients of volatile contamination. These gradients in turn give valuable information on the source, migration pathways and potential harm which spilled organics pose to valuable groundwater resources. Often, with careful work, it is possible to track the path of a groundwater plume using this technique. The data from such a field study is very important in determining the placement of interceptor wells to prevent migration of contaminated groundwater into potable aquifer areas.

2. Field XRF Analysis

The second important field instrument is the XRF analyzer. These tools have been in use in America and elsewhere for many years. However, it is only in the last few years that very compact and simple instruments have been developed. In this category, the instrument developed by the Niton Company of Bedford, MA stands out. The Niton instrument is only about 2x3x8 inches and weights approximately four pounds. The instrument is completely portable, operating on a rechargeable battery pack which provides six hours of continuous operation. Sensitivity is in the range of 100-200 ppm. Precision and accuracy can be improved by longer counting times. Typically lead in soil can be analyzed in about 60 seconds.

2.1 SOIL ANALYSIS

The most common use of this instrument at the present time is for determining the presence of heavy metals in soil. The instrument is factory calibrated and can be periodically checked against a know standard. Analysis is simply a matter of placing a soil sample in a plastic bag and placing the instrument detector window against the sample. A safety switch prevents accidental exposure to the low-level radioactive source. Alternatively, the soil can be measured by placing a thin sheet of plastic directly on the soil and pressing the instrument to the surface. A metal template is supplied which provides the hard surface needed to open the detector window for analysis. Using simple technique, it is possible to profile heavy metal contamination over very large areas in a short period of time. Because the measurements are in real time, each successive measurement can direct the field investigator in the proper direction to obtain a complete overview of the contamination pattern.

The Niton instrument is programmed to quantitate several heavy metals of interest. Also, the instrument has an LED window in which the energy spectrum is displayed while a measurement is in progress. Therefore, there is no doubt about the presence of any metal, even when neighboring elements in the periodic table are present and cannot be completely resolved by the instrument. After analysis, one of the modes of data display brings up a cursor line and displays the element symbol with its energy in Kev as the cursor crosses each peak on the energy display. Some future developments of this instrument will include: a wider range of elements which can be detected through use of more than one x-ray source; better resolution through the use of better detector cooling and technology; and better sensitivity which will be achieved by a combination of detector technology, resolution and varied energy sources.

2.2 WATER ANALYSIS

Water analysis by XRF is an old technology. Researchers have been using this type of instrument for over 20 years to detect low levels of heavy metals dissolved in aqueous media. The commonest technique is to adjust the pH of the sample to optimize capture of the heavy metals on ion exchange resins. The resins are impregnated into filter discs through which the aqueous sample is forced. The heavy metal ions concentrate on the resin impregnated paper and can be analyzed for low levels within minutes of collection. Sensitivity is achieved down into the ppb range and is dependent on the resin capacity, instrument sensitivity and the type of aqueous matrix analyzed.

Though it is obvious, it is perhaps important to point out that the above method is designed only to measure dissolved or ionic metal species. Suspended material can also be measured in water samples by XRF if one simply draws or forces the water through a neutral filter paper which is capable of capturing all suspended particles. Again, the large concentration factor achieved by this method enables the determination of very low levels of suspended particulates.

2.3 OIL AND OTHER LIQUID ANALYSES

Any liquid sample can be analyzed by XRF by simply placing the liquid in a sampling cup which can then be exposed to the x-ray source. Again, sensitivity is limited by the instrument's minimum detection level and the sample matrix. Oils are often analyzed by this technique to determine heavy metal contamination prior to recycle or proper disposal.

2.4. AIR ANALYSIS

Particulate matter in the air can be analyzed by XRF using the simple expedient of collection of a large amount of airborne particulates on some kind of inert filter material. Fiberglass is a medium of choice for this analysis, but is not necessary. Any clean filter medium can be used provided that a sufficient sample of air can be filtered. Calibration is a much greater problem for than for water or oil and other liquids. Many researchers have developed techniques for particulate analysis. The most direct technique is to use collected air samples which have already been analyzed by an independent method (e.g. Atomic Absorption) as secondary standards.

3. Summary and Conclusions

The use of portable instruments to address some of the common environmental contamination problems of the Urals Region is already a reality in the Perm Region. A portable (Photovac 10S50) gas chromatograph and a Niton XL analyzer have been in use for some time now, and other similar instruments will be brought to the region in the coming years. It is the hope of this author that such equipment will provide a better capability for rapid assessment of contamination of air, water and soil as the cities in the Urals Region gear up to deal with many years of environmental contamination. Such an approach can provide real-time data on a very cost-effective basis and will help identify the places where limited funds can be best expended to effect environmental cleanup with maximum benefit to all inhabitants, especially the most susceptible population, Russia's children.

USE OF SMALL MAMMALS FOR MONITORING HEAVY METAL CONTAMINATION IN THE ENVIRONMENT

K. SAWICKA-KAPUSTA and M. ZAKRZEWSKA
Department of Animal Ecology,
Institute of Environmental Biology,
Jagiellonian University, Ingardena 6, 30-060 Krakow, POLAND.

Abstract

Bank vole, yellow necked mouse, and common shrew, three most common and wildly distributed species of small mammals in forest ecosystems, were used as monitors to estimate heavy metal contamination of natural environment in Poland. Animals were caught between 1980 and 1994 in seven national parks, two moderate and one heavily polluted forest sites and two pristine areas. Five elements (Cd, Pb, Cu, Zn, Fe) were determined using AAS method in liver, kidneys, bones and whole bodies of animals. Heavy metal concentrations in rodents were lower than in insectivorous species. Animals from pristine areas (Bialowieza and Borecka Forest) had lower concentrations of heavy metals than animals from polluted areas. Extremely high concentrations of toxic metals were found in small mammals from Olkusz Forest. During fourteen years small mammals using as monitors showed different contamination of investigated areas. Recently they confirmed decreasing pollution trend in Poland. They also showed that still ecosystems located close to industrial sources are highly contaminated.

1. Introduction

Poland was one of the European country suffering from extremely high levels of industrial emissions. Atmospheric pollution in the eighties was among the heaviest in Europe making our country the third larger polluter [1]. Due to emission from sources located in Poland, and because of trans-boundary transport the natural environment in the wide range of the country was devastated. Although industrial emission have rapidly decreased in the recent years, new sources of emission developed, we still need to estimate contamination of natural environment. Mosses and lichens are normally used as bioindicators of air pollution [2]. To estimate heavy metal contamination of forest, meadows or field ecosystems small mammals are very useful monitors [3-6]. They are enough small and easy to catch. They have a territory of limited range and

I. Linkov and R. Wilson (eds.), Air Pollution in the Ural Mountains, 127–133.

fairly short life span. They are closely adjusted to their environment [4]. Concentrations of heavy metals in tissues and bodies of small mammals vary with the degree of contamination [3-5]. In many countries small mammals have been used in assessment of heavy metal contamination around different sources of pollution like smelters, mines, waste disposal sites, dumps, along the highways [5, 7-10]. In Poland the levels of heavy metals in tissues and bodies of some rodents and insectivorous species were analyzed for almost twenty years [6, 11-19]. The purpose of this paper is to summarize results concerning using small mammals as monitors in Poland.

2. Material and Methods

Bank vole - *Clethrionomys glareolus* (Schreber, 1780), yellow necked mouse - *Apodemus flavicollis* (Melchior, 1834) - two species of rodents and common shrew - *Sorex araneus* (Linnaeus, 1758) insectivorous species were collected between 1980 and 1994 from different polluted areas in Poland. Presented data were published in several papers listed in references [6, 11-19]. The study areas were located in the Drawieński (DNP), Wigierski (WNP) National Parks and in Bialowieza (BF) Forest - located in the northern part of Poland; in the Swiêtokrzyski (SwNP) National Park and Kampinos (KF) Forest - situated in the central part of Poland; in the Ojcowski (ONP), Gorczański (GNP) and Pieniñski (PNP) National Parks - located in the southern part of Poland. The animals were caught also from Olkusz (OF) Forest - heavily polluted, from Niepolomice (NF) Forest - moderately polluted areas in southern Poland and from unpolluted area Borecka (BoF) Forest - situated in the north-eastern part of Poland.

Concentrations of cadmium, lead, zinc, iron and copper were determined in the whole bodies of small mammals, from which the stomach and alimentary tract were removed earlier. In animals from Ojcowski National Park, Borecka and Olkusz Forests heavy metal concentrations were also determined in liver, kidneys and bones. Dry samples were wet digested in a 4 : 1 HNO_3 : $HClO_4$ mixture. Heavy metal concentrations were determined by AAS. The Kruskal-Wallis and Mann-Whitney nonparametric tests were used to determine the significance of differences among the study areas in the concentrations of heavy metals in the bodies of the animals collected.

3. Results And Discussions

The study in national parks were conducted in 1990 [11, 12]. Among six investigated national parks the most contaminated were Ojcowski and Gorczański NP (Fig. 1, 2). Levels of toxic metals (Pb and Cd) in small mammals from some national parks were much higher than expected for protected areas. The highest average lead concentrations were found in the bodies of common shrews from Gorczański NP and of bank voles from Ojcowski NP. These concentrations were significantly higher than those in the animals from other investigated national parks (Fig. 2). Cadmium concentrations in common shrew bodies indicate contamination of Ojcowski, Swiêtokrzyski NP and both mountain parks Gorczañski and Pieniñski NP (Fig. 1). No contaminated national parks are Drawieñski and Wigierski NP located in the northern part of Poland.

Bank voles using for assess contamination of Bialowieza Forest were caught in 1980 and 1985 [13, 14]. Lead and cadmium concentrations in livers, kidneys and bones of bank voles showed higher contamination of Bialowieza and Niepolomice Forests in 1985 than in 1980. Levels of toxic metals in animals from Bialowieza Forest were much higher than expected in the control area [14]. According to those findings since 1985 Bialowieza Forest was not treated as a reference area for Poland in any studies on the degree of contamination (Fig. 3).

FIGURE 1. Cd levels in the bodies of bank voles and common shrews from Polish National Parks (1990).

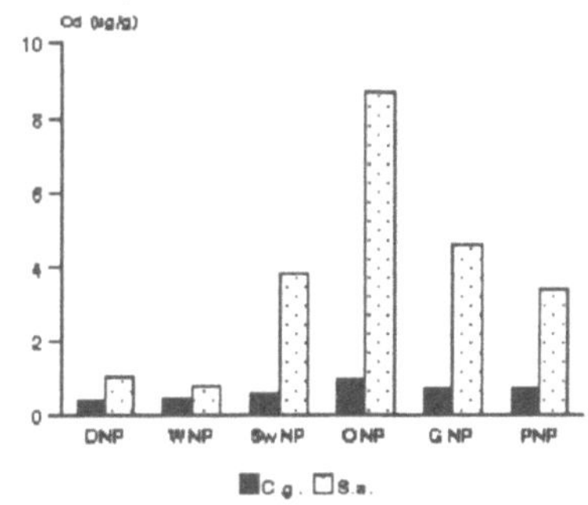

FIGURE 2. Pb levels in the bodies of bank voles and common shrews from Polish National Parks (1990).

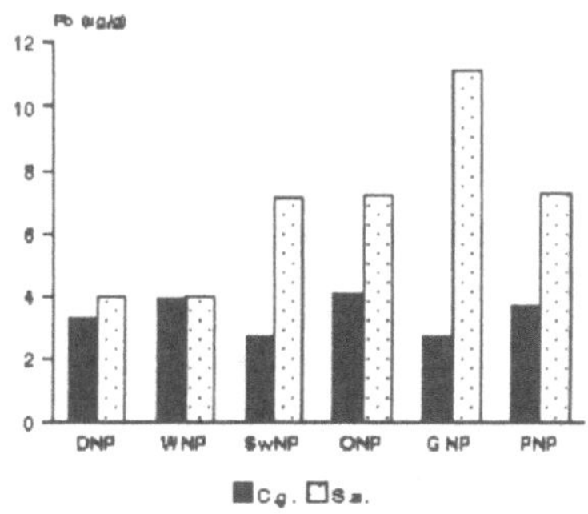

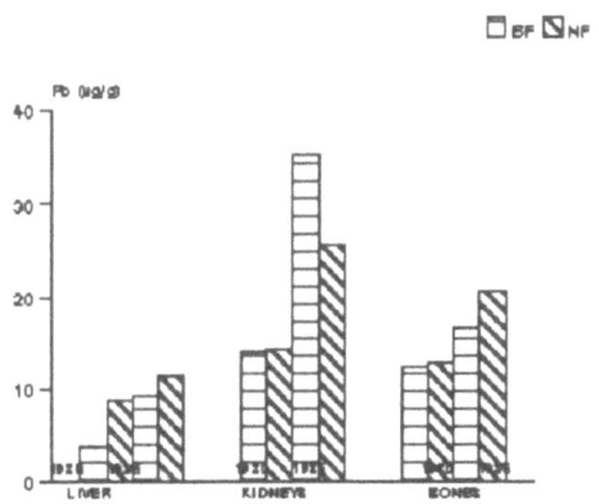

FIGURE 3. Pb concentrations in the tissues of bank voles from various polluted areas in. areas Poland (1980-1985)

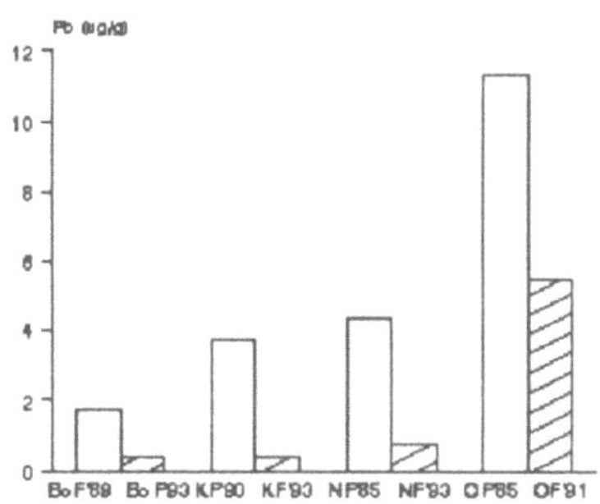

FIGURE 4. Pb concentrations in the bodies of bank voles from the same in different years.

In 1985-1993 in four forest sites in Poland monitoring studies were conducted. Bank voles were caught in heavily polluted Olkusz Forest, in two moderate polluted Niepolomice and Kampinos Forest also in Borecka Forest - situated in the north-eastern part of Poland far from any industrial centers (Fig. 4). Lead and cadmium levels in the bodies of bank voles caught in 1993 in investigated forest were lower than in the previous time [15, 16]. Bank voles used as monitors confirm a decreasing contamination of study forest areas in comparison with contamination found in the same areas a few years earlier. It was connected with decreasing of dust emissions, mainly lead particulates, in Poland in the recent years [1].

In 1991 bank voles and common shrews were caught in three different polluted areas: Olkusz, Niepolomice and Borecka Forest [15, 17]. Cadmium and lead concentrations in the common shrew bodies are statistically higher than those of the bank voles from the same forest area. The highest toxic metal concentrations were found in the animals from Olkusz Forest (Fig. 5, 6). The relatively low levels of both metals in small mammals from Borecka Forest declare this site as the new control area in Poland.

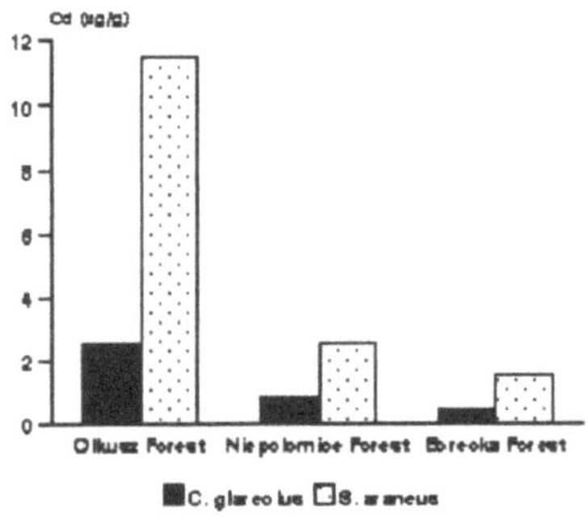

FIGURE 5. Cd concentrations in the bodies of small mammals from different polluted areas (1991).

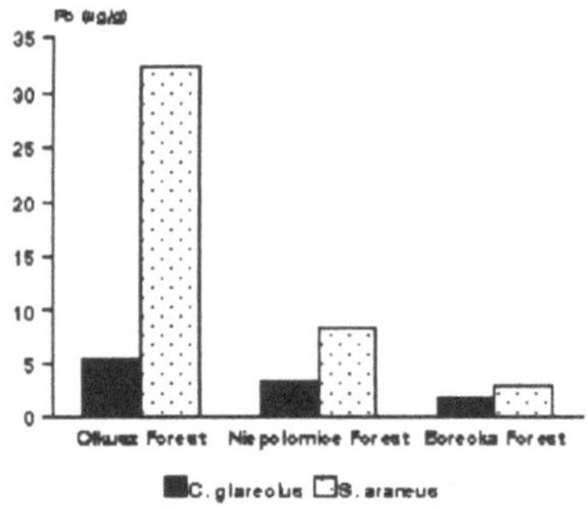

FIGURE 6. Pb concentrations in the bodies of small mammals from different polluted areas (1991).

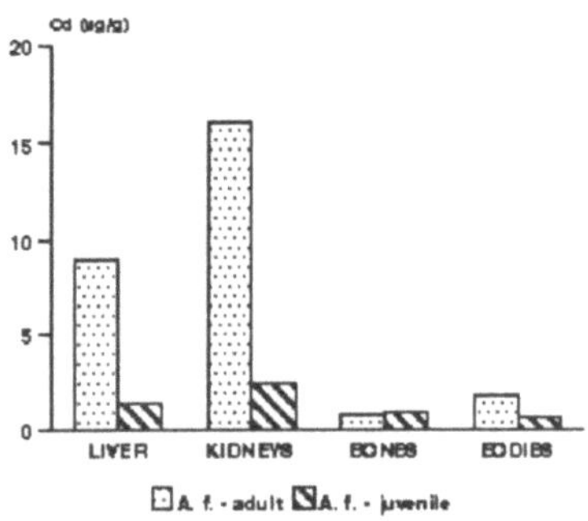

FIGURE 7. Cd concentrations in the tissues tissues of yellow necked mouse *(Apodemus flavicollis)* from Olkusz Forest (1994).

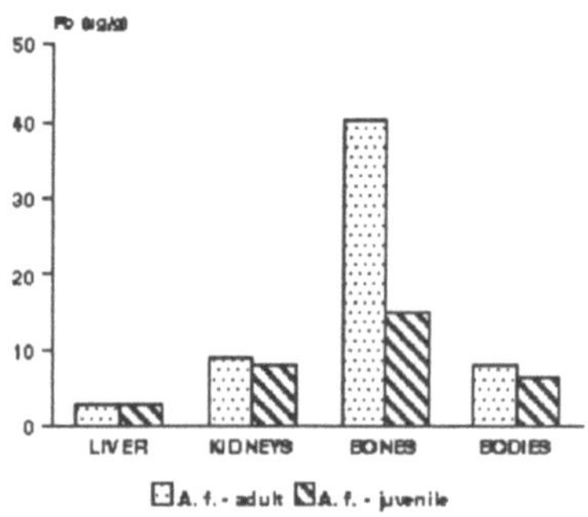

FIGURE 8. Pb concentrations in the tissues of yellow necked mouse *(Apodemus flavicollis)* from Olkusz Forest (1994).

Olkusz Forest is situated 5 km north-east from Pb-Zn smelter in Upper Silesia Industrial Region. In 1994 heavy metal concentration in two different age classes of yellow necked mouse from Olkusz Forest were analyzed [18]. There are statistical differences in cadmium concentrations between two months old and adult animals. In case of lead the juveniles and adults mice have the same concentration in their tissues and bodies with exception for bones (Fig. 7, 8).

The similar lead level which was accumulated in tissues of young mice in very short time indicate high contamination of their food and natural environment. In 1991 and 1994 in Olkusz Forest three species of small mammals (*Apodemus flavicollis, Clethrionomys glareolus* and *Sorex araneus*) were caught and heavy metal concentrations were analyzed in their livers, kidneys and bodies [17-19]. The same cadmium concentrations were found in livers and whole bodies of *A. flavicollis* and *C. glareolus* in 1991 and 1994. *S. araneus* accumulated less cadmium than animals from 1991. Small mammals from both years had the same amount of lead in their bodies (Fig. 9-12).

FIGURE 9. Cd concentrations in the liver of small mammals from Olkusz Forest.

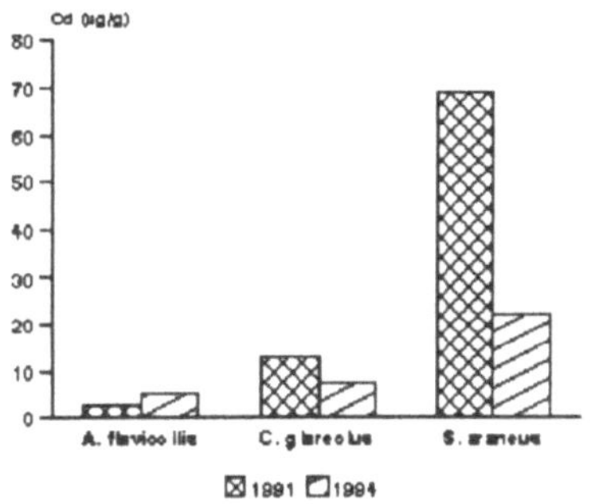

FIGURE 10. Cd concentrations in the bodies of small mammals from Olkusz Forest.

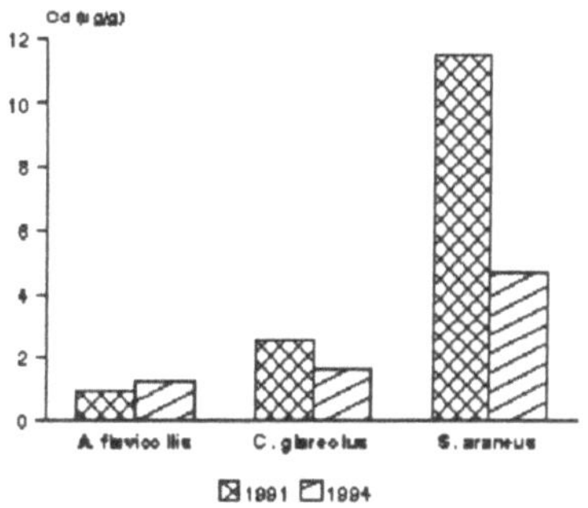

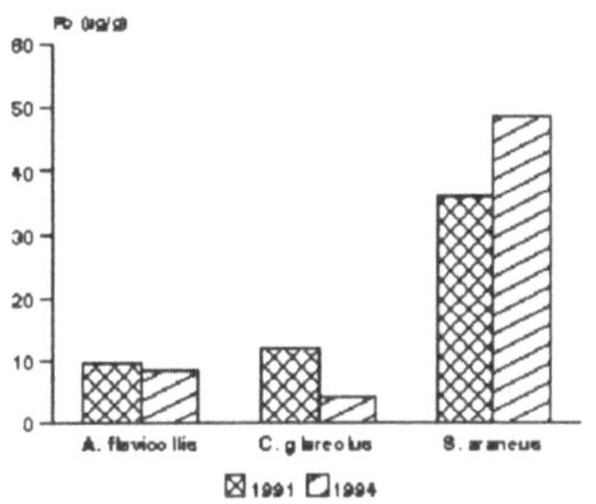

FIGURE 11. Pb concentrations in the kidneys of small mammals from Olkusz Forest.

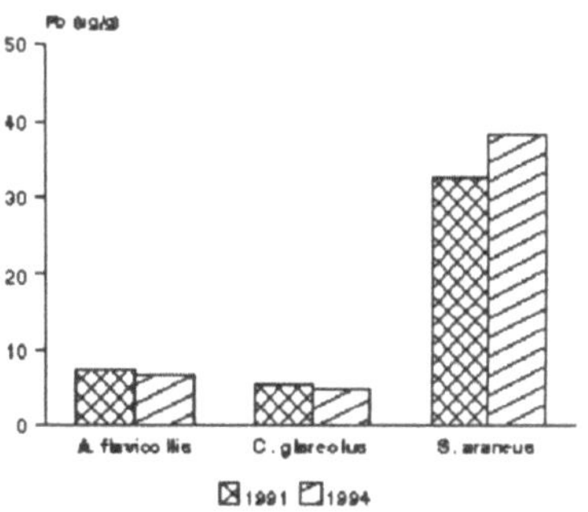

FIGURE 12. Pb concentrations in the bodies of small mammals from Olkusz Forest.

Concentration of toxic metals in small mammals from Olkusz Forest confirmed that areas located close to industrial sources are still contaminated although industrial emission in Poland has generally decreased in the recent years [1].

4. References

1. Environmental Protection. (1996) Information and statistical data, General Statistical Office, Warsaw, Poland.
2. Tyler, G. (1989) Uptake, retention and toxicity of heavy metals in lichens, Water, Air and Soil Pollut. **47**, 321-333.
3. Samiullah, Y. (1990) Biological monitoring of environmental contaminants: animals, MARC, Rep. 37, Monit. Assess. Res. Center, London.
4. Martin, M.H. and Coughtrey, P.J. (1982) Biological monitoring of heavy metals pollution - land and air, Applied Science Publ., London.
5. Talmage, S.S. and Walton, B.T. (1991) Small mammals as monitors of environmental contaminants, Rev. Environ. Contam. Toxicol. **119**, 47-145.
6. Sawicka-Kapusta, K. and Zakrzewska, M. (1994) Small mammals as monitors of environmental contamination, Abstracts Third European Conference on Ecotoxicology - SECOTOX, Zurich, Switzerland, August 28-31.
7. Beyer, W.M., Pattee, O.H., Sileo, L., Hoffman, D.J., and Mulhern, B.M. (1985) Metal contamination in wildlife living near two zinc smelters, Environ. Pollut., Ser. A, **38**, 63-86.
8. Cloutier, N.R., Clulow, F.A., Lim, T.P., and Dave, N.K. (1985) Metal (Cu, Ni, Fe, Co, Zn, Pb) and Ra-226 levels in meadow voles Microtus pennsylvanicus living on nickel and uranium mine tailings in Ontario, Canada: Environmental and tissue levels, Environ. Pollut., Ser. B, **10**, 19-46.
9. Cooke, J.A., Andrews, S.M., and Johnson, M.S. (1990) Lead, zinc, cadmium and fluoride in small mammals from contaminated grassland established on fluorspar tailings, Water, Air, and Soil Pollut. **51**, 43-54.
10. Scanlon, P.F. (1987) Heavy metals in small mammals in roadsite environments: implications for food chains, Sci. Total Environ. **59**, 317-323.
11. Zakrzewska, M., Sawicka-Kapusta, K., Perdenia, A., and Wosik, A. (1993) Heavy metals in bank voles from Polish National Parks. Proc. of the Second European Conference on Ecotoxicology, 11-15 May, Amsterdam, The Netherlands., W. Slooff and H. de Kruijf (eds.), Sci. Total Environ., Supplement, Part **1**, 167-172.
12. Sawicka-Kapusta, K., Zakrzewska, M., and Lenda, B. (1993) Heavy metals in the common shrews from Polish National Parks, in R. J. Allan and J.O. Nriagu (eds.), Heavy metals in the Environment, Intern. Confer. Toronto, vol. **1**, 123-126, CEP Consultants, Edinburgh .
13. Sawicka-Kapusta, K., Górecki, A., and Lange, R. (1987) Heavy metals in rodents from polluted forests in southern Poland, Ekol. pol. **35**, 345-354.

14. Sawicka-Kapusta, K., Swiergosz, R., and Zakrzewska, M. (1990) Bank voles as monitors of environmental contamination by heavy metals. A remote wilderness area in Poland imperilled, Environ. Pollut., **67**, 315-324.
15. Sawicka-Kapusta, K., Zakrzewska, M., Kowalska, A., Lenda, B., and Skrobacz, M. (1995) Heavy metal concentrations in small mammals from Borecka Forest, Arch. Ochr. Œrod. **3-4**, 229-234.
16. Sawicka-Kapusta, K., Zakrzewska, M., and Chelstowska, J. (1994) Heavy metals in small mammals from three different forest areas in Poland, in I. Meszaros (ed.), Toxicity, Hazard, Risk, Proc. of SECOTOX 3rd Regional CCEE, Balatonaliga, Hungary, Acta Biol. Debr. Oecol. Hung. **5**, 277-284.
17. Sawicka-Kapusta, K. and Stachowicz, A. (1993) Heavy metals in small mammals from contaminated forest in Southern Poland, Abstracts First SETAC World Congress Ecotoxicology and Environmental Chemistry - a Global Perspective, Lisbon, Portugal, March 28-31.
18. Borowski, W. (1996) Levels of heavy metals in small mammals from Olkusz Forest, M. Sc. Thesis Jagiellonian University, Kraków.
19. Sawicka-Kapusta, K., Zakrzewska, M., Kowalska, A., and Stachowicz, A. (1995) Heavy metals in small mammals from different forests in Southern Poland, Abstracts 2nd European Congress of Mammalogy, J. Gurnell (ed.), Southampton, England, 27 March - 1 April.

INFLUENCE OF AIR POLLUTANTS IN THE ^{7}BE SIZE DISTRIBUTION OF ATMOSPHERIC AEROSOLS

C. PAPASTEFANOU and A. IOANNIDOU
Nuclear Physics Department
Aristotle University of Thessaloniki, Thessaloniki 54006, GREECE

Abstract

The atmospheric behavior of ^{7}Be aerosols was studied by using 1-ACFM cascade impactors. The activity distribution of ^{7}Be measured by gamma spectrometry (E_γ=477 keV), was largely associated with submicron aerosols in the accumulation mode (0.4-2.0 μm). The activity median aerodynamic diameter, AMAD ranged from 0.62 to 1.0μm (average 0.8 μm). The geometric standard deviation σ_g ranged from 1.87 to 2.50 (average 2.22). Low AMAD of ^{7}Be aerosols have been observed at locations characterized with relatively low pollution. Some dependency of AMADs on height has been also observed. In near marine environment the ^{7}Be activity size distribution was observed in higher size range of aerosol particles (AMAD 0.82 μm).

1. Introduction

Beryllium-7 is a relatively short-lived ($T_{1/2}$=53.3 d, $\tau = 1/\lambda$ =7.7d) naturally occurring radionuclide of cosmogenic origin which is formed by spallation processes of light atmospheric nuclei, such as carbon, nitrogen and oxygen, when they absorb protons and even neutrons of the primary component of cosmic rays [1-3].

Once ^{7}Be (inorganic ion) is formed in the troposphere, it rapidly associates primarily with submicron-sized aerosol particles [4]. Beryllium-7 in these fine aerosols may subsequently enter the marine as well as the terrestrial environment and vegetation via wet or dry depositional events. Following deposition, ^{7}Be will tend to associate with particulate material (particle-reactive element).

Beryllium-7 has come to bc recognized as a potential tool in studying the description of environmental processes such as aerosol transit in the troposphere [5], aerosol deposition [6], and aerosol trapping by above ground vegetation [7].

Bcryllium-7 as well as other natural radionuclides like $^{212+210}$Pb, ^{22}Na, ^{35}S and $^{32+33}$P participates in the formation and growth of the accumulation mode aerosols (0.07 to 2 μm diameter), which is a major reservoir of pollutants in the atmosphere. Following its production by gas-phase nuclear transformation, this isotope condenses on the aerosol population growing by condensation of non-radioactive species (e.g.,

I. Linkov and R. Wilson (eds.), Air Pollution in the Ural Mountains, 135–140.

H_2SO_4 or organics) and the fate of ^{7}Be will become the fate of the carrier aerosols [8]. On the behavior of ^{7}Be with atmospheric aerosols it was concluded from early aerosol studies that considerable coagulation occurred during migration from the stratosphere and upper troposphere to ground level [9-10].

This paper summarizes results of an investigation regarding the aerodynamic size distribution of atmospheric aerosols in the context of ^{7}Be distributions in order to find out any influence of air pollutants on ambient aerosols.

2. Instrumentation

The aerodynamic size distribution of ^{7}Be in atmospheric aerosols was obtained by using Andersen 1-ACFM cascade impactors. This 1-ACFM impactor involves a flow rate of 1.7 m^3h^{-1} (28.3 L $min^{-}\mu$ or 1 ft^3 min^{-1}) and eight atmospheric pressure stages for collecting aerosols above 0.4 μm. The effective cutoff diameters (ECDs) of this impactor were as follows: 0.4, 0.7, 1.1, 2.1, 3.3, 4.7, 5.8., and 9.0 μm. Stainless-steel plates were used for aerosol collection. Glass fiber filters were used as a back up filter to collect particles below the 0.4-μm collection plate.

The length of each collection period was 1 week. The samples were collected at different sites (locations) in the suburban area of the city of Thessaloniki, Greece (40° 38'N, 22° 56'E), including rural areas, industrial areas, the airport area, high elevations, and marine environment (Fig. 1). In this plan view, the direction of stream-lines of transferred air masses by the local wind blown, named Vardaris, are also shown.

The stainless-steel collecting plates of the impactor as well as the backup filters used as plane radioactive sources were measured for ^{7}Be activity (E_γ=477 keV) by γ-spectrometry using a high resolution (1.9 keV at 1.33 MeV ^{60}Co), high efficiency (42%), low background HPGe detector. Uncertainties in counting ^{7}Be activity varied from 10% up to 20%. Less than 3 mg of particulate material at any stage was collected in each experiment, thus overloading ($\geq$10 mg at any stage) was excluded.

3. Results

A typical plot of the activity size distribution of ^{7}Be versus aerodynamic diameter (D_p) is represented in Fig. 2. This distribution was selected from 12 measurements made over an almost 1-year period (26 August 1993 to 5 August 1994). Atmospheric aerosol size distribution appears to follow a trimodal distribution expected for condensation-derived aerosols with the following mode ranges: the Aitken nuclei mode ranges from 0.003 to 0.07 μm (average 0.015 μm), the accumulation mode ranges from 0.07 to 2 μm (average 0.3 μm) and the coarse mode ranges from 3 to 36 μm (average 10 μm) [11]. Beryllium- 7 aerosols are accumulation mode aerosols as is evident from the plot of Fig. 2.

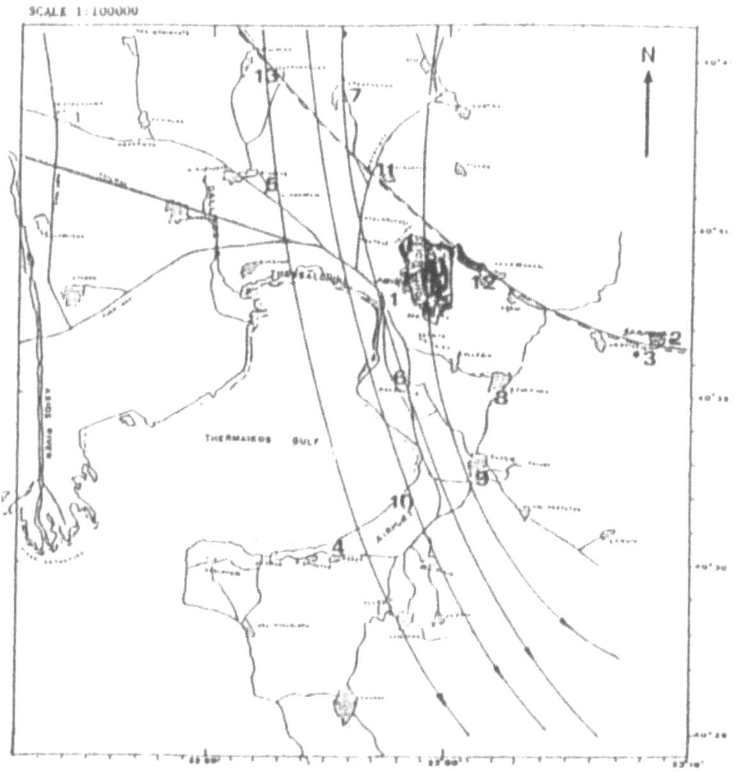

FIGURE 1. Map of the city of Thessaloniki and suburban areas.
The aerosols sampling location are shown and numbered according to Table 1.

A summary of experimental data regarding the ^{7}Be size distrbution of atmospheric aerosols is presented in Table 1 for 12 measurements carried out from August 1993 through August 1994. The activity median aerodynamic diameter (AMAD) varied from 0.62 to 1μm (average 0.80 μm) and the geometric standard deviation (σ_g) varied from 1.87 to 2.50 (average 2.22). The AMAD and σ_g calculations were made by plotting the cumulative distribution on log-normal probability paper.

4. Discussion

In a rural region, height 250 m, 50 km far from the city of Thessaloniki, beyond Mt. Hortiatis (site No. 2, Fig. 1), under quite clear (without significant pollution) atmospheric conditions and a relatively dry (precipitation free) climate, AMAD of ^{7}Be aerosol was found to be 0.68 μm, i.e., significantly lower than the mean AMAD (0.90 μm) as determined by 11 measurements carried out at the center of the city

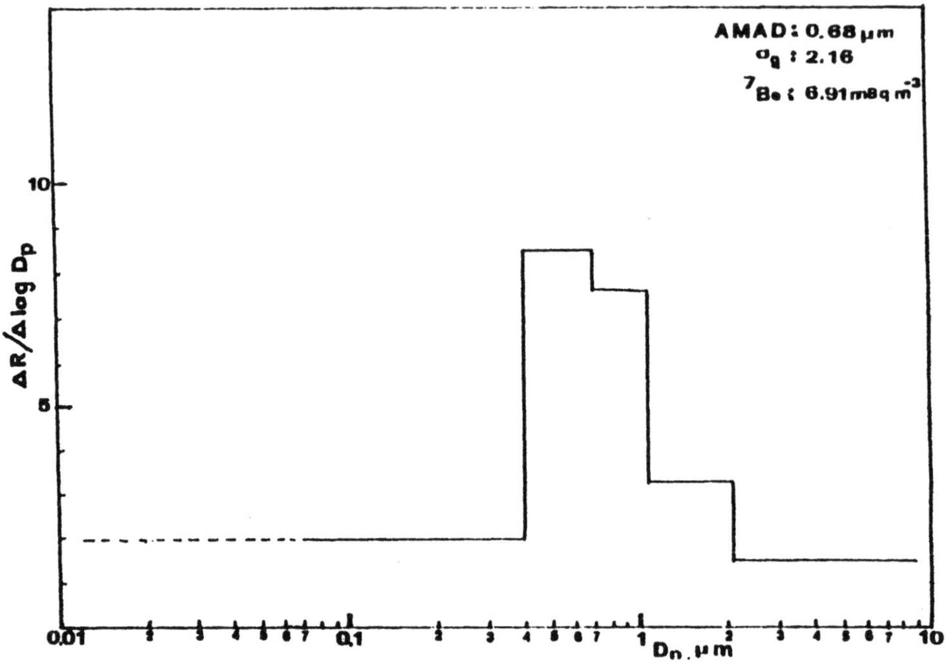

FIGURE 2. Aerodynamic size distribution of ^{7}Be aerosols (Sarakina region (site 2), 26 August 1993)

TABLE 1. Experimental data for ^{7}Be aerosols in different locations.

Date	Location	R (MBq/m^3)	Size Range (μm)					AMAD (μm)	σ_g
1993									
8/26	2 Sarakina	6.91	22.15	29.93	21.46	13.26	13.20	0.68	2.16
9/6	3 Mt. Hortiatis	7.38	27.98	26.91	30.40	8.47	6.24	0.62	2.24
10/13	4 Perea	9.10	12.12	29.44	25.29	14.18	18.98	0.82	1.88
12/15	5 Industrial Area	4.04	22.05	16.24	25.32	20.05	16.34	0.84	2.50
1994									
2/3	6 Kalamaria	6.16	13.28	19.17	25.02	22.35	20.18	1.00	2.31
2/23	7 Oreokastro	3.98	14.96	16.32	28.94	22.4	17.88	0.97	2.24
3/18	8 Panorama	5.92	19.18	24.19	16.16	25.19	15.28	0.97	2.34
4/12	9 Thermi	10.55	19.09	19.89	18.12	30.28	12.62	0.87	2.27
5/6	10 Airport	8.47	12.63	25.95	24.27	23.41	13.72	0.90	2.09
6/13	11 Efkarpia	7.78	23.45	24.53	20.20	20.20	11.63	0.74	2.37
7/1	12 Asvestohori	9.44	32.02	19.74	22.85	17.28	8.11	0.63	2.39
8/5	13 Neohorouda	10.38	23.30	31.16	26.70	18.84	–	0.64	1.87

of Thessaloniki (site 1, Fig. 1). The AMAD of ^{7}Be aerosol at that area was varied between 0.80 and 1.20 μm [12].

In a highland, height 1000 m, 50 km far from the city of Thessaloniki (site No. 3, Fig. 1), the AMAD of ^{7}Be aerosol was found to be 0.62 μm. This is comparable to the rural area (site 2, Fig. 1), reflecting clear atmospheric conditions.

In a coastal site, near the Thermaikos Gulf, 25 km far from the city of Thessaloniki (site 4, Fig. 1), under atmospheric conditions of the marine environment, the AMAD of ^{7}Be aerosol was found to be 0.82 μm, that is, it was shifted to higher particle size range.

Sites where the AMAD of ^{7}Be aerosol varied between 0.62 and 0.74 μm were located on a line with direction from West to East (see the dotted line in Fig. 1). That line crosses the direction of the streamlines of air masses that transfer pollutants from North-West (Industrial zone) to South-East. Below the dotted line, the AMAD of ^{7}Be aerosol varied between 0.82 and 1.00 μm, which is higher than that observed above the dotted line.

It is concluded that in the activity size distribution of ambient aerosols ^{7}Be is shifted to large particle sizes in the presence of pollutants. Removal of small aerosol particles in the submicron size range of the activity size distribution either by scavenging or by deposition of particles on any surface may result in a depletion of small particles in the activity size distribution. Subsequent ^{7}Be condensation on all aerosols effectively enriches large particles in the activity size distribution.

The radioactive aerosol is only generated by attachment and there is no nucleation, in contrast to the sulfate aerosol (Nucleation + attachment) [13]. Freshly produced ^{7}Be should attach to an existing aerosol or coagulate with other nuclei during its lifetime, as mean attachment half-lives are of the order of a minute less [14].

Furthermore, the ^{7}Be activity size distribution was associated with a small size range of aerosol particles with AMAD 0.68 μm at height 250 m and with AMAD 0.62 μm at height 1000 m, showing a dependency of the AMAD of ^{7}Be aerosol on the height. In marine environments, the ^{7}Be activity size distribution was associated with a relatively high size range of aerosol particles (sea spray) with AMAD 0.82 μm.

5. Acknowledgement

This research was sponsored by the National Foundation of Scholarships (IKY) of Greece under contract No. 6466 (25 May 1993).

6. References

1. Bruninx, E. (1961). *High-Energy Nuclear Reactions Cross-Sections.* I. Rep. CERN 61-1. Geneva, Switzerland.

2. Bruninx, E. (1964). *High-Energy Nuclear Reactions Cross-Sections.* III. Rep. CERN 64-17. Geneva, Switzerland.
3. Silberberg, R. and Tsao, C.H.. (1973). *Cross Sections Of Proton-Nucleus Interactions At High Energies.* Rep. NRL. 7593. Naval Research Laboratory Washington, D.C.
4. Bondietti, E.A., Papastefanou, C. and Rangarajan, C. (1987). *Radon And Its Decay Products: Occurrence, Properties, And Health Effects.* American Chemical Society, Washington, D.C., ACS Symposium Series 331:377-397.
5. Martell, E.A. (1970). *Radionuclides in the Environment.* American Chemical Series, Washington, D.C., Adv. Chem. Ser. 93:138-157.
6. Young, J.A. and Silker, W.B. (1980). *Earth. Planet. Sci. Lett.* 50:92-104.
7. Bondietti, E.A., Hoffmann, F.O. and Larsen, I.L. (1984). *J. Environ. Radioactivity* 1:5-27.
8. Bondietti, E.A., Brantley, .J.N. and Rangarajan, C. (1988). *J. Environ. Radioactivity* 6:99-120.
9. Lokhart, L.B. Jr., Patterson, R.L. Jr., Saunders, A.W. Jr. (1965). *J. Geophys. Res.* 7:6033-6041.
10. Gaziev, Y.I., Malakhov, S.G., Nazarov, L.E. and Silantiev, A.N. (1966). *Tellus* 18:474-485.
11. NRC, National Research Council (1979). *Airborne Particles.* University Park Press, Baltimore.
12. Papastefanou, C. and Ioannidou, A. (1995). *J.Environ. Radioactivity* 26 :273-282.
13. Hopke P.K. (1991). The initial atmospheric behavior of radon decay products. In: *Indoor Air Pollution.* Editors .J.K. Kay, G.E. Keller and .J.F. Miller, Lewis Publishers, Chelsea, pp.141-169.
14. Porstendorfer, J. and Mercer, T. (1980). *Natural Radiation Environment III.* National Technical Information Service, Springfield, Virginia 1:281-293.

ENVIRONMENTAL CONDITIONS IN THE AREAS EXPOSED TO EMISSION FROM THE GAS PROCESSING INDUSTRY IN SOUTH URAL

A.Z. KALIYEV
Orenburg State Agrarian University
18 Chelyuskintsev Street, Orenburg 460795, RUSSIA

Destabilization of the environmental condition in agriculture is linked with intensive use of ecologically unsafe technologies, forest destruction and water contamination. Contamination by toxic pollutants from industrial wastes, pesticides, radioactive substances and heavy metals currently has a global scale. These substances enter the environment in the form of industrial wastes and emission from the gas processing industry. The operation of the Orenburg Gas Processing Complex (OGPC) results in the emission of highly hazardous gases that are products of unrefined gas incineration. Some examples include nitric oxides (NO_2, NO), sulphareous anhybride (SO_2), hydrocarbons, mixtures of natural mercapants and heavy metals. Moreover the gas-processing enterprises produce annually about 4 millions m^3 of waste water. The main purpose of our research is to identify the impact of air pollutants emission and sewage from the gas-processing enterprises on the environmental condition.

We have studied the effects of pollutant emission within 56 km from the emission source. We considered meteorological conditions and population density in the area. Table 1 provides data on heavy metals concentration in soil at different distances from OGPC.

TABLE 1. Soil contamination by heavy metals at different distances from the Orenburg Gas Production Complex.

Distance	Element Concentration (mg/kg)									
	Cu	Zn	Pb	Ni	Co	Cr	V	Ti	Mo	Mn
Background	25	56	43	40	20	600	100	7100	4	1210
5-10 km	45	100	40	70	17	300	90	4500	2	800
15-20 km	91	136	36	460	32	1840	272	6340	5	1380
30-35 km	40	100	20	300	20	2000	150	5000	4	1100

The result of the study indicate that the maximal levels of heavy metals both in snow cover and in soil are highest in the 15-20 km zone, and the total index of chemical contamination in the given area exceed the limits for permissible concentration.

I. Linkov and R. Wilson (eds.), Air Pollution in the Ural Mountains, 141–143.

The observed land contamination 20 km far from the source is consistent with the earlier reported hypothesis that heavy metals are released into the atmosphere in the form of aerosols. The residence time of 1um particles is about 5 days, while smaller particles may stay in the atmosphere for more than 3 or 4 weeks.

Concentration of heavy metals in plants of the studied sites was compared to the biological standards set by the State Sanitary and Epidemidogical Department (GOSSANEPIDUPRAVLENIYE, GSEU) (Table 2).

TABLE 2. Concentration (in mg/kg)of common heavy metals and Maximum Allowed Concentration (MAC) for food crops..

Distance		Cu	Zn	Pb	Ni	Co	Cr	V	Ti	Mo	Sr	Mn	Fe
Background		10	5	0.5	3	18	5	0.1	0.1	1.2	620	20	0
MAC		30	50	5	3	1	0.5					2K	20
5-10 km	min	1.2	7.5	0.1	0.5	0	0.7	0.2	2	0.2	0	15	10
	max	3.6	18	5	0.9	7	5	5	364	0.3	36	73	30
15-20 km	min	2.5	13	0.5	0.5	0	0.1	0.1	1	0.1	0	14	14
	max	7.5	6	1.6	1.2	7	4	0.4	5	1	7.5	40	35
30-35 km	min	1.4	12	0.2	0.1	0	0.1	0	1.7	0.1	0	20	75
	max	7	5	4.2	2	0.1	8	6	4	5	7.5	25	

The high level of soil contamination by heavy metals leads to their concentration in plants. The highest plant contamination was also detected about 20 km from the source. The concentration of Zinc, Nickel, and Cobalt exceed the permissible standards of GSEU by 2, 3, and 7 times respectively. The background values are exceeded for chromium, vanadium and titanium. Among the fodder crops, the highest accumulation of metals occurs in alfalfa grass that was irrigated by sewage. Almost all plants have elevated levels of lead, vanadium and titanium. An increased bioaccumulation of Zinc was observed in such crops as-alfalfa, corn, Sudan grass (*Sorghum halepense*). Grain crops such as spring wheat and barley are also being contaminated by Zinc, nickel, chromium, strontium, vanadium and titanium in concentrations several times exceeding the background values.

The total concentration of contaminants (Zn, Mn, Pb, Cr, Cu, Mo, Ni) in plants correlates with their concentration in soil (R = 0.83). We do not recommend cultivation of grain crops used for food on the cultivated lands in the vicinity of the OGPC. Instead, perennial crops should be cultivated in accordance with existing recommendations on reducing the level of mobile forms of heavy metals in contaminated soil.

Because sewage is often used locally for irrigation of farm crops, water wastes have been tested for mutagencity. The "Ames" Salmonella test using TA 98 and TA 100 strains as well as chromosome aberration in Allium Sepa onion rootlets tests were used. Results of the tests show low mutagenicity and uncertain promutagenic activity in samples taken from the water tank for seasonal irrigation. Samples taken from the sewage running into the water tank did not show any mutagenic activity. Result of these and other bioassays indicate that chemicals present in the OGPC waste waters are weak mutagens. They induces gene mutation in bacterial cells and in the generative cells of animals, gene mutations of the chromosome indivergency type are induced in the somatic cells of plants and animals. Weak mutagens still stimulate mitotic activity in plant tissues.

To prevent further environmental damage by industrial contaminants, it is necessary to establish special monitoring services aimed to carry out comprehensive control over the soil and other ecosystem compartments. A monitoring strategy and regulatory procedures should be designed. These measures will facilitate forecasting of the current and future environmental changes due to the industrial activities.

STUDY OF THE ENVIRONMENTAL CONTAMINATION IN THE MAJOR INDUSTRIAL AREAS OF THE URALS WITH FOCUS ON REMEDIATION OF HUMAN HEALTH PROBLEMS

Report of the Working Group on Environmental Contaminants[1]

T.C. HUTCHINSON
Environmental and Resource Studies Program
Trent University, Peterborough, Ontario, CANADA

M.S. SYMINGTON
Canadian Heritage and Development Studies Graduate Program
Trent University, Peterborough, Ontario, CANADA

1. Goals

The major goal of the group was to identify the methods of documenting impacts of pollution in the Ural Region. This may include attempting to obtain funding, plan a proposal, and organize a steering committee to implement the work.

Pollutants of concern are causing both ecosystem damage and human health effects. It is the opinion of the group that human health effects are more important. Therefore decisions on environmental studies to be conducted should focus on the long-term goal of reducing adverse health impacts (with particular emphasis on the health of children).

Our group recommends working directly with the human health group as these pollutants have a significant impact on *both* health and environment. Questions posed to the group: what should be monitored and measured. In particalar, what are the *Russian* priorities?

A key challenge is to minimize the cost of reducing environmental effects, since money for both conducting environmental studies and for reducing emissions are limited. Consideration of benefit-cost ratios for environmental studies at different industries and for different pollutants is therefore warranted. Examples of *criteria for identifying industries and pollutants for study* include:

1. Focus on industries that have the most severe environmental impacts, including human health.

[1] Members of Working Group: Chairman: T.C. Hutchinson, Rapporteur: M.S. Symington (Canada), M. Frontasyeva, E. Genikhovieh, O. Pononiazeva, I. Gyelxfyenbuym (Russia), E. Steinnes (Norway), F. Papastefanov (Portugal), A. Dzhabasov (Kazakstan), G. Petersen (Germany), C.I. Davidson (U.S.A.), and K. Sawicka-Kapusta (Poland).

I. Linkov and R. Wilson (eds.), Air Pollution in the Ural Mountains, 145–150.

2. Choose industries where information is already available, such as pollutant emissions, meteorology, and history of emissions.
3. Choose pollutants that are responsible for human health impacts, or are indicators of other pollutants responsible for health impacts.
4. Choose towns, industries and pollutants where it is likely that progress can be made in reducing environmental impacts and where local authorities are supportive.

Priorities may differ within the Urals region, in that there are different pollutants throughout this area (see paper by Chukanov and Korobitsin for information on pollutant emmissions in the Ural Region). Priority measurements by pollution assessments may include:

- chemistry of suspended particles
- chemistry of total dustfall
- measurement of gases
- soil chemistry
- chemical analysis of food raised and eaten in the area
- drinking water analyses
- survey/mapping

Gaseous emissions were felt to be the most detrimental to human health. However, there was not agreement among the group in choosing the most important gas to monitor. The top two choices were SO_2 and HC's, with the third choice being NO_x. It was agreed that O_3 was not a priority.

It was suggested that a practical approach was to begin by making a map and grid of this area. The suggestion is to identify "hot spots" around point source areas for assessing the impact of pollution both on the people and the ecosystem We are dealing with a large area of approximately 800 x 500 km.

To organize the monitoring you need to be sure about the sources of pollutant industries. Around emission sources the monitoring grid must be more intense and for other places it would be more spacious. Several media should be studied: air, soil and vegetation. These allow the study of both the history of pollution and on-going inputs into the ecosystem. Soil, for example, is an accumulator of many pollutants. Looking at air contamination in comparison to the density of population will also allow us to recommend the most suitable places for living.

It was felt by some Russian participants that we must distinguish between science and charity, in that while some communities desperately need help, the NATO Workshop was designed to focus on the scientific approach to remediation of the air pollution problems and their health consequences, not to bring large scale aid from outside the Russian republic, welcome as this would be.

Biological, ecological, instrumental, chemical, remote sensing, GIS. Focus needs to be on simple methods, on high priority items, and low cost. Remote sensing for the Urals can be obtained from outside Russia. However, this should not replace on-the-spot measurements.

2. Developing a Proposal for a Focused Scientific Assessment of the Problems

For a particular industry or urban centre we suggest looking at the impact, both historical and on-going, of pollution along transects of perhaps 50 km centred on the industrial site(s). We suggest using line transects located at right angles based on prevailing wind directions. Components to measure include:

1. *Chemistry of precipitation, (i.e. Heavy metals, organics, sulphate, nitrate, ammonium, etc.).* Suspended particles are more difficult to measure than dustfall, so begin by initially measuring dustfall. Airborne concentration samplers with battery operated pumps could be used. Never put these collectors on the ground. They must always be at least 1-2 meters off the ground. Wind direction information should be incorporated with the data collected from the dustfall buckets. This method is very inexpensive as a single collecting bucket costs only one dollar. Pumps for suspended particles are $100-200. Reports from USA EPA on protocols and problems for dustfall buckets, as well as information on problems with interpretation should be included. Have dustfall collectors every few km along the 50 km transect. The area in which they are located should be covered with vegetation, rather than an unvegetated area which will give inputs of local suspended particulate and blowing soil.
2. *Soil.* Soil samples should be collected and analyzed from each point along the transect for levels of heavy metals, pH, organic content.
3. *Water.* Drinking water sources should be analyzed. The majority of the residents in the Ural region do not drink bottled water. They boil water (it still contains heavy metals). Therefore, there is a need to sample water from the local taps and wells.
4. *Food.* Another major area of concern is the food sources for people (Figure 1). A diet survey should be completed to estimate exposures. Food should be prepared (both plant and animals) as if it were to be eaten and analyzed for its chemical content. Perhaps a chemical analysis of food baskets of what is actually eaten could be carried out, which focuses on what is grown and/or raised in the area as well as food that is brought into the region. Local diets include berries, apples, mutton, pork, beef, poultry, and lamb.

With data generated from such a monitoring program, local authorities could create a map of the of the pollution occurring so that the people of this area can decide where to 1) live and 2) where to grow their food.

3. Possible Approaches

It appears, then, that there could be two approaches in setting up a study such as this:

1. Long-term large-scale program including dust fall analysis, soil analysis, vegetational analysis, food analysis could be undertaken in each community, drinking water analysis and moss monitoring.

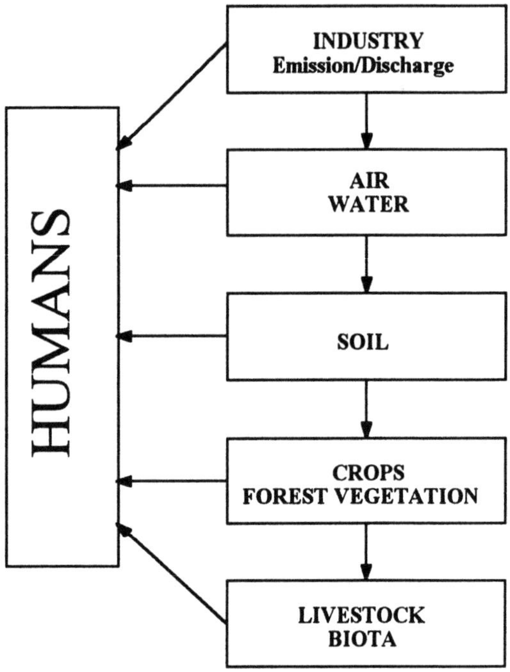

FIGURE 1. Pathway Conceptualization

2. Small components of larger investigation in 1. The most important components are to collect and analyse dust fall and to look at soil concentrations- since that is an integrator of the food supply.

The study can be as large or as small as funds are available. A lot of this type of monitoring has already been performed in certain areas of the Urals, therefore it is essential to collect existing data and expand on it in order to thoroughly document levels of pollution, both current and historical. This mapping of pollution is necessary in order to provide information to help both city planners and residents decide where farming and housing should take place.

There are regular meteorological stations in Russia with information (mostly wind data) on 50 x 50 km grid. Many thousands of these stations exist. Some use balloons to get profiles of wind and temperature. There is a central databank for these data. Also precipitation, temperature, and pressure data are available. Every major enterprise in the USSR was required to do dispersion modeling, provide emissions inventory, etc. This information was put into a report called "Volume of the maximum permissible emissions for this enterprise." There are airborne concentrations obtained, but sometimes this information is not publicly available. It is important that this information be made available.

It was suggested that the chemical analysis of stack emissions be done. Most of the source emissions data available are not from stack measurements but rather from mass balances. There are large uncertainties in these mass balances. But it is very expensive to do stack sampling. Perhaps a limited amount of stack sampling could be conducted if funding is available.

All of the data could be entered into a geographical database using GIS. Portugal is currently putting information on world wide web. Again, all of this information must be publicly available, preferably all in the same place. Currently you have to pay large sums of money to get this information and this is inappropriate for a region urgently in need of asssistance.

Data to be collected and approaches available include geochemical, biological, ecological, instrumental, chemical, computational (modeling), remote sensing (from outside the country). Other working groups want all sites to have radionuclides, as well as meteorological conditions information.

There are different philosophies of environmental study design to be considered. One idea is to conduct extensive studies over broad regions with a limited number of pollutants and auxiliary information (e.g., emissions, meteorology, etc.). Another idea is to conduct intensive studies where more pollutants are monitored more and auxiliary information is obtained for a limited number of locations.

Another aspect of study design is to consider how the data will be used. For example, in the USA, one goal of environmental sampling is sometimes to determine source-receptor relationships. This can be done using chemical element balance techniques where source-to-sink matrices are developed and applied to characteristics of source emissions. In Russia, it is not clear that source-receptor relationships are important since many of the problems involve contamination surrounding a large source complex where the source of pollution is obvious and well known. Subtle detective approaches are not needed.

We considered a specific proposal from Eiliv Stiennes of Norway to use moss sampling as a surrogate of environmental exposure to pollutants. This moss proposal consists of monitoring of heavy metal and radionuclides in the Ural region. This proposal could be used in addition to the fundamental pollution mapping of the area. It could be part of the European network program based on moss analysis. This program also takes surface soil samples. The atmospheric deposition of heavy metals in this region is from local point sources. It is probable there is also long range transport from Siberia. Deposition pattern should be mapped in the whole region. The moss techniques could be used here also. One should take samples of surface soil and samples of the sub soil. This can begin to look at the natural source as well as the anthropogenic sources. The soil samples could also be looked at for radionuclides. So far, studies in this region have concentrated on hot spot areas. The regional extent of this contamination has not been systematically studied. Peat and sediment samples could also be looked for the history of pollution in this region. This would be helpful to both the health and epidemiological research. Mosses would not grow along the lower plains, so one would be restricted to higher areas for the Ural region. This is to look at regional trends, and if other things need to be looked at in more detail, this should also be done. The next European wide elemental monitoring using mosses is in the year 2000. This program and data are described in "Atmospheric Heavy Metal Deposition in Europe" edited by A. Ruhling, Pub. Nord, 1994.

4. Criteria for Study Sites

How many cities do we start with? There is a need to work closely with health experts from the Urals, as they can help in identifying cities. Sites should be selected based upon:

1. Best available supporting data, best mass balances, known health data.
2. Largest population, largest number of sick children.
3. Cooperation with local authorities. Difficult to team up. Must find locations where local authorities care about the local people. Cooperative medical personnel.

Start with one or more cities, but no more than five or six. These six cities could be in close vicinity, or well separated. A city could be picked with no background information or where the contamination is known to be severe and where significant health problems are reported. It would be useful to have a pilot study to start to document potential problems and to arrange logistics.

Local centers of pollution exist throughout the Urals. There cannot be just one group undertaking the project, rather it would be better to have a set of local studies with a central coordinating office.

5. Potential Funding Sources

Potential sources of funding include the Russian Meteorological Service, Ministry of Health, Committee for Nature Protection, local agencies, NATO and the European Union in Brussels. Cooperation between European countries and the former Soviet Union is also a possibility. Partners from former Soviet countries get about 70-80% of funding. Need to include economic aspects of the project.

The Proceedings from this workshop should be distributed as widely as possible, especially within the Russian Republic and neighboring republic. These proceedings will be in Russian and in English and will include all of the Russian papers given at the workshop.

We agreed to keep our Working Group in existence on a voluntary basis after the May Workshop and to help appropriate persons in a) analyzing information, and b) develop a research proposal aimed at documentation and remediation.

Part 2

Health Consequences of Air Pollution

Ore reduction plant in Magnitogorsk
Photo: R. Wilson

ASSESSMENT AND COMPARATIVE ANALYSIS OF HEALTH RISKS CAUSED BY AIR POLLUTION IN CITIES OF RUSSIA

B.A. REVICH
Centre of Demography and Human Ecology, Institute of Forecasting of Russian Academy of Sciences, 32 Krasikova ul., Moscow, 117418, RUSSIA.

A.A. BYKOV
Plekhanov Economic Academy, Stremiannyi per. 28, Moscow, 113054, RUSSIA.

Abstract

The paper deals with the analysis of the antropogenic impacts of air pollutions in Russian cities, the pollution levels of ambient air and health risks for the Russian population. It includes the assessments of toxic substances emissions, the analysis of air pollution monitoring in Russia. Basing upon the data of stationary monitoring posts, and values of health risks from certain air pollutants we made the interval and mean assessments of the health risks orders for populations in industrial cities in Russia. The results of the calculations are presented as absolute numbers of deaths caused by the air pollution in Russian cities with populations of certain size under exposure, and the relative contribution of pollution effects to the corresponding mortality rates. The potential contribution to overall mortality is approximated by few percents. Using a classification of risk orders, a comparative analysis was made of danger from air pollution versus main causes of death and natural hazards. This analysis indicates the relatively dangerous pollutants which should be further investigated and included in cost-benefit analysis so as the optimal risk reduction decision could be suggested.

1. Introduction

Although ambient air pollution is known to affect humans, quantitative measures of the impact are quite uncertain. One of the most successful approaches to evaluate adverse health consequences caused by actual concentrations of pollutants in ambient air, is risk assessment. The quantitative values of lifetime risk units (RU) recommended by international and some national organizations are often used in

I. Linkov and R. Wilson (eds.), Air Pollution in the Ural Mountains, 153–180.

assessing the health risk from different sources (chemical pollutants, radioactive exposure, electromagnetic waves etc.). Recommendations of the World Health Organization (WHO) [1,2] and the United States Environmental Protection Agency (U.S.EPA) [3] in their integrated risk data base (IRIS) were taken for a basis in calculating health risks caused by chemically polluted ambient air in Russian cities. We also used well established approaches [4-7] for assessing health risks from these substances not clearly defined in terms of risk units in the mentioned recommendations. In all cases, expert values of risks have been averaged, and the uncertainty range reflects the differences in the risk recommendations and the inaccuracies in methods used for measuring air pollutants.

The air pollutants can be classified either carcinogenic or non-carcinrogenic according to their impacts and consequences for human health. Carcinogens increase the incidence of cancer diseases. The group of non-carcinogens includes 'classical' pollutants contributing significantly to the health hazard; chlororganic and inorganic (Cu, Zn, Pb etc.) substances.

The first stage of the risk assessment procedure includes the identification of major hazards, their epidemiological assessment, and estimation of exposed population sizes. Evaluation actual concentrations of ambient air pollution in cities is compared with maximal allowed concentrations (MAC - threshold limits) or with safe levels of exposure (SLE - specific Russian standard). Russia has established threshold values (MAC, SLE) for more than 400 chemicals [8].

2. Assessment Of Toxic Chemical Emissions Into Ambient Air

In Russia, data on 47 substances emitted by industrial enterprises and other stationary sources are collected annually by regional statistical agencies. The State Committee for Statistics issues annual report [9] at the federal level. Data on emissions in 1991 - 1994 are presented in Table 1. During this period total amount of emission decreased by 10 000 tons (30%). Unfortunately, the information available on emissions of most hazardous chemicals is incomplete. For instance, data on emissions of such carcinogenic chemicals as asbestos, vinyl chloride, beryllium, acetaldehyde, tetrachlorethelene, dimethyl-formamin and some others are lacking although their concentrations are monitored by the state authorities. Nevertheless, data on some chemicals of the third and fourth groups of toxicity are collected even though these chemicals are less dangerous for human health and are less abundant in environment.

3. Monitoring Ambient Air Pollution

The Federal Service for Hydrometeorology and Environmental Monitoring controls concentrations of chemical pollutants in ambient air according to the Operational Manual for Air Pollution Control [10]. Air samples are collected 3-4 times per day during 20 min. each, in 239 cities at 671 meteorological stations (data for 1994).

TABLE 1. Dynamics of toxic chemical emissions into ambient air from stationary sources in cities of Russia, tons per year [9]

Chemical	1991	1992	1993	1994
Total, thousand tons	31830	28127	25126	21929
Suspended particles, thousand tons	6452	5530	4709	3870
Sulfur dioxide	9187	8148	7222	6512
Carbon monoxide	7590	6774	5896	5140
Nitrogen oxides	3024	2715	2456	2085
Hydrocarbons	3122			
VOC	1831	1766	1525	1238
V_2O_5	6452	5679	4957	3364
Manganese	2008	1959	1486	945
Copper	7509	6959	5347	4943
Nickel	4946	4299	3767	3553
Mercury	7.9	6.3	7.8	6.6
Lead	1503	1404	1149	876
Chromium	463	400	299	187
Nitrogen acid	1620	1054	839	622
Ammonia	71017	70983	56287	46913
HCl	5050	3963	3003	2612
Hydrogen cyanide	1428	1036	2902	2745
Sulfuric acid	41784	38070	33349	32001
Arsenic	1357	1034	1007	1046
Soot, thousand tons	109.8	107.0	98.3	79.9
Se dioxide	28.6	22.0	4.3	3.6
Hydrogen sulfide	29911		20402	14476
Carbon bisulfide	27331	17304	16480	11021
Compounds of fluoride	16018	12456	12762	11619
Cadmium	11.8	5.5	6.2	5.0
Chlorine	1886	1641	1380	1101
Cyclogectane	958	930	965	545
Benzene	24168	17967	16348	12922
Xylene	50174	41913	36465	26977
Styrene	2372	1938	1866	1351
Toluene	64647	52687	44892	32472
Benzo(a)pyrene	156	89.4	54.3	61.4
Naphthalene	3106	3018	3056	2752
Dichloroethane	10772	13823	3985	3640
CCl_4	803	961	433	303
Isopropyl spirit	5327	5499	3057	1816
Methyl spirit	7604	6071	5317	3708
Phenol	4502	7449	3056	2440.7
Butyl acetate	12535	11015	9026	6009.3
Ethyl acetate	16562	12709	10071	6797
Akrolein	837	684	721	630
Formaldehyde	4799	3981	2543	3439
Acetone	38477	29803	23535	14515

Ftaliy anhydride	685	601	420	288
Caprolactam	1044	961	598	333
Acetic acid	4706	4528	2960	1782
Methyl mercaptan	3389	2190		960
Protein - vitamin forage	829	61.7	28	36
Petroleum, thousand tons	3869.9	562.4	593.9	467.8
Others, thousand tons	279	71	1162.7	990.5

In addition, samples of air are taken during one month to determine concentrations of metals and benzo(a)pyrene. Concentrations of toxic chemicals are determined in regional laboratories of the Service, but some chemicals such as metals and benzo(a)pyrene are determined in specialized laboratories at the Institute for Experimental Meteorology (city of Obninsk) or in other institutes. Data on the concentration of toxic chemicals are reviewed in 27 regional centers and summarized in the Department of Environmental Monitoring of the Main Geophysical Observatory (St.-Petersburg) for distribution over the whole country. This department prepares annual reports on ambient air contamination in Russian cities [11], including some data provided by Sanitary and Epidemiology (Sanepid) centers. Usually regional Sanepid centers control air quality only in the sanitary zones near the industrial enterprises. The number of air quality measurements is being reduced. The results of air quality investigations are analyzed and presented in annual reports. The reports give the percentage of air samples with concentrations of pollutants exceeding allowed limits. During recent years, monitoring networks have been created in a number of Russian cities (Cheliabinsk, Dzerzinsk of Nizniy Novgorod region) by local governments. The data are kept locally without being transferred to the federal level. In most cases, laboratories of industrial enterprises implement monitoring for only several chemical compounds mainly in the air of sanitary zones.

4. Ambient Air Quality Standards

Ambient air quality standards (MACs - threshold limits) have been established by the Department for Sanitary and Epidemiological Inspection, Ministry of Health as a part of sanitary control system. These values were introduced using the Methodological Recommendations [8]. Air pollutants are divided to three groups depending on their toxicity: mostly reflex, mostly resobtive, reflex-resobtive toxicity. Single threshold limit of short-term exposure is established for the first group of chemicals with reflex toxicity. For the second group, two standards are recommended: for daily exposure and for short-term exposure at 99 % probability level of its detection in a long lasting experiment. In the third group of substances, the short-term exposure standard is established for reflex toxicity, and daily exposure standard for resobtive toxicity. According to this document, epidemiological studies to assess air pollution impacts on human health are not required In Table 2 various national and international standards of air quality are summarized. One can compare Russian standard with that recommended by WHO or with some other national standards.

TABLE 2. Maximal allowable concentrations (MAC) used in Russia for chemical substances in ambient air, WHO recommendations and national standards of other countries, μg/m³

Chemical substance	Averaging time period for standard	Russia	WHO recom mendations for European countries	Other countries	
				Country	Standard
A) "Classical" pollutants					
Carbon monoxide, mg/m³	30 min.	5	60		
	1 h	no	30	USA	40
	8 h,	no	10	USA	10
	24 h	3	no		
Nitrogen dioxide	30 min.	85	no		
	1 h	no	200		
	24 h	40	no		
	1 year	no	40	USA	100
				Germany	80
Sulfur dioxide	10 min.	no	500		
	30 min.	500	no		
	24 h	50	125	USA	365
	1 year	no	50	USA	80
Suspended particles (TSP)	30 min.	500	no		
	24 h	150	no	USA	PM_{10} - 150 $PM_{2.5}$ - 65
	1 year	no	no	USA	PM_{10} - 65 $PM_{2.5}$ - 20
Ozone	30 min.	160	no		
	1 h	no	no	USA	235
	8 h	no	120		
	24 h	30	no		
B). Inonorganic substances					
Ammonia	30 min.	200	no		
	24 h	40	no		
Arsenic	24 h	3.0	no		
	1 year		no	USA	0.0002 - 3.3
	Unit risk/lifetime		$1.5 \cdot 10^{-3}$ per 1 μg/m³		
Asbestos, fibers/m³	?	0.06	no		
	Unit risk/lifetime		1 000*)		

Cadmium	24 h	0.2	no	USA	0.005 - 8.0
	1 year	no	no	USA	0.0006-1.67
	Unit risk/lifetime		$1.8*10^{-3}$ per 1 μg/m³		
Chromium (Cr^{+6})	24 h	1.0	no	USA	0.12 - 0.39
	1 year	no	no		0.00008-0.68
	Unit risk/lifetime		$4*10^{-2}$ per 1 μg/m³		
Hydrogen chloride	30 min.	200	no		
	24 h	200	no		
Hydrogen fluoride	30 min.	20	no	Germany	1
	24 h	5	no	USA	2.8 - 4.0
	1 year		no		8.3
Hydrogen sulfide	30 min.	8.0	no	USA, other countries	60
	24 h	8.0	no	France	80
	1 year		no	Italy	3
				Germany	5
				Bulgary	
				Hungary	
				Czech R.	
				Yugoslv.	8
				Spain	4
Lead	24 h	0.3	no	USA	1.5
	1 year	no	0.5	USA	0.14 - 2.50
Manganese	24 h	1.0	no		
	1 year	no	0.15		
Mercury	24 h	0.3	no		
	1 year	no	1.0		
Nickel - soluble compounds	24 h	0.2	no		
Nickel - metal compounds	24 h	1.0	no		
	1 year	no	no	USA	0.24 - 3.3
	Unit risk/lifetime		$3.8*10^{-4}$ per 1 μg/m³		
Vanadium	24 h	2.0	no		
	Unit risk/lifetime (leukemia)		$6*10^{-6}$ per 1 μg/m³		

C) Organic substances					
Benzene	30 min.	1 500	no		
	24 h	100	no		
	1 year		no	USA	0.1 - 76.7
Benzo(a)pyrene, ng/m^3	24 h	1.0	no		
	1 year		no	USA	0.03 - 0.3
	Unit risk/lifetime		$8.7*10^{-5}$ per 1 $\mu g/m^3$		
Carbon disulfide	30 min.	30	no		
	24 h	5	no		
Carbon tetrachloride	30 min.	4.0	no		
	24 h	0.7	no		
Chlorine	30 min.	100	no		
	24 h	30	no		
Dioxins, pg/m^3	24 h	0.5	no		
Formaldehyde	24 h	3	no	Bulgary	
	30 min.	35	100	Hungary	
				Yugoslv.	12
Methil mercaptan	30 min.	0.1	No		
Phenol	30 min.	10.0	no		
	24 h	3.0	no		
Styrene	30 min.	4.0	no		
	24 h	2.0	no		
	1 week	no	260		
Toluene	30 min.	600	no		
	24 h	600	no		
	1 week	no	260		
Vinil chloride	ESEL**	5	no		
	Unit risk/lifetime		$1*10^{-6}$ per 1 $\mu g/m^3$		
Xylene	30 min.	200	no		
	24 h	200	no		

*) Population with 30% smokers, Lung cancer: 10^{-6} - 10^{-5}, mezotelioma: 10^{-5} - 10^{-4}.
**)Estimated Safe Exposure Level: Russian specific standard, approximation of MAC.

5. Concentrations of Chemical Pollutants in Ambient Air

5.1. CARCINOGENIC SUBSTANCES

The International Agency for Research on Cancer (IARC) adopted a classification scheme for carcinogenic substances dividing them to four groups. The most hazardous chemicals which have been proved by epidemiological studies to cause cancer are included in the first group (1A). Eleven chemicals produce carcinogenic effect through inhalation.

5.1.1. Asbestos

Although three Russian cities have enterprises of asbestos industry, the data on asbestos emissions are not available. Potential sources of asbestos emission into the ambient air are asbestos processing plants and manufacturers of materials which contain asbestos (City of Asbest, Sverdlovsk region, town of Ak-Dovurac, the Republic of Tuva). Data on asbestos concentrations in air can be obtained only from the Report of Sverdlovsk Regional Center of Sanepid, 1994 [12]. Average annual concentrations of dust in the city of Asbest vary from 60 to 300 $\mu g/m^3$. Asbestos fibers form 12 % of the mass (limit is 0.06 $fibers/m^3$), including 2 % of respirable fibers[1].

5.1.2. Benzene

Oil processing and petrochemical plants are the major sources of benzene emission. 13 000 to 24 000 tons of benzene are annually released into the air. Concentrations of benzene in the air of most industrial cities with large petrochemical plants (Gubakha, Ishimbay, Kstovo, Omsk, Salavat, Samara, Toliatty, Usolie-Sibirskoye) are between 20 and 60 $\mu g/m^3$ (limit for 24 h exposure is 100 $\mu g/m^3$). High level of benzene pollution can be expected also in some other cities with oil processing plants, where emissions are essential, but air quality is not systematically monitored. These are: cities of Angarsk, Volgograd, Kirishi, Lipetsk, Niznekamsk, Novokuybishevsk, Riazan and Yaroslavl.

5.1.3. Vinyl chloride (VC)

Approximate recommended limit is 5 $\mu g/m^3$. Sources of emissions of VC into air are manufacturers of this compound and factories which use it in their technological processes. No data are available emissions of VC into air. VC is produced by five enterprises in Dzerzinsk, Usolie-Sibirskoye, Sterlitamak, Volgograd, and Zima. Ambient air concentrations of VC are not regularly monitored, but some mesurements have been conducted in a few special studies. For instance, according to computer model estimations, concentrations of VC in north-western district of Usolie-Sibirskoye, where a factory is located, may reach 25 $\mu g/m^3$, that is 5 times higher than the recommended limit. In another city with chemical plants (Dzerzinsk) increased concentrations of VC spread over 12 kilometers from the factory [13]. Air pollution levels in Dzerzinsk and Usolie-Sibirskoye are significantly higher than those in foreign cities with chemical plants.

---

[1] *Editorial Note. The fibers from the mine of Asbest are chrysotile uncontaminated by tremolite which may be less important than comparable fiber concentrations in the world*

5.1.4. Arsenic

More than 1000 tons of arsenic is emitted annually in the ambient air. It is released by copper smelters and other metallurgical plants in the following cities: Karabash, Krasnouralsk, Mednogorsk, Verkhnii Ufaley, Gorniak, Dalnegorsk, Novosibirsk, Vladikavkaz, Riazan. No data on arsenic concentrations in the air are available, but very high levels of soil pollution have been detected in a number of cities, suggesting possible air pollution.

5.1.5. Nickel

Nickel is mostly emitted into the air by nickel smelters at Kolskiy Peninsula (cities of Monchegorsk, Nickel, Zapoliarniy) and in Norilsk, Orsk, Zlatoust, Vladikavkaz, Electrostal, Cheliabinsk, Saratov, Kursk. The data on concentrations of nickel in the ambient air of these cities are limited. The highest level of *Ni* in the air of Norilsk is 23 $\mu g/m^3$, that is 23 times higher than MAC. 58 % of all samples from Norilsk have concentrations of Ni above threshold limit.

5.1.6. Chromium

From 200 to 460 tons of *Cr* are emitted into the ambient air annually. Six-valent chromium and its compounds have carcinogenic properties, but usually only total concentrations of chromium are mesured in air. Concentrations of the six-valent chromium are measured in nine cities, and six-valent chromium concentrations exceeding the MAC level (1.5 $\mu g/m^3$ for 24 h exposure) were recorded only in the city of Vladimir.

The group 2A of carcinogenic substances includes chemicals for which carcinogenic effect is proved with high degree of certainty.

5.1.7. Benzo(a)pyrene

Benzo(a)pyrene (BP) is released annually in amount of 61 to 156 tons as a result of combustion of organic fuels (coal, oil, benzine). Iron smelters, aluminum producing and petroleum refining plants release BP as well. Increased concentration levels of BP in the air can be found practically in each town where coal is used as a fuel for boiler houses. Great emissions of BP are detected at aluminum smelters of Krasnoyarsk, Bratsk, Novokuznetsk. During past years concentrations of BP in atmosphere decreased due to downfall of plants, but considering delayed effect of exposure to carcinogens we can expect increased rate of lung cancer during 15 to 20 years in the future. The highest BP levels (6-15 ng/m^3; limit for 24 hs exposure is 1 ng/m^3) are found in cities with world largest aluminum and iron smelters: Shelekhovo, Novokuznetsk, Bratsk, Magnitogorsk, Nizniy Tagil, Petrovsk-Zabaykalskiy, Krasnoyarsk, Cheliabinsk, Lipetsk. In these cities concentrations of BP exceed 3 ng/m^3 almost at all air quality control stations. Long-term exposure to 3 ng/m^3 concentration of BP increases lung cancer incidence in the general population. High concentrations of BP are detected in cities located near large electric power stations: Gubaha, Kansk, Nazarovo, Novochercassk, Cheremhovo; and in cities with many boiler houses which use coal as a fuel: Abakan, Biysk, Zeya, Zima, Irkutsk, Chita and some others.

Concentrations of BP in the air of these cities range from 1.5 to 15.6 ng/m^3. Large oil processing plants are other sources of BP emission, they cause increased concentrations of BP in air (2 to 3 ng/m^3) which are detected at all stations of air quality control.

5.1.8. Cadmium

Cadmium emissions are produced mainly by metallurgic plants in Belovo, Vladikavkaz, Dalnegorsk, Karabash, Kishtim, Kirovograd, Krasnouralsk, Mednogorsk and some other settlements, by cadmium-containing battery plant in Kursk, by cadmium-containing paint plant in the town of Duliovo, Moscow region. Total cadmium emission ranges from 5.0 to 6.1 tons annually. We estimated that approximately 10 tons of cadmium per year are emitted as a result of oil and coal combustion. For comparison, 33 tons of cadmium were emitted by nonferrous metallurgy and 6.5 tons - by oil and coal combustion in the countries of European Union in the year 1980. Systematic control of the ambient air for cadmium was undertaken in 1992-1994 in 52 cities at 94 control stations. Average concentrations during this period covered the range 0.005-0.01 μg/m^3, and were much lower than MAC (0.3 μg/m^3). Some special environmental researches found significantly higher concentrations of cadmium, for instance cadmium concentrations in the ambient air near metallurgical plants in Vladikavkaz and Belovo were above MAC [14,15].

5.1.9. Formaldehyde

Formaldehyde is emitted into the air by chemical and metallurgical plants as well as manufacturers of construction materials and polymeres, furniture and internal combustion engines. Average annual concentrations of formaldehyde in the ambient air of many cities ranges from 3 to 12 μg/m^3 (MAC for 24 h exposure - 3 μg/m^3).

5.2. CLASSICAL POLLUTANTS

5.2.1. Suspended particles

Only the total concentration of suspended particles (TSP) is determined in the air of most large cities. In only a few cities hazardous respirable fraction of suspended particles (PM_{10}) has been measured. For many years high concentrations of TSP have been recorded in the air of 50 Russian cities. Among them, are cities with aluminum industry and with cement plants (Dalnegorsk, Kamensk-Uralskiy, Komsomolsk-on-Amur, Krasnoyarsk, Lipetsk, Magnitogorsk, Novokuznetsk, Novotroitsk, Orsk, Old Oskol, Cheliabinsk, Cherepovets) and northern cities, where coal is used as a furnace fuel (Barnaul, Vorkuta, Ulan-Ude, Khabarovsk, Yakutsk and other cities). Elevated concentrations of TSP are commonly detected at all monitoring stations of these cities, i.e. practically the whole urban population is under increased exposure of SP. Mean SP concentration in the air of most polluted cities in 1992 was 275 μg/m^3, that is about two times higher than MAC for daily exposure (150 μg/m^3).

5.2.2. Nitrogen dioxide

Average concentrations of NO_2 in air were 42-44 μg/m^3 in 1992-1994. Concentrations were higher than MAC for daily exposure (40 μg/m^3). Elevated concentrations of NO_2 (more than 60 μg/m^3) were detected by more than two control stations in 15 cities, where transport exhaust fumes inputs were about 50-70 % of total air pollution (Moscow, St. Petersburg, Saratov, Ulianovsk, Vladivostok) or large metallurgical plants (Bratsk, Lipetsk, Magnitogorsk, Mednogorsk, Electrostal) or enterprises of chemical industry (Berezniki, Volzkiy, Sterlitamak, Usolie-Sibirskoye) existed.

5.2.3. Sulfur dioxide

This substance is a leading air pollutant in the classification based on total masses of emissions. High concentrations of SO_2 were detected in ambient air before the year 1989 (mean concentration for Russia in 1984 was 83 μg/m^3, in 1988 - 30 μg/m^3), but since 1989, laboratories of Roshydromet switched to new express method based on film sorbent use. As a result, concentration fell down promptly to 10 -12 μg/m^3, that is much lower than concentrations in the air of industrial cities of other countries. The highest air levels of SO_2 concentrations (more than 50 μg/m^3 - daily MAC) have been detected in 7 Russian cities: the city of Norilsk - a major source of SO_2 emission (2.1 million tons per year), cities with nickel smelters (such as the city of Nickel) and with copper smelters (Mednogorsk).

5.2.4. Carbon monoxide

This is the third major air pollutant preceded by suspended particles and sulfur dioxide, according to classification based on masses of emission. 5.9 to 7.6 million. tons of CO are released every year. Annual mean concentration of carbon dioxide varies from 1.54 to 1.57 mg/m^3 while the limit equals to 3.0 mg/m^3 for daily exposure. In the air of most Russian cities, concentration of CO is below the limit. Even in large cities like Moscow and St.-Petersburg, elevated concentrations of CO were detected only by very few stations. The highest levels of CO pollution are detected in Vladivostok by six monitoring stations, and also in cities with iron smelters such as Komsomolsk-on-Amur, Lipetsk and Khabarovsk.

Average annual concentrations of 3 classical pollutants: suspended particles, SO_2 and NO_2 - in the ambient air of a number of Russian and large foreign cities are presented in Table 3.

Concentration of suspended particles in the air of large cities of the European part of Russia is approximately the same as in cities of Western Europe. Concentrations of NO_2 are lower in Russia which is understandable considering heavier traffic in Western countries. Concentrations of SO_2 are much lower in Russia. On our opinion, analytical methods used for collecting and analyzing air samples in Russia require an examination.

5.3. INORGANIC SUBSTANCES

5.3.1. Lead

From 0.9 to 1.4 thousand tons of lead are released in air by industrial enterprises annually. In recent years, emissions of lead have decreased substantially due to reduction in manufacturing. Special researches recorded high concentrations of lead in the ambient air. For instance, practically all the inhabitants of such small towns as Belovo, Karabash, Dalnegorsk are exposed to high concentrations of lead

TABLE 3. Average annual concentrations of SO_2, NO_2 and suspended particles in ambient air of central districts in some cities of the world, 1975-1991, μg/m^3.

City, source of information		Suspended particles	SO_2	NO_2
Moscow	[16]	100	1	80
St. Petersburg	[16]	160	5	80
Yekaterinburg	[16]	70	9	40
Nizniy Novgorod	[16]	180	9	40
Samara	[16]	160	10	50
Calcutta	[17]	270 to 550	50 to 60	
Chicago	[17]	80 to 90	30	
Lisbon	[17]	90 to 160	20 to 60	
Sidney	[17]	90 to 150	30 to 70	
Warsaw	[17]	no data	20 to 40	
Athens	[18]	no data	74	194
Barcelona	[18]	216	42	122
Kheln	[18]	82	61	134
Lion	[18]	no data	86	85
Paris	[18]	no data	38	84

ranging from 15 to 36 μg/m^3 (MAC for 24 hs exposure is 0.3 μg/m^3). Increased concentrations of lead in the air of residential zones were detected near battery plants in St.-Petersburg, Kursk and other cities [19]. It may be assumed that local areas of lead contamination exist in some other large industrial cities, which are not controlled by Russian Committee for Meteorology (Roscomhydromet).

5.3.2. Copper (oxide)

Industrial enterprises release from 5 to 7 thousand tons of copper per year. Largest sources of emission are metallurgical plants in Revda and Krasnouralsk of Sverdlovsk region and Karabash of Cheliabinsk region. Mean annual concentration is 0.08-0.09 μg/m^3 (less than MAC for 24 hs exposure - 2.0 μg/m^3).

5.3.3. Mercury

More than 6 tons of mercury are emitted in ambient air annually mainly by factories producing mercury containing apparatus: major sources are factories in the town of Klin in Moscow region, city of Saransk in Mordovia, and enterprises which use mercury for chlorine chemistry (Sterlitamak, Usolie-Sibirskoye and others). Scientific research organizations in a number of cities have studied concentration of mercury in ambient air for many years. Local pollution has been identified in the areas near mercury storage facilities and in some other places.

5.3.4. Hydrogen sulfur

Main sources of hydrogen sulfur emissions are paper plants, coal chemical plants of metallurgical factories, gas and petroleum processing plants, chemical factories, synthetic fiber manufacturing facilities. Concentrations exceeding MAC (8 μg/m3, that is a level of smell detection by humans) were recorded in 70% of cities where monitoring was performed. Concentrations of hydrogen sulfur often exceeded MAC in the air of cities with paper factories (Segeza -79% of samples, Amursk - 14.4% of samples, Bratsk - 6%, Selenginsk - 5.5%); in cities with petroleum refining industry (Berezniki - 25% of samples, Oha - 26 % of samples, Gubaha and Sizran - both 8 %) and in cities with synthetic fiber manufacturing plants (Krasnoyarsk - 7 % and Tver - 6 %).

5.3.5. Carbon disulphide

Annual amount of emissions ranges from 11 to 27 thousand tons. Synthetic fiber industry and coking coal chemical plants are major sources of CS_2 emissions into air. There are 26 synthetic fiber manufacturing enterprises in Russia. Highest concentrations of CS_2 were observed in cities with paper industry: Arkhangelsk, Baikalsk, Bratsk, Kaliningrad, Novodvinsk, Selenginsk; in cities with synthetic fiber manufacturing enterprises: Balakovo, Kemerovo, Tver; and in cities with chemical industry: Berezniki, Volgograd. High level of CS_2 pollution possibly exists in Riazan, but CS_2 concentration has not been systematically monitored.

5.3.6. Fluoride compounds. Hydrogen fluoride

Sources of emissions: aluminum smelters, fertilizer manufacturing plants. Annual emission of hydrogen fluoride ranges from 12 to 16 tons. Median annual concentrations of hydrogen fluoride exceed MAC for 24 hs exposure (5 $\mu g/m^3$) approximately twice in cities where aluminum smelters are located. The ambient air concentration of *HF* in the residential zone of Novokuznetsk near the aluminum smelter exceeds MAC 16 to 18 times, i.e. it reaches 80 to 90 $\mu g/m^3$.

5.4. CHLORINE COMPOUNDS

5.4.1. Hydrogen chloride

Annual emission to the atmosphere ranges from 3.6 to 10.8 thousand tons. Ambient air concentrations are measured in 32 cities. Concentrations exceeding MAC (for 24 hs exposure - 200 $\mu g/m^3$) for short-term exposure were detected in cities with large chemical plants: Saratov, Syzran, Samara, Perm, Nalchik, Zavolzie, Dzerzinsk, Volgograd.

5.4.2. Dioxins

Enterprises releasing chlorine compounds represent potential sources of such super toxicants as dioxins. 60 enterprises can be included in the list of potential sources of this chemical: 23 factories producing organic chemical compounds for agricultural plants protection, 30 paper factories, 3 condenser manufacturing factories, and 4 chemical-metallurgical plants. Environmental pollution levels in locations of dioxin emitting enterprises are very high. In the past, information was availably only on surface water and soils pollutions. In recent years, data on ambient air pollution appeared as well. MAC of 0.5 pg/m^3 for dioxin concentration in air was set up in 1994. At that time, concentration in the air of Chapayevsk was 15 pg/m^3 [20].

5.5. OTHER CHEMICAL SUBSTANCES

5.5.1. Ammonia

With the amounts of annual emission ranging from 47 to 71 tons, this chemical is at the top in the list of specific pollutants. Ammonia is emitted into the air by metallurgical plants, fertilizer manufacturing factories and by various enterprises of chemical industry. Measurements of ambient air concentrations of ammonia were conducted in more than 70 cities. Concentrations exceeded MAC (for 24 hs exposure - 40 μg/m^3) in cities with fertilizer manufacturing enterprises (Belgorod, Voskresensk, Toliatty) and in cities with large chemical plants (Dzerzinsk, Kemerovo, Omsk, Samara, Solikamsk, Tomsk). Increased concentrations of ammonia are observed on the whole territories of such cities as Kemerovo, Omsk and Dzerzinsk.

5.5.2. Methyl merkaptan

Annual emissions vary from 1 to 1.34 tons. Major sources of emission are provided by paper industry. In Amursk, Arkhangelsk, Bratsk, Dolinsk, Koriazma, Novodvinsk, Poronaysk, Selenginsk, Siktivkar, Ust-Ilimsk, Uhta, ambient air concentrations exceed MAC (0.1 μg/m^3 for 30 min. exposure) in 20 to 98 % of samples.

5.5.3. Styrene

This synthetic chemical compound is released in the ambient air annually in amounts up to 2 thousand tons. Major sources of emission are manufacturing of synthetic rubber, some other petrochemical enterprises and exhaust fumes of motor cars. In 1992, mean annual concentration of this chemical was 4 μg/m^3 (MAC for 24 hs exposure is 2 μg/m^3). The highest concentrations are recorded in the air of Usolie-Sibirskoye (10 μg/m^3), Gubaha, Yekaterinburg, Cheliabinsk (5 to 6 μg/m^3). Styrene control in ambient air is not implemented in a majority of cities with styrene, polystyrene and synthetic rubber manufacturing enterprises, including the cities with large petrochemical plants (Niznekamsk, Omsk, Voronezh, Yaroslavl, Salavat, Angarsk, Krasnoyarsk and others).

5.5.4. Phenol

Ambient air concentrations are monitored in more than 90 cities at 230 to 235 stations. Mean annual concentration range is 2 - 3 μg/m^3 (MAC for 24 hs exposure is 3 μg/m^3). Concentrations of phenol exceeded MAC at least in 1% of samples in 84 - 85 % of monitored cities, at least in 5 % of samples in 15 - 25 % of monitored cities and at least in 10 % of samples in 4 - 10 % of monitored cities. Such a high percentage of cities with increased concentrations of phenol periodically recorded in ambient air is due to the location of air quality control stations near large streets or highways. Mean annual concentrations of phenol exceed MAC for 24 hs exposure in Moscow, in cities with metallurgical industry (Komsomolsk-on-Amur, Magnitogorsk, Volgograd, Nizniy Tagil, Novokuznetsk, Norilsk, Lipetsk, Riazan), and in those with chemical and petrochemical plants (Perm, Gubaha, Kirovo-Chepetsk, Tiumen, Sizran, Berezniki, Saratov, Usolie-Sibirskoye, Dzerzinsk).

In Table 4 we attempted to summarize the results decribed above.

TABLE 4. Comparisons of certain contaminant concentrations detected in urban air in Russia and other countries, μg/m^3

Substance and	Russia			Other countries	
allowable normative level in air of suttlements in Russia	Concentrations in cities averaged over 1990-1994 years	Tendencies in last years	Concentrations in the most contaminated cities *μg/m^3*	Cities	Concentrations
Suspended particles (TSP) MAC, 24 h - 150; MAC, 30 min. - 500.	150 - 170	13 % decrease	In air of about 100 cities - from 160 to 450	Europe USA Lisbon Sydney Calcutta	10 - 150 40 - 90 90 - 160 90 - 150 270 - 550
Nitrogen dioxide MAC, 24 h - 40; MAC, 30 min. - 85.	42 - 44	6% increase	60-100 (15 cities)	Europe Athens Barcelona Keln London Paris Lyon	20 - 90 194 122 134 121 85 85
Sulfur dioxide MAC, 24 h - 50; MAC, 30 min. - 500.	Data in evaluation	Factor of 2 decrease	135 highest level in Orel city (nickel plant), Orenburg oblast	Athens Barcelona Keln London Lyon Milan Paris Warsaw Lisbon	74 42 61 41 86 99 37 20 - 40 20 - 60

				Calcutta Chicago Sydney	50 - 60 30 30 - 70
Carbon monoxide, mg/m^3 MAC, 24 h - 3; MAC, 30 min. - 5.	1.5 - 1.6	5.4% increase	Vladivostok, Komsomolsk-on-Amur, Lipetsk, Khabarovsk, Saratov, Irkutsk	Europe	< 20
Ammonia MAC, 24 h - 40; MAC, 30 min. - 200.	39 - 44	28%. decrease	Moscow St.-Petersburg, Voskresensk, Samara, Omsk, Kemerovo, Solikamsk et.al.	Europe Japan Canada Japan (max)	4 - 26 to 210
Asbestos, fibres/m^3		no data			100 - 1000
Benzo(a)pyrene, ng/m^3 MAC, 24 h - 1.0	2.1 - 3.0	35% decrease	3.0 - 15.6 ng/m^3 in air of about 25 industrial cities; >10 ng/m^3 of 10 cities	USA Europe (in 60th) West Germany Current	1 – 5 <100 decreased to 10 1-10
Benzene MAC, 24 h - 100; MAC, 30 min. -1500	90 - 93		200 - 280 Moscow region, St.-Petersburg Samara, Toliatti, Beloretsk	Big cities Near oil refinery plant	3 - 160 few hundreds
Vanadium MAC, 24 h - 2.0.	0.06 - 0.09		no data	Near steel smelting plants	0.07
Vinil chloride ESEL$^{*)}$ - 5	no data		up to 25-100 Dzerdzinsk, Usolie-Sibirskoye	Europe Near VC (PVC) productio n plants	0.1 - 0.5 10, but may reach up to 100
Hydrogen fluoride MAC, 24 h - 5; MAC, 30 min. - 20	6 - 8	5% increase	>5 In 20 cities	USA Germany (industr. Centers)	0.02 - 2.00 0.5 - 3.8 (1.3 aver.)
Hydrogen chloride MAC, 24 h - 200; MAC, 30 min. - 200.	86 - 95		1200 - 2800 Berezniki, Biisk, Dzerjinsk, Krasnoiarsk, Omsk, Perm, Tomsk, Solikams Vladivostok, Volgograd		

Cadmium MAC, 24 h - 0.2; MAC, 30 min. - no.	0.005 - 0.08		0.28 - 1.96 Belovo, Vladikavkas	Europe	0.005-0.05 usually <0.02
Xylene MAC, 24 h - 200; MAC, 30 min. - 200.	51 - 82		>200 2400 – 3400 max in more than 10 cities.	no data	
Manganese MAC, 24 h - 1.0	0.06 - 0.09		no data	USA, industrial USA, max	0.2 - 0.5 10.0
Methil mercaptan MAC, 24 h - no; MAC, 30 min. - 1.0.	0.047- 0.055		0.05 11 cities with paper production	no data	
Arsenic MAC, 24 h - 3.0		no data		Europe USA max	1 - 10 2.6 - 11.0 1,000
Mercury MAC, 24 h - 0.3	lower 0.1		0.1 - 0.3 Amursk Usolie- Sibirskoe, Sterlitamak	Europe Near plants	0.5 - 20.0 to 10 000
Lead MAC, 24 h - 0.3	0.06 - 0.10		0.05-0.3 100 max near Pb-smelters	Europe	0.5 - 3.0
Nickel (metal compounds) MAC, 24 h - 1.0	0.06 - 0.10		To 23 $\mu g/m^3$ - Norilsk	Canada, nickel plants	0.5 - 0.7
Hydrogen sulfide MAC, 24 h - 8.0; MAC, 30 min. - 8.0.	2.0	unchang ed (last 5 years)	50 - 100 max in 15 cities with cellulose-paper, oil, coke, chemical and fiber industries	California Belgium (cellulose plants)	170 - 200
Carbon disulfide MAC, 24 h - 5; MAC, 30 min. - 30.	10 - 16	12.5% decrease	6 - 30 in 20 cities with synthetic fibre production*), chemical or coke production**), paper production: Baikalsk, Bratsk, Novodvinsk et.al.	Austria (viscose product.)	< 157

Styrene MAC, 24 h - 2; MAC, 30 min. - 40	4 - 10		2-10 1240 max in 20 cities with chemical production: Perm, Berezniki, Gubaha, Solikamsk, et al	Europe Industrial cities Chemical plants	0.3 20 20-30
Toluene MAC, 24 h - 600; MAC, 30 min. - 600.	100 - 111		>600 Moscow region 1000 - 3000 cities with chemical plants	Europe Near plants	0.1 - 204.0 100-20,000
Phenol MAC, 24 h - 3.0; MAC, 30 min. - 10.0	2 - 3		3-12 in control; >60 in 17 cities with chemical production ; 70-123 in 30 cities with metallurgical, or oil industry	Paris Osaka Portland Paris, Osaka, Portland Near plants	0.7 - 0.8 1.0 - 4.0 0.2 - 0.4 0.2 - 4.0 13 - 26 190 max
Formaldehyde MAC, 30 min. - 35MAC, 24 h - 3;	3 - 12	8.2% decrease	8-27 In more than 20 cities with chemical and oil production	Europe Paris (roads) Rotterdam Germany Italy Los-Angel Tokyo industrial plants	5 - 10 41 – 120 7 7-12 7.1 18.5 8.5 60 - 90
Chlorine MAC, 24 h - 30; MAC, 30 min. - 100.	13 - 17		30 - 40 in 5 cities with chemical and metallurgy: Berezniki, Norilsk, Saratov, Solicamsk and Electrostal	Europe	1.0 - 3.7
Chromium (Cr^{+6}) MAC, 24 h - 1.0	0.08 - 0.10		Probably increased Cr^{+6} In Pervouralsk and Vladimir	USA cities near	0.005-0.525 1.0 max

*) Barnaul, Balakovo, Voljskii, Kemerovo, Krasnoiarsk, Mytischi, Tver;

**) Berezniki, Volgograd, Magnitogorsk, Nijnii Tagil, Cherepovets.

6. Approximate Size of Population Exposed to Increased Levels of Contamination of Ambient Air

Evaluation of population size under exposure (Fig.1) was based on ambient air pollution data for different chemicals relating to more than 100 Russian cities. Among pollutants, suspended particles occupy the first place followed by benzo(a)pyrene. More than 15 million. people are exposed to increased concentrations of suspended particles and approximately 14 million. - of benzo(a)pyrene. Elevated concentration of suspended particles in the ambient air may contribute to some increase in mortality caused by respiratory diseases, high levels of benzo(a)pyrene determines elevated risk of cancer in 24 cities with steel and aluminum smelters and in 30 cities with petroleum

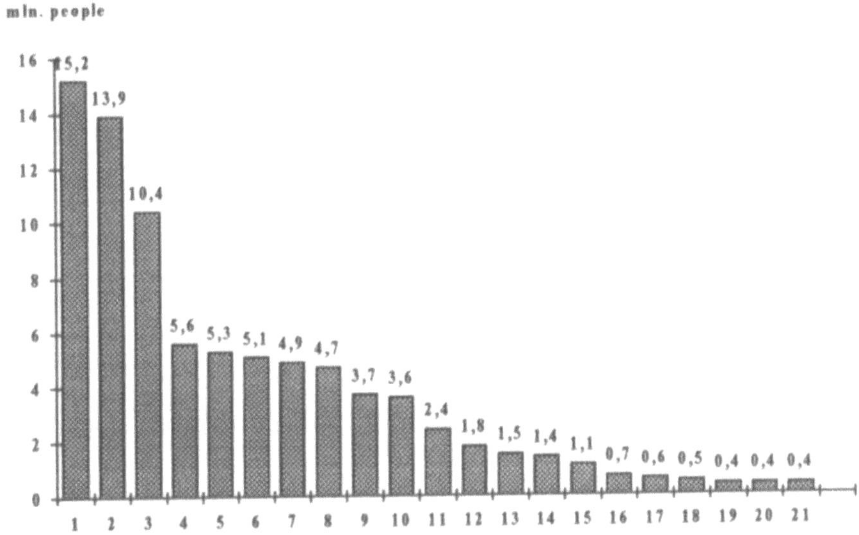

1. Suspended particles	8. CO	15. Methyl mercaptan
2. Benzo(a)pyrene	9. Ammonia	16.HCl
3. Phenol	10. Styrene	17. Ni
4. NO_2	11. Pb	18. As
5. HF	12. Benzene	19. Cd
6. Carbon disulfide	13. NO	20. Vinyl chloride
7. Formaldehyde	14. Hydrogen sulfide	21. Hg

FIGURE 1. Approximate sizes of population exposed to high levels of certain air pollutants

refining industry or large thermoelectric power stations. Phenol is in the third place with 10.4 million. people exposed; then nitrogen dioxide follows (5.6 million.). Increased concentrations of these substances are measured in the air of most cities with intensive traffic. More than 5 million. people live in areas with elevated concentrations of sulfur fluoride, carbon disulfide, formaldehyde and carbon monoxide; In the cities with big chemical plants, more than 3 million. are exposed to increased concentrations of ammonia and styrene. A significant number of people (more than 1 million) are affected by elevated concentrations of lead, benzene, nitrogen oxide, hydrogen sulfide and methyl mercaptan. High levels of toxic heavy metals and vinyl chloride are detected in ambient air of several cities with over 1 million inhabitants.

7. Assessments Of Health Risks Caused By Chemically Contaminated Air In Russian Cities

Knowing the pollution levels in the ambient air and the size of the population under exposure we can assess and compare the health risks from air pollutions for populations in Russian cities. The aim is to make clear the contribution of existing atmospheric air pollution to mortality rates. For this purpose we have:

- analysed the national (U.S.EPA), international (WHO) and other approaches to health risk assessment [1-7],
- combined their recommendations, and
- used averaged values for point assessments with uncertainty range reflecting the variance in recommendations.

The scheme of our analysis is presented in Fig.2. Some results are showed in Table 5.

TABLE 5. Assessments of health risks from air pollutions in Russian cities

Substances	Annual risk of death, 10^{-6} ($y^{-1}/\mu g/m^3$)	Refe rence	Averaged annual concentration ($\mu g/m^3$)	Absolute annual health risk (deaths per year)	Population size (mln.)
Suspended particles	~4 (0.8-17)	[3-5]	265	16100	15.2
Nitrogen dioxide	<3	[3]	75	<1200*)	5.6
Benzo(a) pyrene	~660 (25 - 1300)	[1,2]	0.005	45	13.9
Vinyl chloride	~0.6 (0.01-1.2)	[1,2]	50	12	0.4
Benzene	~0.09 (0.06-0.12)	[1,2]	37.6	9	2.6
Formaldehyde	~0.09	[2]	18.7	8.2	4.9
Arsenic	~60	[1]	0.6 **)	18	0.5
Cadmium	26	[1]	1.2	12.5	0.4
Nickel	~5.3 (4.9-5.7)	[1,2]	2.6	8.3	0.6

*) Upper bound. **) Expert assessment

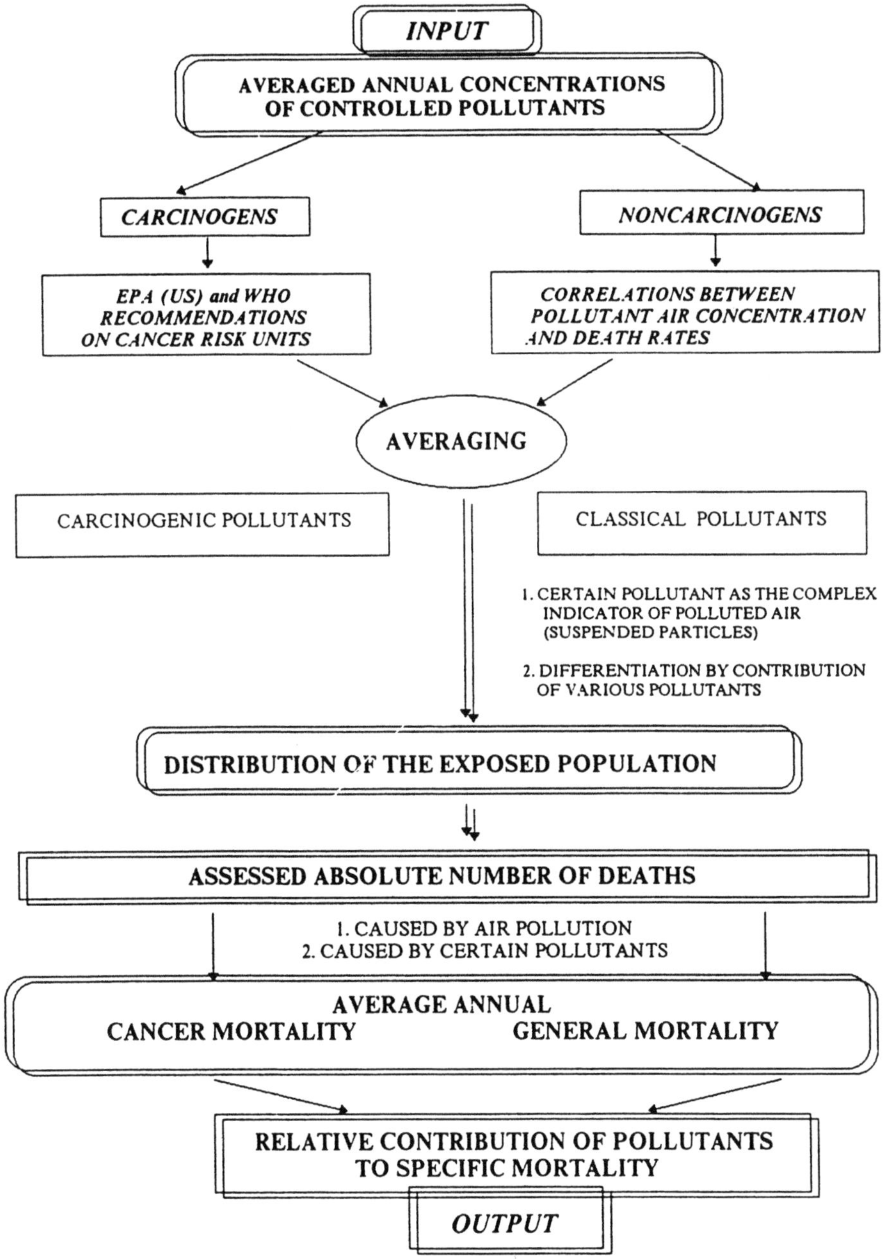

FIGURE 2. Methodological procedure used in the paper for health risk assessments.

The choice of pollutants, presented in Table 5, is determined by the Russian monitoring standards and analytical methods of air pollution control. Only about 20 substances are being constantly measured at stationary monitoring posts. For the purpose of this study, the pollutants were chosen referred to as potentially hazardous in the range of measured concentrations in urban air. Results of our calculations are presented in the third column of Table 5 in absolute numbers of deaths caused by the atmosphere pollution in Russian cities. Last column of Table 5 indicates the size of the population under exposure of corresponding pollutants.

All pollutants included in the tables can be divided in two groups (classes) according to their health effects pattern:

1) Pollutants increasing the frequency of overall mortality. These are: suspended particles and NO_2.

   It should be noted that this class of pollutants also contribute to respiratory tract diseases, but these calculations were excluded from the paper, because of the complications in evaluating air pollution impact on morbidity in Russian population.

2) Carcinogenic pollutants that increase the incidence of specific cancer diseases: leukemia and respiratory tract cancers. This group comprises all the pollutants except those in the first class.

The total number of annual deaths from the exposure to suspended particles is approximately 16.000 for 15 million population of Russian cities. That gives a relative contribution to overall annual mortality of about 5%. It is important to note here that the data of health risk from suspended particles were obtained by the regressive analysis showing the correlation between the mortality rates and the air pollution levels. That's why these figures characterise the total contribution of polluted air to the annual mortality more than the certain contribution of the suspended particles, which is lower.

Nitrogen dioxide and sulphur dioxide are relatively important factors in death risk. Although the assessment of health risk caused by the nitrogen dioxide has a wide range of uncertainty, it obviously plays an essential part in mortality caused by ambient air pollution.

A number of environmental epidemiological studies show the correlation between high concentrations of three major pollutants - suspended particles, SO_2 and NO_2 and excess mortality rates. For instance, excess mortality rates due to respiratory diseases are recorded in such cities as London, Athens, Lyon, Marseilles, Krakow, Milan and Katovice in Europe; Santa-Clara, Donora, New York, Atlanta, Houston, St. Louis, Chicago, Minneapolis, San-Francisco, Los-Angeles, Philadelphia, Steubenville in the USA; Mexico city and in some other cities of the world. Strong correlation has been found between mortality rate and concentration of suspended particles above threshold limit established in the USA of 150 $\mu g/m^3$. For example, elevation of SP concentration in air by 100 $\mu g/m^3$ caused an increase in daily mortality by 3.8 % in Steubenville, by 4 % in London, and by 3 % in New York [21].

Among the classical pollutants sulphur dioxide has been thoroughly investigated in terms of health consequences. No evidence of important contribution of sulphur dioxide to the death risk was found in this study. This fact can probably be explained by the inaccuracy of measurements of this substance in Russia. It is obvious that methods of analytical control for sulphur dioxide currently used in Russia have to be improved. Such conclusion is proved by a comparative analysis of sulphur dioxide concentrations in big cities of other countries. They may be at least ten times high, as it is shown in Table 2.

The contribution of noncarcinogenic pollutants to all deaths caused by air contamination is beyond 50%. Carcinogenic pollutants (second class): benzene, benz(a)pyrene, vinylchloride, arsenic, cadmium, nickel, formaldehyde. The contribution of these substances to the mortality from cancer diseases is from 1 to 3 %.

The total contribution of carcinogenic pollutants to all deaths caused by air pollution is less than 10%. The contribution of pollutants studied elsewhere to all deaths caused by air contamination is about 40%. Further researches are necessary.

8. Comparative Analysis Of Risk Orders

The aim on this stage of our study was to make a comparative analysis with other kinds of risks for better understanding of the hazards posed by air pollutions (Fig.3). To this purpose, we used a rough scale of risk orders (Fig.4).
The introduction of an order of hazard scale facilitates:

- the comparisons of quite uncertain health risk assessments;
- the process of decision making on the stage of risk management.

Basing upon this scale of fatal risk orders for population we have made a comparative analysis of fatal risk orders from air pollutants versus main causes of death and natural hazards. The results are presented in Table 6 and Table7.

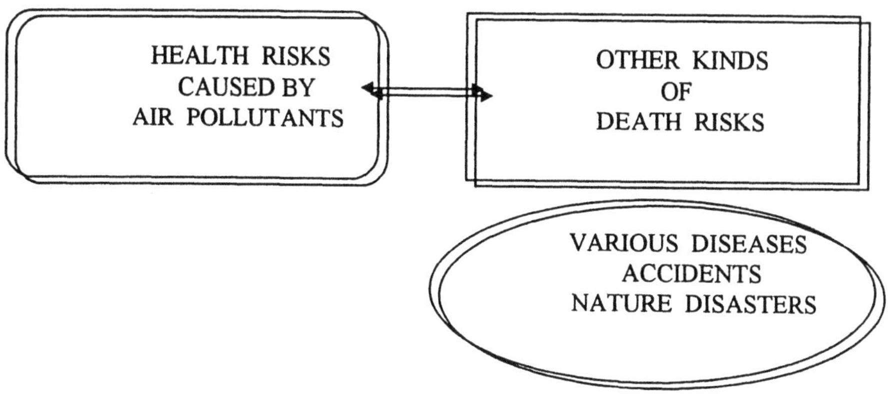

FIGURE 3. The scheme of comparative death risk analysis

TABLE 6. Comparative analysis of population death risk orders

Health hazards	***Range of annual death risk***						
	$<10^{-7}$	10^{-6}	10^{-5}	10^{-4}	10^{-3}		$10^{-2}>$
Ambient air pollutions:							
Suspended particles				I----	----*--	I	
Nitrogen dioxide				I---*	I		
Arsenic				*			
Cadmium				*			
Vinylchloride			I--	*--I			
Nickel			*				
Benzene			I-*--	I			
Benzo(a)pyrene		I-----	---*--I				
Formaldehyde		*					
Fatal diseases:							x
Heart diseases						x	
Cancer diseases						xx	
Marrow vessels diseases					x	x	
Pneumonia					xx		
Diabetes					x		
Arteriosclerosis				x	x		
Suicides					x		
Murders in Moscow					x		
Accidents:					xxx		
automototransport					xx		
falls				xx	x		
drawned				x			
fires				x			
others				x	x		
Nature disasters:			o	o			
Flood, tsunami			o				
Earthquake			o				
Typhoon,cyclone,storm			o				
Thunder storms		o					
Hurricane, tornado	o						

A rough scale guide of annual risk orders								
Low			Middle			High		
neglec-tive		relative-ly low		relative-ly high		vital	extreme	
<	10^{-8}	10^{-7}	10^{-6}	10^{-5}	10^{-4}	10^{-3}	10^{-2}	>
A rough guide for hazard scale								
8	7	6	5	4	3	2	1	

FIGURE 4. Classification of death risk orders for population

TABLE 7. Comparative analysis of various death risks using the hazard scale

Hazard marker	Hazard class number	Air pollutants	Fatal diseases, accidents etc.	Natural hazards
	1		All diseases	
	2		Heart diseases	
	2		Cancer diseases	
	2-3		Marrow vessels diseases	
	3	Suspended particles	Pneumonia	
	3		Diabetes	
	3		Suicides	
	3		Murders (Moscow)	
	3		All accidents	
	3		Transport accidents	
	3-4	Nitrogen dioxide	Arteriosclerosis	
	3-4		All accidents, except transport	
	4	Sulphur dioxide	Falls	
	4	Arsenic	Drowing	
	4	Cadmium	Fires	
	4	Vinylchloride		
	4-5	Benzene		Natural disasters
	4-5	Nickel		
	5	Benzo(a)pyrene		Flood, tsunami, earthquake, storm, typhoon, cyclone
	6	Formaldehyde		Thunder storms

The comparison of health risks from air pollutions in Russian cities and other kinds of risks, i.e. a major group of fatal diseases and natural disasters are showed in Table 7. We may conclude that the death risks from:

1) air pollution (defined through suspended particles) equivalent to the death risks from such diseases as pneumonia and diabetes; or all risks from road accidents; or risks from murders and suicides;
2) nitrogen dioxide is close to death risk due to arteriosclerosis - the disease included in the list of leading causes of death;
3) vinylchloride, arsenic and cadmium can be compared with such causes of accidential deaths as falls, melts, fires and burns, and others (exluding road accidents);
4) nickel is compatible to the risks from all nature disasters and catastrophes;
5) benzene, benzo(a)pyrene, formaldehyde are equivalent to the natural risks, such as flood, tsunami, earthquakes, typhoons, cyclones or storms.

9. Conclusion

The analysis shows the following general results:

1) Increased concentrations of the following chemicals are recorded in ambient air of Russian cities:
   - ammonia - up to 60 - 200 $\mu g/m^3$ in about 20 cities;
   - benzo(a)pyrene - over 10 ng/m^3 in more than 10 cities;
   - benzene - over 200 $\mu g/m^3$ in more than 10 cities;
   - vinyl chloride - 25-100 $\mu g/m^3$ (estimation); exact number of cities is unknown;
   - xylene and toluene - in more than 20 cities;
   - nickel - to 23 $\mu g/m^3$ in Norilsk, the exact number of cities is unknown;
   - carbon disulfide - 6 - 30 $\mu g/m^3$ in about 20 cities;
   - styrene - 2 - 10 $\mu g/m^3$ in more than 10 cities;
   - phenol - 3 - 12 $\mu g/m^3$ in practically all controlled cities;
   - hydrogen fluoride - over 5 $\mu g/m^3$ in about 20 cities;
   - chlorine and hydrogene chlorine in cities with chlorine chemical plants;
2) Concentrations of sulfur dioxide is underestimated;
3) Concentrations of nitrogen dioxide, suspended particles and formaldehyde are comaprable.
4) There are local problems of air contamination by hydrogen sulfide, carbon disulfide, methyl mercaptan, cadmium and lead;
5) Data on air contamination by asbestos, arsenic, chromium (IV) are not available or very limited.

The risk assessments and comparative risk analysis provide evidence of the health hazards for the urban population in Russia associated with increased levels in air of such pollutans (reviewed in this study) as suspended particles, nitrogen dioxide, benzene, vinylchloride, arsenic, cadmium, nickel, benzo(a)pyrene. Further investigation is required including both the analysis of health risks from other pollutants and elaboration of the presented results.

10. References

1. Air Quality Guidelines for Europe. WHO (1987). Regional Publications, European Series, No 23.
2. Rolaf van Leeuwen F.X. (1997). WHO Air Quality Guidelines for Europe. European EpiMarker, April 1997.
3. U.S.Environmental Protection Agency (1995). Integrated Risk Information System - IRIS.
4. Introduction To Cost-Effectiveness Analysis Of Risk Reduction Measures In Energy Systems. (1986). IAEA-TECDOC-383, Vienna.
5. R.Mendelson (1980) An economic analysis of air pollution control from coal-fired power plants, Journal of Environmental Economics and Management, vol.7.
6. Hamilton L.H. (1979) Health effects of electricity generation. Presented at the conference on health effects of energy production, sponsored by Chalk River Nuclear Laboratories, Ontario, Canada, 1979.
7. Bykov A.A., Murzin N.V. (1997). The Problems Of Individual, Societial and Environmental Safety Analysis. Nauka, St.-Petersburg. (in Russian).
8. Temporary methodological recommendations for updating threshold limit values MAC of ambient air pollutants in residential zones. (1988). Ministry for Public Health of the USSR # 4681-88 of July 15. (in Russian).
9. General report on ambient air quality protection in 1991 to 1994. (1992, 1993, 1994, 1995). Goscomstat of RF. (in Russian).
10. Operational Manual for Air Pollution Control. (1990). Leningrad. (in Russian).
11. Annual reports on ambient air quality in Russian cities for 1992, 1993, 1994 years. (1993 - 1995). Main Geophysical Observatory, St. Petersburg. (in Russian).
12. Report of Sverdlovsk Regional Center of Sanepid of the year 1994. (1995). Yekaterinburg. (in Russian).
13. Grachiova M.N., Tikhomirov Yu.P. (1993). Sanitary assessment of vynilchlorine and polyvinilchlorine manufacturing as sources of ambient air pollution. Hygiene and Sanitary, pp. 11-12. (in Russian).
14. Yarushkin V.Yu. (1991) Environmental hygiene and children health in the regions with metallurgical zinc plants. Ph.D. thesis, Irkutsk (in Russian).
15. Bushtueva K.A., Revich B.A., Bezpalko L.E. (1992). Hygienic assessment of cadmium in environment. Zinc and cadmium in environment. Nauka. (in Russian).
16. Annual report on ambient air pollution in Russian cities and regions, 1992. (1993) vol.2, Attachment, St. Petersburg. (in Russian).
17. City Air; Quality Trends; Global Environmental Monitoring; (1995) Monitoring System, V. 3, UNEP, WHO.

18. Romien I. (1995). WHO Pan-American Health Organization; Report at 7th Conference of International Society of Environmental Epidemiologists, August 30, 1995. The Netherlands.
19. Revich B.A. (1997). Lead in environment and childrenís health in Russian cities. Occupational Medicine and Industrial ecology.(in Russsian, to be published)
20. Revich B., Aksel E., Dvoirin V., Kolbeneva L., Pervunina R. (1996). Dioxins in the environment and population health of Chapaevsk (Russia). 16th Symp.on Chlorinated Dioxins and Related Compounds, V.30, Amsterdam, pp.350-353.
21. Schwartz J. and Dockery D.W. (1992). Particulate Air Pollution and Daily Mortality in Steubenville, Ohio. American Journal of Epidemiology, vol. 135, no1.8

LEAD CONTAMINATION OF RUSSIAN CITIES: ASSESSMENT OF RISK TO CHILDREN'S HEALTH

A.A. BYKOV
Plekhanov Economic Academy, Stremiannyi per. 28, Moscow, 113054, RUSSIA.

B.A. REVICH
Centre of Demography and Human Ecology, Institute of the Forecasting of Russian Academy of Sciences, 32 Krasikova ul., Moscow, 117418, RUSSIA.

Abstract

Annual industrial lead releases into the air of Russian cities varies from 1.1 to 1.6 thousand tons. In spite of the substantial decrease of lead emissions in recent years, high levels of lead in the environment (PbE) are still measured. They create increased lead levels in human organisms and can cause various undesirable health consequences. Children are especially sensitive to lead contamination. Blood lead level (PbB) is often used as an indicator of the influence of lead on a child's health, and the acceptable level is 10 ug of Pb per 100ml of blood. The probability of exceeding this level of PbB defines the health risk. This paper presents general results of a health risk assessment of lead exposure in children living in Russian cities. The Integrated Exposure Uptake Biokinetic Model of the US EPA was used to predict total lead intake with pathway differentiation. The health risk of elevated PbB is based on monitoring data of PbE. Predicted model curves relating PbB to PbE can be used by decision-makers in forming environmental policy. Calculated PbB of children living in the cities with relatively low and moderate levels of PbE is close to the standard (10 ug/dl). In highly contaminated cities this standard can be exceeded by 100%. Estimated corresponding risk values are 50% and 85%. Analysis of the different contributing pathways for total lead exposure shows that the principal contributors are: first, food products; then soil/dust; with the lowest contributions from air and drinking water. An assessment of health consequences shows the following approximate distribution of hazard: two million children in Russian cities can have problems in behavior and learning; 400 thousand will need treatment; and the health of 10 thousand children is in danger while 400 children will need immediate hospitalization. These results lay a foundation for planning future studies (owing to the weakness of existing data), building short- and long-term environmental policy, and defining the most effective means to health risk reduction.

I. Linkov and R. Wilson (eds.), Air Pollution in the Ural Mountains, 181–193.

1. Introduction

Industrial enterprises release 1.1 to 1.6 thousand tons of Pb per year into the air in Russian cities. While these levels have decreased in recent years because of reduced production, high Pb concentrations are still measured in the environment (PbE). These high concentrations can result in elevated lead levels in human organisms and can produce undesirable health consequences, especially in children. Blood lead (PbB) is often the indicator of the impact of lead on children' health. In recent years the acceptable level of PbB in the USA has decreased from 20 to 10 ug/dl of lead in blood. The probability of exceeding this level of PbB defines health risk. Health hazards caused by different levels of PbB in children [1] are shown in Figure 1. This paper presents general results of health risk assessment of lead exposure in children living in Russian cities and some results of special risk analysis have been made for the city of Belovo.

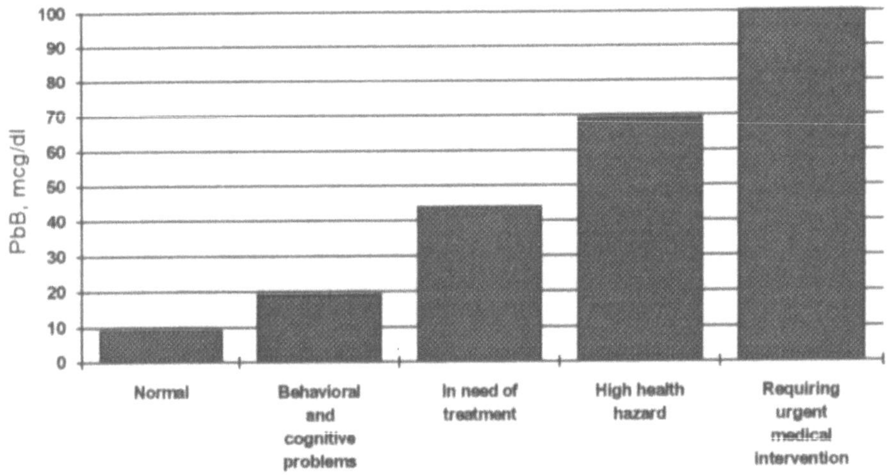

FIGURE 1. Hazard assessment by index of lead in blood (PbB)

2. Mathematical Model for Health Risk Assessment of Lead Exposure in Children

Mathematical models are widely used in health risk assessment in recent years. We used the Integrated Exposure Uptake Biokinetic (IEUBK) Model of the US EPA [2]. This model allows predicting: 1) total Pb intake with pathway differentiation; 2) PbB of children under 7 years of age; 3) health risk as the probability that PbB exceeds the control level. The model curves of PbB dependence on PbE in the environmental compartments can also be used by decision-makers in forming environmental policy.

3. Input Data Analysis for Assessing Risk of Increased PbB of Children Living In Russian Cities

To assess the health risk from increased PbB of children resulting from high levels of Pb in the environment, one must use a set of input data, among which lead from food, soil, dust, air and drinking water are all important sources. Risk assessment is less sensitive to the data on Pb in maternal blood. We proposed that lead intake from painted materials is negligible.

Our experience in data collection for health risk calculations shows that, unfortunately, there are no reliable date for the whole set of necessary parameters in Russian cities. The most adequate and complete data we have found are for the Ural region and a few cities (e.g. Belovo) where special studies have been carried out. The distribution of lead in the environment, which is strongly dependant on the distance from an emission source, is well characterized for this case and can be used for more thorough analysis. Data on environmental lead concentration elsewhere is insufficient for making a multilevel ranking of Russian cities based on the magnitude of environmental lead pollution. Hence, in our risk assessments we used only two categories of cities: with relatively low and medium lead contamination and with high lead contamination. The last category is formed by the cities with industrial plants having high levels of lead emissions to the environment.

3.1. LEAD CONCENTRATIONS IN AMBIENT AIR OF RUSSIAN CITIES

The data on PbC in ambient air of living districts in Russian cities are the most representative. Existing data are obtained for more than 130 Russian cities with a total population exceeding 56 million people. Unfortunately, for the majority of data obtained, we have no information about the monitoring locations. This can lead to significant deviations from the mean values and misrepresents the real picture of air contamination in the city. Samples taken near heavily traveled roadways will show PbC in air. In particular, high levels of PbC measured in such cities as Vladimir, Makhachkala, Tobolsk, Tumen and Yaroslavl could be explained by this factor.

The level of PbC in air of most Russian cities (more than 80) varies from 0.01-0.1 ug/m3. This range reflects low levels of air contamination. PbC's in the air of living districts of industrial centers (about 30), with typical production processes having high lead emissions exceed Russian air standards [maximum acceptable concentration (MAC) is 0.3 ug/m3]. These levels of air lead contamination in industrial centers are defined as relatively high. PbC's in air of the remaining cities (about 25) fall to an intermediate range of 0.1-0.3 ug/m3.

3.2 LEAD CONCENTRATIONS IN SOIL AND DUST OF RUSSIAN CITIES

The data on PbC in soils of living districts in Russian cities are less representative in comparison with the data on PbC in ambient air. During the last two decades laboratories of the Russian State Committee for Meteorology and Water Resources controlled and studied Pb (and other heavy metal) contamination of soils around more than 110 Russian cities and settlements. These data are inadequate for the purposes of risk assessment:

1) These data show the levels of PbC in soil for a 5-km zone around the settlement, and they do not directly reflect soil contamination of living districts. These data can be very useful for the determination of the Pb contamination of food production in the region and were used for the following assessment of dietary intake. Risk assessment based on the IEUBK model requires the data on PbC in soil of living districts as the input parameter. These concentrations would be greater than those measured in the 5-km zones.

2) The data on PbC in dust are practically absent, so we equated dust concentrations with soil concentrations in health risk calculations.

3) The set of cities and settlements where soil contamination has been controlled overlap the corresponding set of cities with measured PbC's in ambient air. The analysis of air and soil contamination data for coinciding cities in both sets shows that for the cities with PbC in air less than MAC, PbC in soil slightly correlates to PbC in ambient air, being practically constant in the range 0.01-0.3 ug/m3 of PbC in air. PbC in soils of living districts of such cities varies from 30-150 mg/kg with an average value close to 100 mg/kg. This number has been used in risk assessment as the input parameter of PbC in soil and dust for all Russian cities where PbC in ambient air is less than the MAC. For the 30 industrial centers with high levels of PbC in ambient air we used the corresponding value of 250 mg/kg (the range for different cities was 50-450 mg/kg).

3.3. LEAD CONCENTRATIONS IN DRINKING WATER OF RUSSIAN CITIES

PbC in drinking water is less than 20 ug/l in most European countries. In Russia, the data on PbC in drinking water are very poor. The Russian standard for lead in drinking water, the maximum acceptable concentration (MAC) is 30 ug/l. PbC in Moscow drinking water varies from 0.7-4 ug/l. This range (mean value is 2.0 ug/l) serves as the basis for risk assessment for Russian cities with low and medium lead contamination of the environment.

It is quite possible that problems of drinking water contamination can arise in regions with smelters or industrial waste storage which is contaminated with lead. In particular, PbC has been measured in some samples of drinking water from metallurgical cities (Verkhniaia, Pyshma, Krasnouralsk, Revda at close to the MAC. Therefore we use a level of PbC in drinking water of 10 ug/l (with a range of 1.0-30.0 ug/l) for risk assessment in highly contaminated cities.

3.4. DIETARY LEAD INTAKE IN RUSSIAN CITIES

Data on PbC in food products and dietary lead consumption are scarce and contradictory. The variance of published data on PbC in food products covers a few orders of magnitude. Some papers present averaged data for PbC in food products consumed by Russian children. The daily lead intake with food has the following age distribution: children 1-3 years old consume 13.7 ug/d; 4-6 year- olds, 64 ug/d; 7-14 year-olds, 67.9 ug/d; 14-17 year-olds, 87.2 ug/d [6]. Thus lead intake from food for children younger than 7 years of age falls in a range from 14-68 ug/d. These data correlate closely with other published work [7]. We used these data as the basic parameters for risk calculation for low and medium contaminated Russian cities. The data of Shurikhin [5] show higher levels of Pb intake with food by children under 7 years of age. This daily intake is in a range of 50-160 ug/d which is close to the corresponding WHO limit for children (1000 ug/wk). Therefore this level of intake was used only for cities with high lead contamination of the environment.

3.5. LEAD CONCENTRATION IN MATERNAL BLOOD

As stated above, this parameter has little influence on the risk assessment calculations. We used the following values; 20 ug/dl for high contaminated Russian cities and 10 ug/dl for others. Note that these values are very close to predicted concentrations of blood lead in children. Generalized data on Pb concentration in different parts of the environment and intake with food products are presented in Table 1. These data were used in health risk assessments of lead exposure for children in Russian Cities.

TABLE 1. Generalized and averaged data on PbE of Russian cities and Pt intake from food products

Rank of cities	Number of cities	Population size, mln.	Preschool children, mln	PbC in air, $mcg/m^{\text{Á}}$	PbC in drinking water, mcg/l	PbC in soil and dust, mcg/g	Pb intake with food products, mcg/d
Relatively low and middle levels of environmental lead contamination	over 100	over 53	4.3	0.01-0.3 (0.15)	0.7-4.0 (2.0)	30-150 (100)	14-68
Relatively high contamination level (cities with industrial sources of lead emission)	about 30	2.5	0.2	over 0.3 (0.6)	1.0-30.0 (10.0)	50-450 (250)	48-163
Total	over 130	over 56	about 4.5				

4. Risk Assessments of Increased PbB of Children Living in Russian Cities

Using the generalized and averaged data of Table 1, we calculated mean values of PbB of children under 7 years of age, accounting for the contributions of different pathways of lead intake and the risk of increased lead concentrations in blood.

4.1. ASSESSMENT OF MEAN PbB OF CHILDREN

Predicting the average lead concentration in blood is one of the main benefits of using the IEUBK model. It is also possible to estimate the contribution of different pathways to the lead exposure and blood lead level of children. Note here that the IEUBK model belongs to the class of steady-state models, since it is based upon equilibrium processes of lead intake, exchange between organs and elimination from the organism. This principal feature of the model restricts its predictive power, but if we need assessments of large populations, averaged over a time interval of about a year, the model is fruitful and its predicted results are quite adequate for application to real situations.

However, in order to draw correct conclusions we must have reliable input data. In our case, input data are very uncertain, in fact often contradictory. It therefore is necessary to improve both the accuracy and reliability of our data. Thus we must point out that all assessments have been made with data of rather poor quality. This factor should be kept in mind when interpreting results and drawing conclusions based on these findings.

Calculations of mean values together with the range of age variation of mean PbB of children living in Russia are presented in Figure 2. It can be seen from Figure 2 that the predicted blood lead of children living in cities with relatively low and medium levels of lead contamination of the environment is very close to the MCA of 10 ug/dl. In the highly contaminated cities this standard is exceeded by 100%.

Analysis of the contributions of different pathways to total lead exposure of children shows that the main contribution to PbB is from food products first, then soil/dust and least of all from air and drinking water. It must be pointed out that the model is very sensitive to the data from food intake and these data are not sufficiently reliable for the present study. Using the known literature data on Pb intake from food in economically developed countries will significantly decrease this contribution to total lead exposure. At the same time, the contribution from drinking water can increase when more accurate and precise data on lead concentrations in Russian cities become available.

4.2. RISK ASSESSMENT OF INCREASED Pb CONCENTRATIONS IN BLOOD

The results of health risk assessment are presented in Figure 3. We also estimated the number of children with increased PbB levels. This histogram is drawn in Figure 4. and shows that approximately 2 million children in Russian cities can have problems in behavior and learning. About 400 thousand will need treatment; the health of about 10 thousand children is in danger and about 400 children need immediate hospitalization (see hazard scale in Figure 1.). And it is also seen that all these consequences are caused by lead exposure.

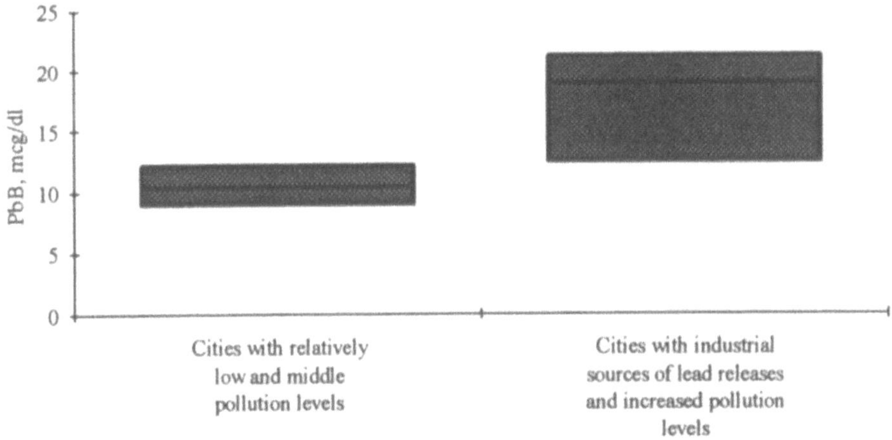

FIGURE 2. Assessments of mean values of PbB in Russian preschool children.

5. Risk Assessments of Increased PbB of Children Living in the City of Belovo

For the city of Belovo, where special studies have been done, there is a complete data base from which one can assess the risk for different exposure scenarios which are dependant on distance from an emission source. Since 1930 a zinc plant has operated in Belovo which releases to the air not only lead but also zinc, cadmium, sulphur dioxide, oxides of nitrogen and other pollutants. In recent years the plant has operated in three stages: up to 1990, normal operation; decreased power during the 1994-96 period; temporary halt to operations in the spring of 1996. These three stages of operation provide three scenarios of lead exposure for children. For each scenario we used the corresponding concentrations of Pb in the environment as reported in previous publications [3,4,7] and new data measurements made in 1996. We also used three zones of contamination corresponding to distance from the plant: 0.5 km, zone 1; 1-2 km, zone 2; 3-4 km, zone 3.

For defined scenarios and zones:

At the first stage, we assessed lead intake in children and contribution of pathways to total exposure. The main pathways contributing to blood lead levels of children living in Belovo are the following: food products, soil/dust and air. Contaminated drinking water does not play an essential role. In fact for any scenario the water pathway contribution is less than 1%. Contaminated air in the normal operation scenario, and for the closest zone, is the main factor influencing PbB in children. For other zones, the contribution from air does not exceed 20%. The soil/dust pathway contribution varies from 3%to 13%. In scenario 2 (partial operation) or scenario 3 (interrupted operation, the air pathway is quickly eliminated and soil/dust pathway contribution grows to approximately 20%. In all scenarios of Pb

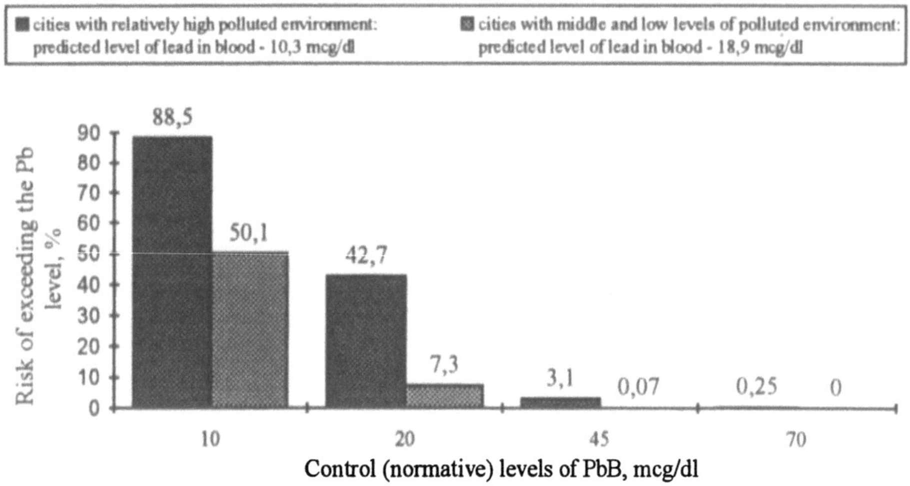

FIGURE 3. Risk distribution of PbB for preschool children in Russia.

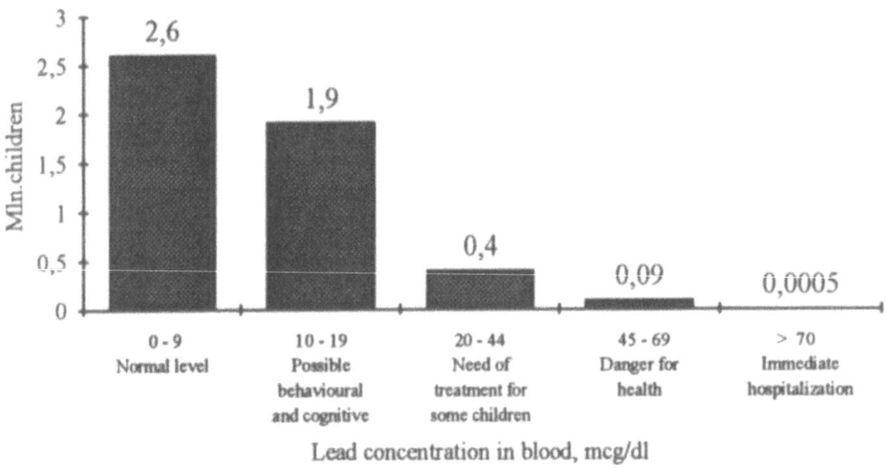

FIGURE 4. Distribution of children population according to health hazards

exposure, with the exception of the first zone, in the first scenario, food is the principal pathway (about 75-80%). Contaminated air is the main hazard factor for children's health in the first zone under normal operation.

At the second stage, we assessed the mean values of PbB in children. The mean value depends upon the age of the children. Examples of zonal distribution of mean values of PbB are given in Figure 5 for the first and third scenarios of lead exposure.

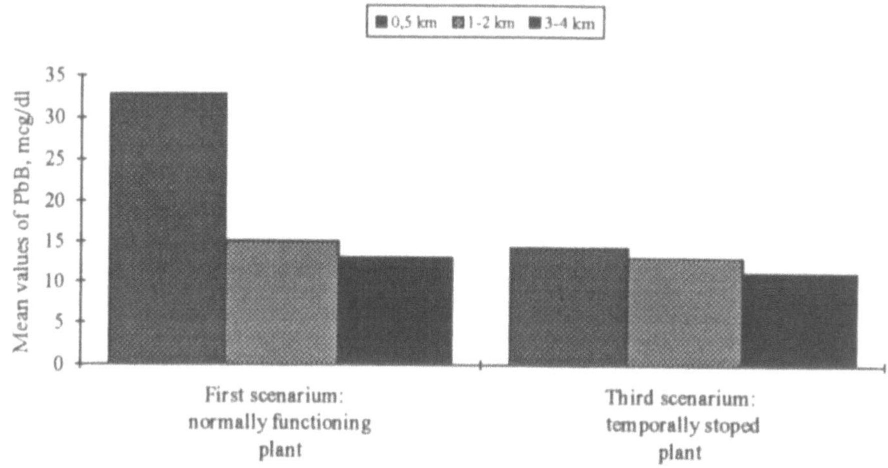

FIGURE 5. PbB mean values under different scenarios for the city of Belovo

Note that predicted mean value of PbB of children from 6-7 years old (Figure 2) varies from 9-11.7 ug/dl. As measured in May 1996, the blood level of 7-year-olds is equal to 9.8 ug/dl, a remarkable agreement with predicted levels. Under normal plant operating conditions the range of mean concentration for different zones is from 9.9-36.1 ug/dl. For comparison, known data from measurement of Pb in maternal blood is 23.4 ug/dl, and in newborns, 25.3 ug/dl, both values falling in the middle of the predicted range. This is another confirmation of agreement between calculated and measured PbB.

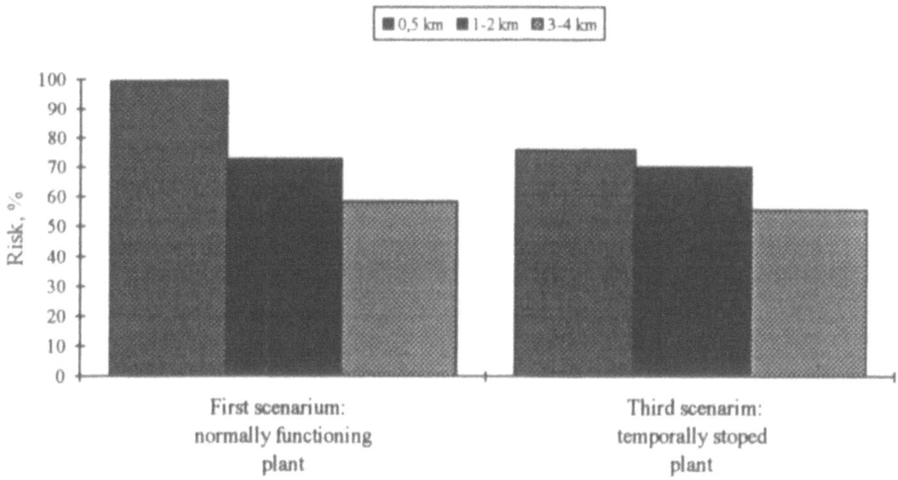

FIGURE 6. Risk-probability that PbB exceeds the control level of 10 ug/dl in children in the age range 0-7 years in Belovo.

At the third stage, we assessed the probabilities that PbB exceeds the normative level (see Figure 1). Risk assessments obtained for the first and third scenarios are presented in Figure 6. The chosen control level is the MAC of 10 ug/dl. These calculations show the very high risk levels to children's health independently of the living district or distance from the plant. When the plant stops operation, the risk drops by 75%, but in other zones one sees very small risk reduction. This is explained by the fact that for normal operations, in the first zone the essential part of the risk is from air contamination. This source of contamination is rapidly eliminated after plant closing. The role of other environmental components in the total risk changes very little or not at all because the time for self purging of soil, dust and food is much longer than for air.

Predicted risk values characterizing the probabilities of exceeding different control levels (Figure 1) of PbB and assessments of hazard to children's health under normal plant operation in the city of Belovo are shown in Figure 7. If we take the risk assessments for zone 2 as an approximation of spatially averaged values, we can make the following conclusion. Under normal plant operating conditions 20% of the children in Belovo require treatment, 50% of the children will experience behavior and learning problems and 30% of the children experience no effect from lead.

At the fourth stage, using the slopes of curves relating blood lead of children to PbE, one can determine the effectiveness of reducing PbB. It is measured in percent of decreasing PbB per 1% reduction of PbE. This type of calculation is demonstrated in Figure 8. From these data one can conclude that the most effective reduction of Pb risk to health can be achieved through decreasing Pb in food products. Decreasing lead in air is about one sixth as effective, and the same holds for lead in soil and dust. The least effective action is reducing lead in drinking water. These results provide the framework for building short-term environmental policy, for they provide definition of the most effective ways to reduce lead exposure in children, and hence, essential health risk reduction.

6. Conclusions, Discussion and Prospective Directions for Future Studies.

The IEUBK model is a steady-state model, and, as an equilibrium model, cannot account for variable environmental processes. The dynamic variation of parameters used in the model results in large uncertainties in the absolute values of lead concentrations in children's blood and the health risks calculated by this model. Nevertheless, by using in the analysis relative quantities obtained from the model assessments, one can achieve some positive results. To be more exact, the model allows us to:

1) make the relative hazard ranking of different pathways of lead exposure in children, and thereby

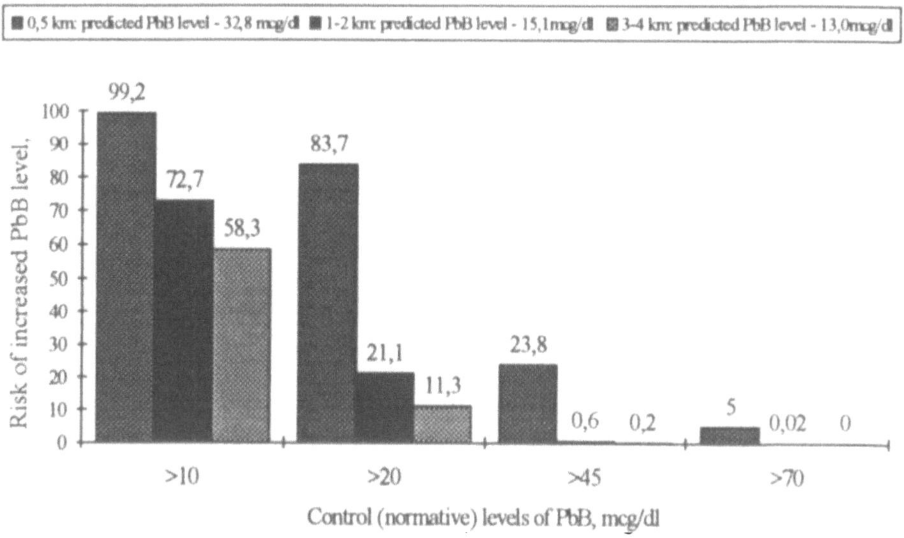

FIGURE 7. Risk-probabilities that PbB exceeds the control level for children from 0-7 years of age - with normal plant operations in Belovo.

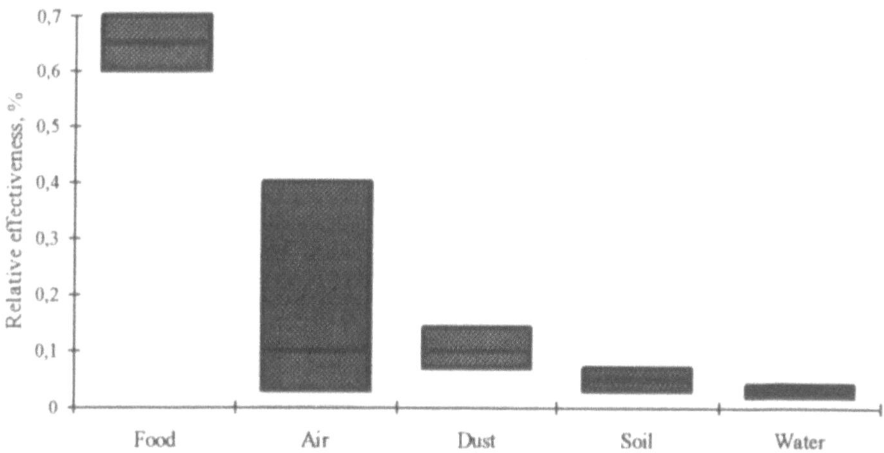

FIGURE 8. Relative effectiveness (%) of PbB reduction per 1% reduction of environmental contamination for preschool children in Belovo.

2) bring out the main hazards to health from different parts of the contaminated environment. The results of relative hazard analysis of different pathways of lead exposure in children living in Russian cities show up the dominant role of lead contaminated food. This pathway can contribute up to 85% of total lead burden. However, food contamination data are very contradictory and unreliable in terms of risk assessment, owing apparently to insufficient attention being paid to this

problem in Russia. It is necessary to continue this analysis because, if it confirms the relatively high levels of PbC in food products consumed in Russia, the actions to decrease lead contamination of food will have priority among the practical measures to reduce lead exposure in children. At the same time, it should be noted that Pb contamination of food in economically developed countries is about ten times less than the values (based on data from national studies) which we used in these calculations. This is an additional argument for more thorough analysis of both methods for measuring PbC in food and the resulting data obtained.

3) determine the effective actions for reducing real lead hazards to human health. Analyzing the contributions of different pathways in total lead intake and the risk to children's health, the slopes of curves relating PbB to PbE or Pb intake with food, makes it possible to determine the most effective steps to be taken to reduce lead exposure in children and suggests the optimum action plan to achieve this goal. This kind of analysis has been done for Belovo city and shows that the most effective reduction of lead risk to health can be achieved through decreasing Pb in food products. Less effective are measures to decrease lead in air, soil and dust, and least effective is reduction of lead in drinking water.

4) show up "weak" links in the system of lead contamination monitoring, data collection and work. For example, the information about PbC in drinking water is practically absent. In this absence, we have had to use data from Moscow drinking water analysis for our risk assessments and this data is not typical of other Russian cities and their drinking water lead levels. These levels, which are lower than for other European countries, could lead to a possible increase in the importance of drinking water when correct data from other Russian cities are obtained and analyzed.

5) define the priorities and directions of activity in planning and realizing environmental policy. Health risk assessments of lead exposure in children living in Russian cities and analysis of the results allow us to determine the following directions of long-term future studies:

- Development of monitoring systems equipped for complex data collection.
- Development of environmental lead standards.

Providing mutually acceptable standards on the basis of acceptable risk principles can be a long-term goal, which could be realized by the following approach:

1) Establish an acceptable risk level in agreement with basic standards for lead concentration in children's blood.
2) Distribute the risk into environmental compartments.
3) Introduce environmental standards for lead in the different environmental compartments on the basis of corresponding compartment risk calculations. Short-term directions in ecological planning, as was noted above, can be formed from the analysis of the effectiveness of risk reduction by actions which decrease lead intake from different pathways.

Supplementary economic analysis including assessment of economic population health damage cause by environmental lead contamination, and generalized economic cost-effectiveness and cost-benefit analyses allow us to make decisions on lead hazard reduction which are optimized from the social, ecological and economic points of view.

7. References

1. *CDC Report*: (1991) Preventing Lead Poisoning in Young Children. A Statement By The Centers For Disease Control. U.S. Department of Health & Human Services/Public Health Service.
2. *EPA Publication* number 9285.7-15-1 (1994). Guidance Manual For The Integrated Exposure Uptake Biokinetic Model For Lead In Children.
3. *Il'in V.B.* (1991) Heavy metals in market-garden soils and foods of the city Belovo. Report of the Institute of Soil Science and Agrochemistry of the Siberian Department of Russian Academy of Sciences, Novosibirsk (in Russian).
4. *Iljin V.B.* (1993) Studies of ceolites use for heavy metal inactivation in industrially contaminated soils for ecologically pure food production. Report of the Institute of Soil Science and Agrochemistry of the Siberian Department of Russian Academy of Sciences, Novosibirsk (in Russian).
5. *Skurikhin I.M.* (1988) Methods of food products analysis. Moscow: "Nauka", pp.132-152 (in Russian).
6. *Smoliar V.I.* (1987) Hipo- and hiper- microelementoses. Kiev, pp.92-108 (in Russian).
7. *Yarushkin V.Yu.* (1991) Environmental hygiene and children health in the regions with metallurgical zinc plants. Ph.D. thesis, Irkutsk (in Russian).

EVALUATION OF EXISTING ENVIRONMENTAL INFORMATION SYSTEMS IN RUSSIA APPLICABLE FOR HUMAN HEALTH EFFECTS ASSESSMENT

H. ÖZKAYNAK, J.D. SPENGLER,
J. J.K. JAAKKOLA, T. FORD, J. XUE,
A. EGOROV, J. SCHWARTZ
Department of Environmental Health, Harvard School of Public Health, Boston, Massachusetts, USA

S. KUZMIN, P. RAKITIN, L. PRIVALOVA,
S. CHEBOTARKOVA, G. ZEMLIANAIA
Environmental Epidemiology Component, Environmental Management Project, Center for Preparation and Implementation for Foreign Projects, Moscow, RUSSIA

1. Introduction

There is growing recognition among the Russian medical and scientific community that environmental conditions are shortening life and contributing to excess morbidity. Experts consider that pollution-related health impacts to be especially high in the industrial centers facing serious ecological problems. Recently, Harvard School of Public Health (HSPH) began a three-year study for the Center for Preparation and Implementation of International Projects on Technical Assistance (CPPI) in Moscow on the topic of environmental epidemiology in the Russian Federation. This project is part of a collaborative effort between the U.S. and Russian scientists which is directed at improving environmental conditions and public health in the Russian Federation.

This paper reports initial findings from our evaluation of data systems existing in Russia applicable for assessment of health risks related to exposures to environmental hazards that would support environmental health decision making. These evaluations emphasized identification of restrictions and limitations of current data and recommendations to improve the current measurement methods. In particular, we have evaluated historic environmental monitoring data and systems in the highly industrialized Sverdlovsk Oblast, and in Cherepovets which is a city in the Upper Volga region. In addition, we have conducted an initial analysis of the relationship between daily air pollution and daily mortality in Nizhni Tagil and Yekaterinburg, two industrialized cities in the Ural region of Russia. This paper describes our initial results from these technical evaluations and offers

I. Linkov and R. Wilson (eds.), Air Pollution in the Ural Mountains, 195–217.

recommendations for improvement of current environmental monitoring systems in Russia. The existing health information systems in the Russian Federation and their use in the assessment of environmental health effects are reviewed in Jaakkola *et al* 1997 presented in this Workshop.

2. Evaluation of Environmental Information

In evaluating the monitoring information, we have considered various criteria: 1) methods and instruments used to measure each type of pollutant; 2) length of operation and completeness of the record; 3) averaging time and frequency of sampling; 4) quality assurance procedures and reliability of data; 5) availability of data in raw, processed or report formats; and 6) availability of statistical analysis or summaries of the data.

Our evaluation to a large extend focused upon data sets previously identified by the Pilot Project summaries and reports performed by CPPI researchers. We examined these selected or short list data sets and visited monitoring sites and institutions responsible for gathering and processing various environmental measurements. Our assessments have been based on review and evaluation of data sets that have been made available to us, and supportive information provided to us by the technical experts at SanEpid, Hydromet, CPPI, and various Federal and local institutions. The study primarily included evaluation of data from three locations in Russia: Cherepovets in Volgogda Oblast, Yekaterinburg and Nizhni Tagil in the Sverdlovsk Oblast.

The city of Cherepovets in Vologda Oblast, has a population of 320,000 and is a major industrial center in European Russia. The modern city was developed in the 1950s around the Northern Steel production facility, the largest iron smelter in Russia and the main employer of about 60,000 people. In the 1960s, two chemical factories producing fertilizers were constructed. Currently there are 20 industrial facilities in the Cherepovets. The city is located on the Sheksna River and the Rybinsk Upper Volga Reservoir.

Yekaterinburg , the capitol of the Sverdlovsk Oblast, is one of the largest cities of Russia. According to 1996 data, there is a population of 1,321,000. Just as in recent years, there still exist negative tendencies in population growth, such as: mortality rate exceeds birth rate 1.63 times; natural net decrease of population is –5.2 per 1,000; and, average life expectancy is now 63.8 years.

Ambient pollution also is a problem of concern. Two hundred seventy-eight industrial enterprises polluting the atmosphere are situated within the territory of Yekaterinburg. There is practically no sanitary zone between apartment houses and industrial sites. The major atmospheric pollution sources in the city are: UralMach stock-joint company, Verkh-Isetsky metallurgical plant, Turbomotor Plant, Ural Chemical Machine-Building Plant Machine-Building Plant named after Kalinin, Novo-Sverdljvskaya, Sverdlovskaya and some other thermal power stations, rubber plant, and optical instruments plant. Ambient air pollution in Yekaterinburg from stationary sources are due to emissions from the following industries: machine building and metal-working industry (39%); construction materials industry (13%); thermoelectricity (10%); ferrous metallurgy (3%); and chemical and petrochemical industry (3%). Transportation sources also contribute to a major portion of the ambient air pollution.

Nizhni Tagil is an industrial city with a population of 450,000. The city consists of three raions. There are four air monitoring stations in Nizhni Tagil. The mortality rate in 1996 was 15.7 per 1,000 people, birth rate was 7.5 per 1,000 people and this caused a net decline in the population rate of –8.2 per 1,000. One hundred seven industrial enterprises pollute the ambient air of Nazhni Tagil. Major sources of pollution are: Nizhni Tagil metallurgical plant (68% of total emission), and chemical plastics plant (28% of total emissions). Other sources of pollution are: ore enrichment factory, UralTransGas, and Ural railroad cars manufacturing plant (Vagonzavod).

2.1. AIR QUALITY MONITORING DATA: CHEREPOVETS AND SVERDLOVSK OBLAST

Air monitoring in these locations is primarily conducted by Hydromet and SanEpid organizations. Hydromet service operates the basic network of routine monitors using standardized testing procedures and equipment. The equipment, instrument shelters and methods have not been revised or updated recently.

SanEpid has supplemented Hydromet's system by installing "portable" monitoring sites to conduct special studies and/or to simply supplement the fixed site network. Local industry operates sites and provides Hydromet with summary data.

The current system has several admirable attributes that include uniformity, number of pollutants measured, dedicated and knowledgeable staff and an established record keeping system. However, the network was designed to assess the short-term acute ambient exposure to industrial generated pollutants. The current network is deficient with respect to:

1. Providing estimates of chronic exposures to industrial and mobile source air pollutants.
2. Insufficient temporal resolution which will limit time-series analysis of health outcomes.
3. The limit of detection (LOD) for many of the measured contaminants is too high and makes it difficult to identify differences in possible exposures. High LODs of measurements methods and limited number of residential oriented monitoring sites limit the spatial resolution needed to define population exposures. These limitations might severely constrain the ability of health studies to discern and correctly attribute adverse effects at low levels.
4. The locations of air monitoring stations have, for the most part, been directed towards identifying industrial impacts close to the source. Given changes that are taking place in industrial production and the interest in assessing health and ecological impacts, different and/or additional locations should be considered.

2.1.1. *Recommendations for Air Monitoring*

The following recommendations refer to the air monitoring system in general. To offer specific system or even site-specific recommendations requires additional analysis, review of QA/QC procedures and records, and consideration of emissions.

1. Daily, 24-hour, measurements of size fractionated particles, PM_{10} and $PM_{2.5}$, should be established at the current network of sites.
2. Weekly, 168-hour, measurements of size fractionated particles, PM_{10} and $PM_{2.5}$, should be established at new residential urban and background rural sites.
3. Passive diffusion sampler methods for SO_2, NO_2, VOC's, O_3 and aldehydes can be used to inexpensively measure 24-hour to 336-hour (2 week) concentrations. Lower detection limits can be achieved to establish spatial and temporal patterns.
4. A two-year study should be designed to assess the lower-level exposures to particles, metals, inorganic and organic gases. The appropriate integration has to be determined based on the analytical technique, sampling device and temporal resolution desired.
5. To improve for particulate measurements a detection limit in the range of 10 to 20 $\mu g/m^3$ may require more sensitive balance as well as temperature and humidity conditioning of filters.
6. Based on established siting criteria for meteorological equipment reference site(s) for the city could be established. To the extent feasible hourly observations of wind speed, direction, precipitation, temperature, humidity (or dew point), solar insulation, visibility, cloud cover, and barometric pressure should be recorded and computerized.
7. Particulate matter collected on the filters should be analyzed for elements and chemicals. Among the metals having possible chronic health effects are Ni, V, Pb, Cd, Cr, and Hg. The techniques of source apportionment can partition the contributions specific sources or sources classes make to ambient particulate pollution.
8. It is critical the existing data be analyzed using appropriate statistical techniques. Until such analysis is performed, it is difficult to appraise fully the usefulness of the network as it exists to address concerns about both the acute and chronic health effects.

2.2. WATER QUALITY: CITY OF CHEREPOVETS

Water is extracted to the Municipal Water treatment plant from the Šeksna River, part of the constructed waterways of the Volgo-Baltijskij Kanal. Upstream of Cherepovets there are municipal, agricultural and industrial discharges. Agriculture and livestock in particular add coliforms to the river (also *Salmonella* sp. and other pathogens). Source water quality monitored once per month suggests high to extremely high pollution levels. Despite the level of pollution, there are swimming beaches along the banks of the Šeksna River.

2.2.1. *Water Extraction and Treatment*

The city has one water treatment plant that relies on prechlorination, coagulation, single-stage sedimentation/filtration and chlorination. Two 80 cm intake pipes supply water to rotating drum coarse wire mesh prefilters. Demand often exceeds supply. The water treatment capacity is designed to produce 145,000 $m^3 d^{-1}$. However, this can be pushed up to 160,000 $m^3 d^{-1}$ with a resultant decrease in water quality. The basic water treatment system for Cherepovets is insufficient for two primary reasons: 1) The source water quality is very poor, and, 2) The demand exceeds the capacity of the treatment plant. The poor quality of water requires effective coagulation and sedimentation, and a slow filtration rate to maximize removal of contaminants. However, in order to supply sufficient water to try meet demand, filtration rates are being compromised. To provide sufficient water pressure in the distribution system, additional water treatment facilities need to be built. The current practice of increasing flow of water through the sand filters to increase supply is clearly a major threat to water quality in the distribution system, and consequently, population health.

There is concern regarding the production of disinfection by-products (DBPs) through chlorination of source water with high organic loadings. An additional concern is the degree of chlorine demand, which may essentially leave no residual chlorine in the distribution water.

2.2.2. *Water Quality Monitoring*

Monitoring of water quality is carried out by both SanEpid and Vodokanal. Vodokanal have a more extensive sampling system than SanEpid and are responsible for maintaining and operating the treatment plant and distribution system. SanEpid have the capability to monitor most chemicals, coliforms and total plate counts. Hepatitis A is a serious problem in Cherepovets, however, it was unclear whether SanEpid had the capability to monitor for Hepatitis A antigens. There appears to be very little attempt to monitor indicator chemicals of specific industrial wastes or of DBP formation.

Most of the Russian water quality criteria are quite good, and for some parameters they are better, than EPA or WHO standards. However, they do not encompass the now enormous number of chemicals that are regulated by the US-EPA in drinking water, or, to a lesser extent the WHO guidelines. This is primarily due to the advanced technologies required for analysis of many of these compounds. However, there is an increasing body of information on the health consequences of exposure to these chemicals.

2.3. WATER QUALITY: YEKATERINBURG

For Yekaterinburg, there are more than 46 water sources, both ground and surface. However, the main reservoir system covers 80% of the city, other sources supply small districts. Testing and analysis of the water is carried out in various forms.

2.3.1. *Water Treatment*

The different water treatment facilities for Yekaterinburg do not provide sufficient capacity, and attempts to increase capacity, given the poor quality of source water may lead to increases in contamination in the distribution system. The addition of two new filtration beds in the WFS facility will obviously improve capacity. There is considerable concern regarding the production of disinfection by-products (DBPs) through chlorination of source water with high organic loadings. An additional concern is the degree of chlorine demand, which may essentially leave no residual chlorine in the distribution water.

2.3.2. *Water Distribution System*

There are major problems with Yekaterinburg's water distribution system. The major water mains are thought to be adequate, however, they can be easily contaminated by older parts of the system. Many of the older houses have old, corroded water systems resulting in secondary contamination and localized outbreaks of infectious disease. There is almost no residual chlorine maintained in the distribution system. This is thought to be primarily because of high disinfectant demand from the organic loading. Water sources are derived from swamp areas which have high concentrations of dissolved organic carbon, and, presumably, humic acids. In addition, there is also high organic contaminant loading to the surface waters.

SanEpid Center conducts routine examinations of water quality in the treatment plant and distribution systems. There are specific sites where they routinely provide control. Laboratories at SanEpid Central are well equipped. The microbiology laboratories monitor coliforms using membrane filtration (Millipore cellulose, one time, filters). They also use pathogen selective media to isolate *Salmonella, Shigella, Clostridium, V. cholerae* (not found this year). They also monitor for Hepatitis A antigen. At present these samples are sent away to a regional center. Other viruses are not monitored except for coliphages. *Pseudomonas* sp. and *Giardia lamblia* (which is a thought to be problem in Yekaterinburg) are also monitored. The chemical laboratories are very good. They have individual laboratories for water, food, soil etc. (water and soil may use the same lab). They have centralized labs for AA and GC analysis that are used for all media.

Vodokanal and SanEpid Central considered the drinking water supply to be reasonably microbiologically safe, and that waterborne disease issues were far less important than exposure to toxic chemicals, in particular DBPs. A deteriorating distribution system, with many interconnections to suspect piping and no measurable residual chlorine, must pose a serious infectious disease risk. Effective disinfection should not be compromised for what as yet, are poorly established risks from DBPs.

Yekaterinburg may provide an opportunity to test cheap and simple detection systems for pathogens, and could be an ideal city for trials of new products. This would be an opportunity to educate Russian microbiologists in new approaches to monitoring water quality, for example, the adaptation of molecular techniques to monitor specific pathogens.

2.3.3. *Hot Water*

Hot water quality is a major problem. By the time it reaches individual houses the water temperature is about 70^{0}C. Often certain houses have to boost the temperature of the hot water with a separate heater. The distribution system, made of pig iron, corrodes faster than the cold water system. As a result anticorrosive chemicals are added at the heating station, resulting in a build up of chemical contaminants, and, in particular, anti-corrosive chemicals that are thought to create serious dermatological problems. Some districts have a separate water source for their centralized heating system. However, the majority use the municipal water supply.

2.3.4. *Population Health and Water Quality*

Cold water is considered a major source of Hepatitis A. People are advised to boil drinking water but many apparently do not. Other important diseases associated with water in Yekaterinburg are dysentery and gastroenteritis, much of which is presumably undiagnosed. There are some disease differences between sites; however, it was not clear whether this could be used for epidemiological studies. As might be expected, districts at the termination of the distribution system should provide the greatest opportunity for comparative evaluations.

There appear to be many dermatological problems associated with exposure to the hot water, ranging from skin irritation to chronic dermatitis. Existing skin conditions are reported to worsen after showering. Anti-corrosive chemicals in the hot water are thought to cause many of the skin problems. There is no information on infectious disease through contact with hot water, although survival and growth of thermotolerant pathogens should be expected in at least part of the hot water distribution system. We recommend the following investigations:

1. Assessment of the role of drinking water pollution as a risk factor for the health of the Yekaterinburg population. Focusing on: quality and sufficiency of available data; define scope and objectives of additional investigations; try to specify some indicators of special significance.
2. Revealing and ranking priority drinking water pollutants as a whole and for different zones. Compute some integral indices for "water risk factors."
3. Estimate actual damage to population health from water: selection of population health indicators; estimate sufficiency vs. uncertainty of existing health data; Linkage of health indicators to water supply conditions in different zones.
4. Designing, organizing, and supervising epidemiological studies. Ranking the water factor with other risk factors.
5. Participating in determining priority directions in health risk management.

Establishing risk from drinking water is extremely complex. It is unrealistic at present to attempt to calculate population exposure based on the current database and sampling strategies. Most water quality parameters measured in Yekaterinburg relate to treatment efficiency and are not designed to reflect human health. To begin to calculate population exposure, it would be necessary to establish a monitoring program that provides contaminant analyses at a large number of randomly distributed sites within the distribution network, and contaminant analyses would need to be inclusive. For example, the coli index is only a very weak indicator of exposure to human pathogens, and very few if any carcinogens are routinely monitored in the system.

For hot water, risk assessment may need to be related to length of distribution pipe, condition of pipe, individual house plumbing etc., or concentration of anticorrosive additives/corrosion products. Risk assessment per se may be unnecessary and a questionnaire based study on incidence of dermatitis with/exacerbated by showering/ other exposure routes may be sufficient, i.e., commercial launderers, dish washers etc. (may have separate, technical, supply), vs. jobs with less water exposures. It may also be useful to look at vulnerability of exposed populations, both to hot and cold water.

2.4. FEDERAL-LEVEL ENVIRONMENTAL INFORMATION

Federal Level environmental data are received from various local and regional sources. Air data are sent to Federal Service of Hydrometeorology in St. Petersburg. Water data are sent to State Hydrochemical Institute of Rostov and soil data are sent to Typhoon Company at the Kalig region. Data are sent once a year both in hard copy and electronic form. It is not very clear to the regional organizations what types of analyses, aside from trends and summary statistics, the Federal group conducts on these data. At the regional level, baseline pollution levels and relationships are investigated in the prognosis of environmental quality.

The main purpose of the Federal Center is to summarize the pollution conditions across Russia in terms of low to very high pollution categories (5 for air pollution). At times, certain problems are encountered when Federal level analyses are conducted. For example, in summarizing trends at the local or regional level, federal researchers use all available data without consideration of monitor location or purpose. Sometimes data from monitors set-up to investigate situations at complaint or industrial areas are mixed together with urban or residential data. These could lead to artificial biases in the trend analysis if local expertise and input are not routinely solicited.

Another difficulty is the exchange of information between regional and Federal centers. Due to resource limitations and data ownership issues, access to federal and regional level data, even by participating institutions is often hampered. Federal level data provides a useful service of reporting overall surveillance of environmental conditions in different regions of the Russian Federation. Average and extreme tendencies by pollutants, and in terms of summary indices, are mostly useful for trend analysis and geographic comparisons. However, this form of data presentation does not support environmental health investigations. At present, neither financially nor by administrative authority, do federal groups have the ability to function as a central clearing house of raw or original environmental data to health researchers. Unless this situation is modified most of environmental health investigations have to depend on data and personnel support from local and Oblast organizations. Unfortunately, this situation limits proper characterization of environmental health problems and priorities across different regions of the Russian Federation. It is recommended that the responsibilities of the federal institutions in charge of analyzing and disseminating environmental data be clarified so that access to raw data can be allowed to authorized environmental epidemiologists and regional organizations interested in analyzing environmental health problems in Russia.

In addition, the data analyses procedures of the Federal institutions should be re-evaluated to determine if the annual reports provided meet the competing needs of different users and academic institutions. Current presentation of basic summary statistics across different cities may need to be modified to include more description on local sources of pollution; pollutant-specific emissions by different sources categories; information on siting of monitors and number of samples collected in each area, and; reporting of more detailed statistics, such as the 50th and 95th percentile concentration values. Moreover, frequency of exceedances of the current standards (e.g., PDKs) or other acceptable limits need to be supplemented by additional analyses of causes and sources of these violations. The current system is designed to qualitatively identify possible problem areas but does not allow or facilitate an in-depth analysis for developing recommendations on how to solve these critical problems. Respective roles of Federal vs. regional authorities in addressing these nationwide issues need to be defined through a series of meetings organized at the Federal Level. Until these new responsibilities and new types of analyses needs are identified, environmental health research is Russia has to rely most heavily on local or regional level data.

3. Acute Effects of Air Pollution on Daily Mortality in Nizhni Tagil and Yekaterinburg

Using historical data on selected air pollutants (TSP, SO_2, CO, NO_2, formaldehyde and phenols), we performed an initial investigation of the relationship between daily air pollution and daily cause-specific mortality in Nizhni Tagil and Yekaterinburg. The purpose of this analysis was to apply state-of-the-art time-series methods developed over the last 10 years to existing information systems in Russia in order to identify the key air pollutants that may be contributing to daily excess mortality. As discussed earlier, however, this investigation is severely hampered by the quality of the air pollution measurements and also to coding of only the principal causes of mortality. Nevertheless, as presented below, the application of time-series methods to existing pollution and health data are useful in determining the potential role of air pollution in acute mortality in Russia.

3.1. DATA BASES

Five years, (1991, 1993-1996) of daily pollution and mortality in Nizhni Tagil and three years (1994-1996) of daily pollution and mortality in Yekaterinburg were obtained. Pollution data consisted of daily average (mean of three 20-minute samples) TSP, SO_2, CO, NO_x, formaldehyde and phenols from four stations in Nizhni Tagil and two representative stations in Yekaterinburg. Measurements from all available stations were then averaged to calculate a single daily average value for each pollutant. Daily total and cause-specific mortality by major categories of death for the entire city of Nizhni Tagil and Yekaterinburg were obtained for each day of the respective study periods. The following categories of cause-specific mortality categories were used in the analysis of Nizhni Tagil and Yekaterinburg data:

- All (excludes causes listed under the control category)
- Cardiovascular: ICD.9 390-459
- Respiratory: ICD.9 460-519
- Other: infectious diseases, tumors, endocrine systems, mental disorders, nervous system, gastrointestinal diseases
- Control: traumas and poisonings, complications due to pregnancy and delivery, dermal and subdermal tissue diseases, musculoskeletal system, malformations, perinatal diseases, unknown causes.

3.2. METHODS

Daily counts of mortality, temperature and pollution series exhibit seasonal cycles, some of which are common. Unless adjusted for long-term cycles, shared seasonal or monthly cycles among cause-specific mortality and environmental variables could confound results. Adopting the technique used in Kinney and Özkaynak (1991), a weighted 19-day moving average filter developed by Shumway (1988) was used to detrend the pollution and meteorological series from Nizhni Tagil and Yeakterinburg. The method involves subtracting the weighted moving average of each variable (X_t) from itself on each observation. In other words, the variable (X_t) on day t=i is filtered as:

$$X\_F_i = X_i - \sum_{i=-9}^{9} Xsubi\, w_i \tag{1}$$

where W_i are the filter weights are provided in Shumway (1988). This process of filtering removes the long period cycles but not the short-term cycles (*i.e.*, high frequencies). When a linear filter such as this is applied to both the predictor and predicted variables prior to regression analysis, linear regression relationships among variables are preserved and can be estimated without bias. In addition, this filter efficiently removes the autocorrelation in the daily pollution and mortality series. Autocorrelation functions were examined to detect any remaining temporal structure in the filtered data, and none were found.

Descriptive statistics were computed for each of the filtered and unfiltered data. Cross-correlations between filtered mortality and filtered pollution series (*i.e.*, TSP, CO, SO_2, NO_2, formaldehyde and phenol) were calculated to determine the importance of the relationship between cause-specific mortality and same day (or lag 0), previous day (or lag 1) and 2-days prior (or lag 2) particulate (TSP) and gaseous pollutant measurements.

Analysis of the daily mortality counts and pollution data was performed using time-series and regression modeling techniques implemented with the Statistical Analysis System (SAS) (SAS Institute Inc., SAS Campus Drive, Cary, NC). Because of low daily counts for some categories of cause-specific mortality (*e.g.*, respiratory and other causes of mortality), we examined two different methods of modeling the pollution health effect relationships. Both ordinary and Poisson regression models were fitted to filtered mortality, temperature and pollution data. Consistency of results and normality of model residuals were examined. In most cases, results from Poisson and multiple regression models were found to be quite similar. Basic analysis involved fitting multiple regression models to five filtered cause-specific mortality variables, using either filtered or unfiltered pollution (*e.g.*, TSP, phenols, etc.), day of the week, temperature and relative humidity (RH) as explanatory variables. The form of the basic multivariate regression model was:

$$M\_F = \beta_0 + \sum \beta_i X\_F_i + W\_D + E \qquad (2)$$

where; M_F are the filtered daily cause-specific mortality; X_F_i are the filtered same day or previous day daily pollution, temperature and relative humidity measurements, and; W_D is the day-of-the-week dummy variable, and E is the error term.

3.3. RESULTS

3.3.1. Nizhni Tagil

Figure 1 shows the daily average TSP measurements in Nizhni Tagil, Russia. Both the original and filtered TSP series are presented. Table 1 presents the summary statistics for all available pollution and mortality variables in Nizhni Tagil between 1991 and 1996. Figure 2 displays the original and filtered all-cause mortality in Nizhni Tagil. Seasonal trends in either the mortality or the pollution series are not immediately apparent. Cross-correlation function between filtered respiratory mortality and various lags of TSP (*e.g.*. lag 0 same day, lag 1 previous day's TSP, and so on) is presented in Figure 3. Highest correlations are observed between same day TSP and respiratory deaths on that day. Similarly, highest correlations between all-cause mortality and phenol concentrations in Nizhni Tagil , on the same day, are also noted. On the other hand, highest correlations between cardiovascular mortality and (lag 1) previous day's TSP also are observed in the Nizhni Tagil TSP pollution series.

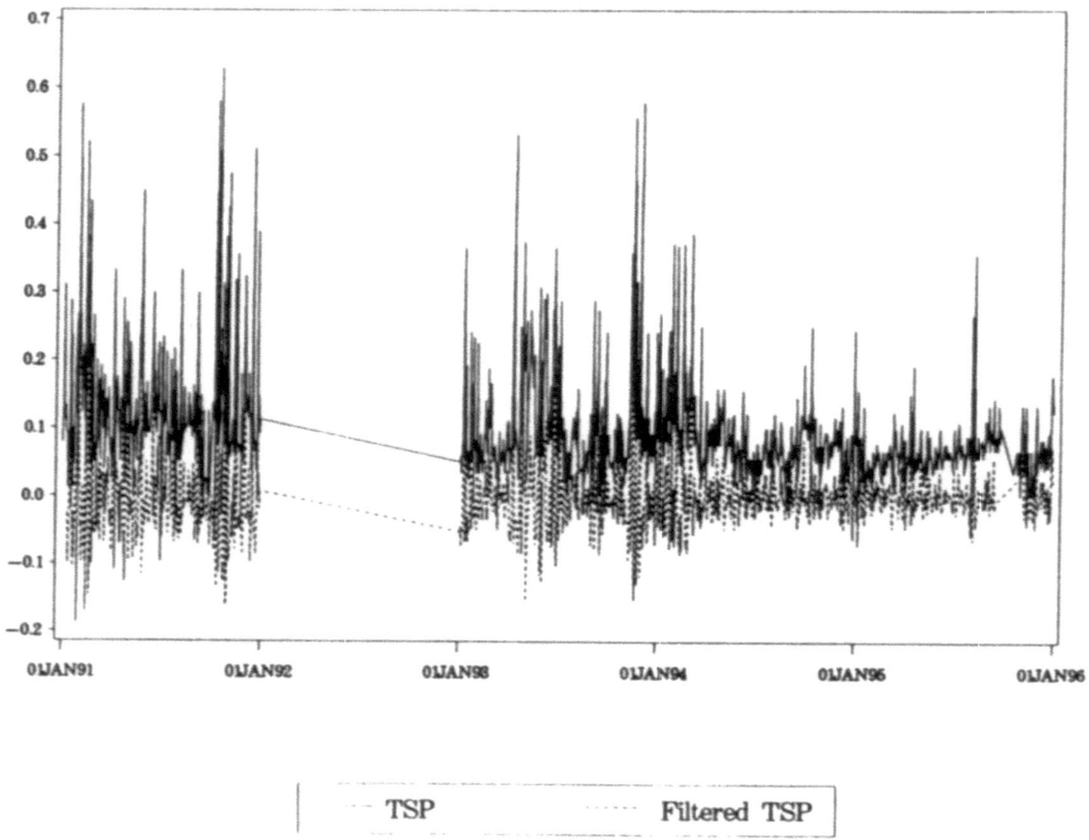

FIGURE 1. Original and filtered TSP <mg/m^3> in Nizhni Tagil: 1991, 1993-1996.

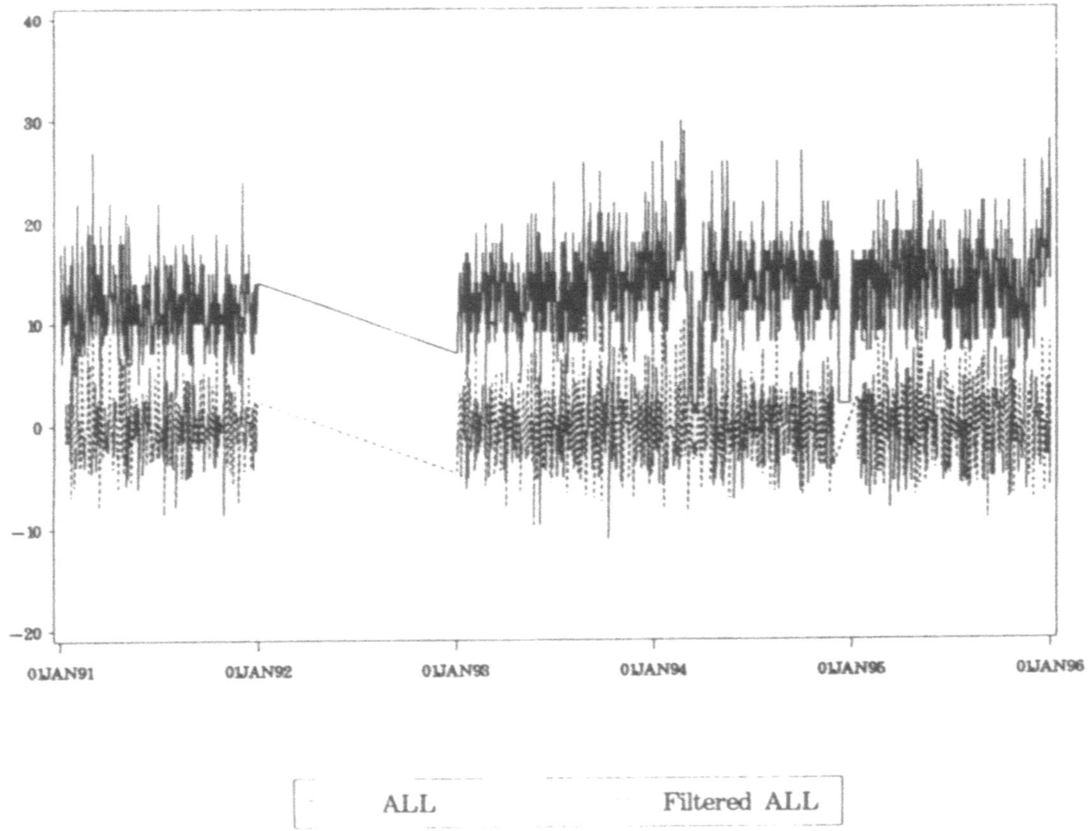

FIGURE 2. Original and filtered all-cause mortality in Nizhni Tagil: 1991, 1993-1996.

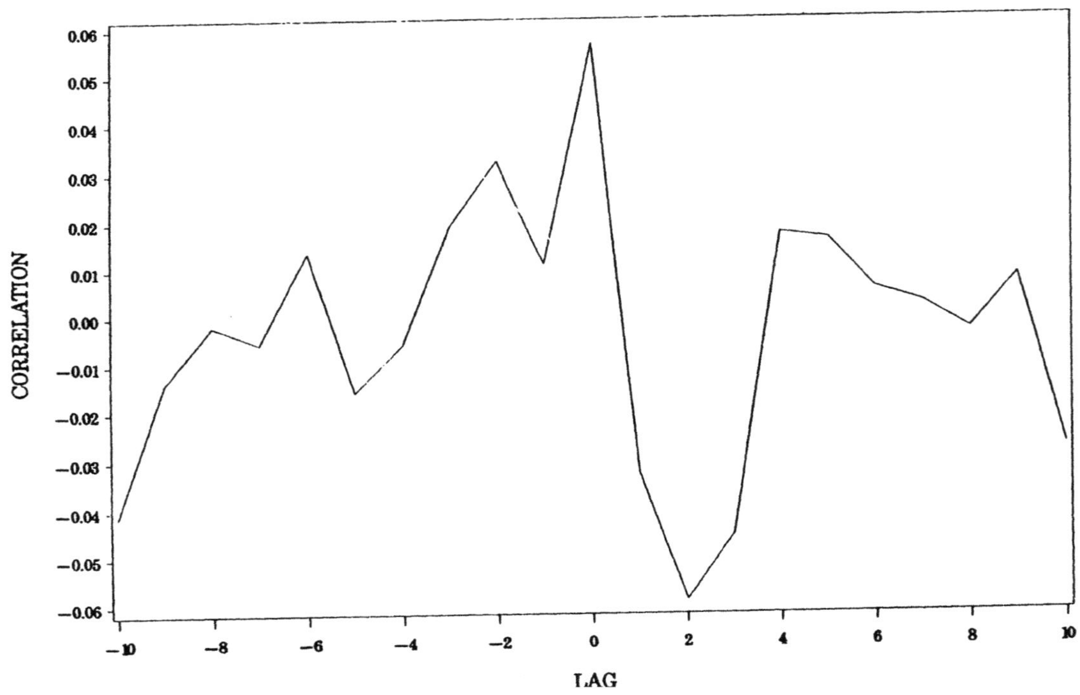

FIGURE 3. Cross correlation between respiratory mortality and TSP in Nizhni Tagil: 1991, 1993-1996.

TABLE 1. Summary statistics for pollution and mortality variables in Nizhni Tagil: 1991, 1993-1996.

Variable	N	Mean	STD	Median	MIN	MAX
Temperature	1529	3.65	12.82	4.04	-30.80	38.68
TSP	1502	0.096	0.080	0.075	0.000	0.627
Phenol	1517	0.004	0.003	0.004	0.000	0.020
NH3	1523	0.058	0.059	0.043	0.000	0.780
Formaldehyde	1495	0.012	0.018	0.008	0.000	0.341
SO_2	1290	0.009	0.013	0.006	0.000	0.187
NO_2	1520	0.025	0.018	0.022	0.000	0.174
HCN	1430	0.001	0.001	0.001	0.000	0.019
NO	1452	0.019	0.108	0.010	0.000	4.013
All	1800	13.25	4.28	13.00	1.00	31.00
Cardiovascular	1800	8.92	3.42	9.00	0.00	24.00
Respiratory	1800	0.85	0.92	1.00	0.00	6.00
Control	1800	3.24	2.09	3.00	0.00	19.00
Other	1800	3.48	1.93	3.00	0.00	11.00

Tables 2-4 present the results from analysis of five years of daily pollution and cause-specific mortality in Nizhni Tagil. Again, two types of basic models were run: Ordinary Least Squares and Poisson. Both these models were run either with filtered or unfiltered mortality in order to evaluate if absolute levels of air pollution or deviations from a local moving average value are associated with the variations in daily mortality. Because of low counts of daily deaths results from the Poisson model formulation are expected to be more representative. The results shown in these tables suggest that TSP and phenol may play a role in contributing to observed daily mortality in Nizhni Tagil. TSP is marginally associated with respiratory mortality (5% to 9% effect corresponding to the mean TSP level of 0.096 mg/m^3); previous day TSP is weakly associated (p=0.06) with cardiovascular mortality (2% to 3% effect corresponding to the mean TSP level), and; same day phenol is marginally associated with all-cause mortality (about 2% predicted effect corresponding to the mean phenol level of 0.004 mg/m^3).

The basic time-series analysis was based on modeling the acute effects of air pollution on cause-specific mortality first by pre-filtering the mortality series. Air pollution variables were either filtered or unfiltered depending on the model selected. However, researchers have also recently used generalized additive models (GAM) by

TABLE 2. Estimated regression coefficients (β) and average percent contribution (e) of daily TSP pollution to daily respiratory mortality in Nizhni Tagil: 1991, 1993-1996

		All	Cardio	Respiratory	Other	Control
OLS Model with filtered pollutants	β	0.00	0.00	0.76*	0.00	0.00
	e(%)	0.00	0.00	8.58	0.00	0.00
Poisson Model with filtered pollutants	β	0.00	0.00	0.88*	0.00	0.00
	e(%)	0.00	0.00	8.82	0.00	0.00
Poisson Model without Filtered pollutants	β	0.00	0.00	0.50	0.00	0.00
	e(%)	0.00	0.00	4.92	0.00	0.00

*: $p < 0.05$

TABLE 3. Estimated regression coefficients (β) and average percent contribution (e) of daily (lag 1) TSP pollution to daily cardiovascular mortality in Nizhni Tagil: 1991, 1993-1996.

		All	Cardio	Respiratory	Other	Control
OLS Model with filtered pollutants	β	1.84	2.47*	0.00	0.00	0.00
	e(%)	1.33	2.66	0.00	0.00	0.00
Poisson Model with filtered pollutants	β	0.15	0.29*	0.00	0.00	0.00
	e(%)	1.45	2.82	0.00	0.00	0.00
Poisson Model without Filtered pollutants	β	0.09	0.22*	0.00	0.00	0.00
	e(%)	0.87	2.13	0.00	0.00	0.00

*: p value is about 0.06

TABLE 4. Estimated regression coefficients (β) and average percent contribution (e) of daily phenol pollution to daily all-cause mortality in Nizhni Tagil: 1991, 1993-1996.

		All	Cardio	Respiratory	Other	Control
OLS Model with	β	88.89*	59.82	0.00	36.41	9.82
filtered pollutants	e(%)	2.66	2.66	0.00	4.14	1.20
Poisson Model with	β	6.15*	6.21	0.00	9.82	7.32
filtered pollutants	e(%)	2.47	2.49	0.00	3.97	2.94
Poisson Model without	β	4.45	4.31	0.00	7.32	5.54
filtered pollutants	e(%)	1.78	1.72	0.00	2.94	2.22

*: $p < 0.05$

locally smoothing the mortality series as a function of time and other key co-variates, such as temperature and RH (cf. Schwartz 1994, 1995), and then regressing them on selected pollution variables. We compared our basic results to those derived from GAM applications. We used two different models. SAS package (Proc Genmod) was used to fit locally weighted moving average cubic spline smooths to the daily mortality series. These were then regressed on pollution and weather variables using the Poisson model formulation. The results obtained were nearly identical to the previously presented basic Poisson model time-series results using the unfiltered pollution variables. Next we used the S-Plus computer software to re-estimate the acute air pollution effects on locally smoothed mortality as a function of time and temperature (with bandwidth 0.8), again through the Poisson model formulation. Results obtained were also very similar to the ones previously obtained using the pre-filtering techniques. An example of an S-Plus model output is shown in Figure 4. Figure 4 displays in three panels the original all-cause mortality series, smoothed mortality as a function of time, and residuals of daily mortality after subtraction of the calculated smooth function of all-cause mortality.

3.3.2. *Yekaterinburg*

Table 5 shows the summary statistics for the pollution and mortality variables used in the analysis of acute mortality effects of air pollution in Yekaterinburg during 1994-1996. Original and filtered all-cause mortality series are shown in Figure 5. Time-series analysis of the relationship between different air pollution variables and cause-specific mortality indicated a statistically significant association between daily (lag 1)

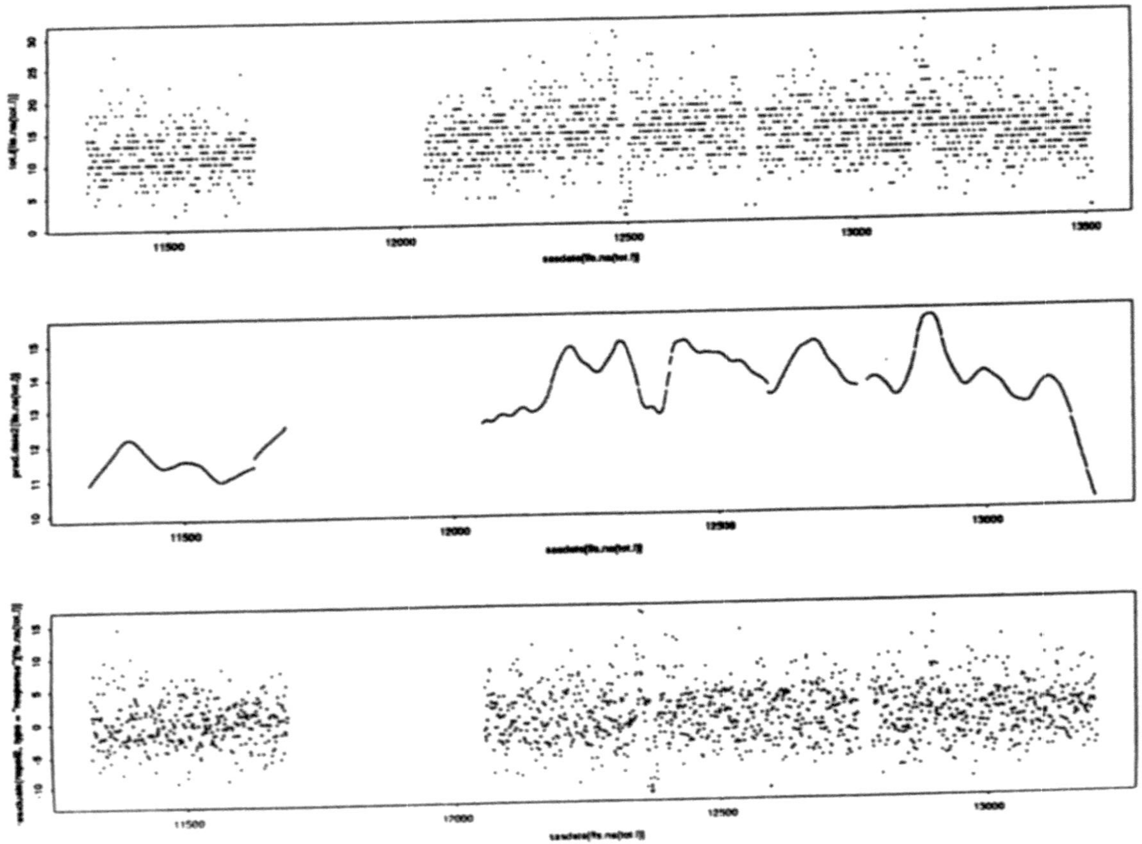

FIGURE 4. All-cause original, smooth and residual mortality in Yekaterinburg: 1994-1996.

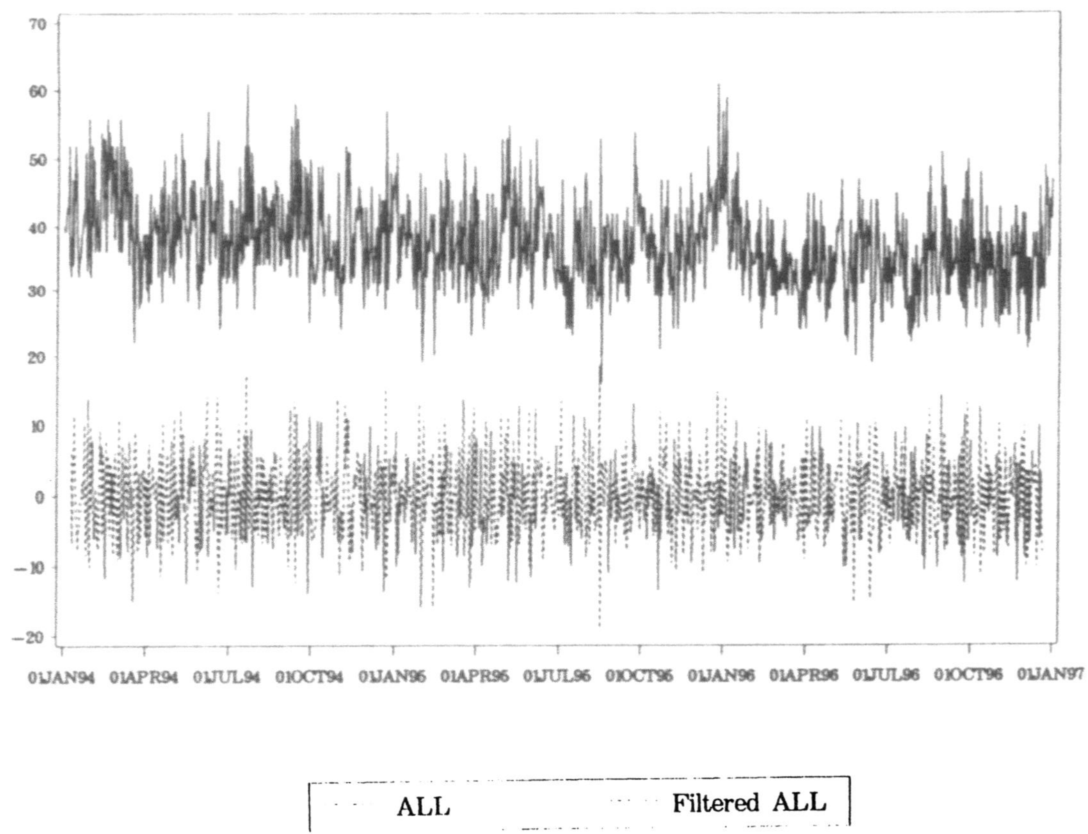

FIGURE 5. Original and filtered all-cause mortality in Yeakterinburg: 1994-1996.

TABLE 5. Summary statistics for pollution and mortality variables in Yekaterinburg: 1994-1996.

Variable	N	Mean	STD	Median	MIN	MAX
Temperature	854	4.155	12.537	4.400	-27.467	28.867
TSP	845	0.122	0.090	0.100	0.000	0.500
Phenol	852	0.002	0.001	0.002	0.000	0.009
SO_2	855	0.009	0.010	0.007	0.000	0.120
NO_2	855	0.080	0.064	0.063	0.000	0.553
NO	849	0.072	0.047	0.063	0.000	0.450
CO	853	2.200	1.292	2.000	0.000	17.000
All	1096	37.43	7.13	37.00	16.00	61.00
Cardiovascular	1096	24.19	5.77	24.00	8.00	42.00
Respiratory	1096	2.19	1.60	2.00	0.00	9.00
Control	1096	13.16	3.85	13.00	3.00	26.00
Other	1096	11.05	3.31	11.00	2.00	24.00

TABLE 6. Estimated regression coefficients (β) and average percent contribution (e) of daily (lag 1) TSP pollution to daily respiratory mortality in Yekaterinburg: 1994-1996.

		All	Cardio	Respiratory	Other	Control
OLS Model with filtered	β	4.70	3.17	2.70**	0.00	0.00
pollutants	e(%)	1.53	1.60	15.04	0.00	0.00
Poisson Model with filtered	β	0.13	0.13	1.27**	0.00	0.00
pollutants	e(%)	1.60	1.60	16.76	0.00	0.00
Poisson Model without	β	0.07	0.04	0.87**	0.00	0.00
Filtered pollutants	e(%)	0.86	0.43	11.20	0.00	0.00

*: $p < 0.05$, **: $p<0.01$, ***: $p<0.001$

previous day's TSP and daily respiratory mortality in Yekaterinburg (Table 6). The predicted effect corresponding to the mean TSP concentration of 0.122 mg/m^3 was in the range of 11% to 16%. No other pollutants were found to be significantly associated with different causes of mortality.

4. Summary and Conclusions

Environmental monitoring activities in the Russian Federation are dispersed among different federal, regional and local organizations. Environmental monitoring includes collection of samples of ambient air, soil, surface, drinking water, food, and snow. Typically, this information is collected by SanEpid, Hydromet, Geological Committee, and the Committee for Environmental Protection. Most of the air pollution data are collected by Hydromet. Location of Hydromet sites are selected and/or approved by SanEpid. SanEpid uses a mobile van or truck to either monitor source-oriented impacts downward of a facility or monitor ambient conditions in a location (*i.e.*, in a stationary sampling mode) not covered by Hydromet. SanEpid also gathers air quality data in non-polluted or control zones.

Meteorological data only are collected by Hydromet in few selected sites in each of the cities monitored by this organization. Sampling is ordinarily done every day of the week, 3 times a day (7 AM, 1 PM, 7 PM). The ambient air monitoring data are gathered in the same way throughout Russia and has various technical limitations. Basically, for particles and some gases, method detection limits are too high and the laboratory procedures could result in poor precision and accuracy. Particle monitoring data especially are of limited use since minimum reported values often are either zero or 0.1 mg/m^3.

The quality and completeness of the air quality data collected throughout varies by pollutant and location. Particulate measurements are considered to be poor and are of limited use for epidemiologic analysis. Additional analysis is needed to determine if sufficient within day variation exists in any of the monitoring sites data. Further examination of gaseous pollution data is also needed to ascertain if levels measured three times a day, over a 20-minute sampling period are representative of 24-hour daily average values. It is unknown whether the current sampling design is sufficient to provide health-relevant air quality measurements (*i.e.*, 8- or 24-hour averages) suitable for the assessment of health effects. The existing monitoring system is designed strictly for compliance monitoring purposes. Namely, indices are generated based on pollutant-specific MACs to determine violations caused by nearby industries. Modification to both instrumentation, laboratory methods, and sampling duration and frequencies are recommended in order to provide reliable data for more sensitive environmental epidemiologic studies.

In summary, current methods routinely used to monitor ambient particle and gaseous pollutant concentrations in Russia are less accurate than those used in Northern America and Western Europe. PM_{10} and $PM_{2.5}$ are not monitored but seem to be about 50% and 20%, respectively of the TSP measured based on limited method comparisons performed by Harvard and CPPI in Cherepovets and Yekaterinburg.

In both Cherepovets and Yekaterinburg the drinking water quality suffers due to poor source water quality, insufficient treatment due to excess demand and problems and leakages in the water distribution system. Centralized hot water systems also pose a concern for public health. The distribution system, made of pig iron, corrodes faster than the cold water system. As a result, anticorrosive chemicals are added at the heating station, resulting in a build up of chemical contaminants, and, in particular, anti-corrosive chemicals that are thought to create serious dermatological problems. Most water quality parameters measured in Cherepovets or Yekaterinburg relate to treatment efficiency and are not designed to reflect human health. The range of potentially toxic chemicals in the water is vast, however, very few are monitored due to technological constraints. In addition, parameters monitored at the treatment plant do not reflect population exposure due to the dramatic changes in water quality within the distribution system. To begin to calculate population exposure, it would be necessary to establish a monitoring program in selected cities in Russia that provides contaminant analyses at a large number of randomly distributed sites within the distribution network, and contaminant analyses would need to be inclusive.

In spite of likely measurement errors, we conducted a preliminary time-series analysis of existing daily air pollution and cause-specific mortality data from two large cities in the Sverdlovsk Oblast. Interestingly, these analyses showed statistically significant associations between particulate pollution and both respiratory and cardiovascular mortality. Specifically, we found that:

- 11% to 16% of acute respiratory mortality is associated with previous day's TSP in Yekaterinburg.
- Around 5% to 9% of acute respiratory mortality is associated with same day TSP in Nizhni Tagil.
- 2% of acute cardiovascular mortality is associated with pervious day TSP levels in Nizhni Tagil.

These findings are consistent with previous results from many different investigations on epidemiology of acute particle health effects conducted in US, Europe, and other countries (cf. Wilson & Spengler 1996; Dockery & Pope 1994). The contribution of particle pollution to acute respiratory or cardiovascular mortality in Russia seems to be comparable to effects predicted in other countries with somewhat lower air pollution levels. However, our analysis also suggests the possibility of potential health effects of pollutants other than just the TSP. We have detected a statistically significant association at the 2% level between same day phenol pollution and all-cause mortality in Nizhni Tagil. Likely sources of phenol in Nizhni Tagil include emissions from coke oven batteries and the Ural Chimplast chemical factory. Due to limitations and the quality of existing pollution measurements it is yet unknown whether other gaseous criteria pollutants, VOCs or aldehydes also contribute to excess mortality or morbidity in Russian cities.

In conclusion, these initial findings suggest that emissions from metallurgical, chemical and mining industry could be contributing to daily acute mortality in industrial cities of Russia. However, in order to quantify better the morbidity and mortality effects of environmental pollutants new epidemiologic studies with more accurate pollution monitoring techniques are needed. Over the next two to three years a number of such new epidemiologic studies will be implemented in the Sverdlovsk Oblast and Cherepovets under the framework of the Harvard/CPPI Environmental Epidemiology project in Russia. These studies will examine the acute and chronic health effects of air and water pollution on different health indicators through various cohort and panel studies of adults, pregnant women, infants and young children. With this new information, we hope to be able to investigate more reliably the role of environmental contaminants on public health in various Russian cities or regions.

5. References

1. Dockery, D. W. and Pope, C. A. (1994) Acute respiratory effects of particulate air pollution, *Ann. Revs. Public Health* 15, 107-132.
2. Jaakkola, J.J.K., Özkaynak, H., Cherniack, M., Wojtyniak, B., Egorov, A., Rakitin, P, Kuzmin, S. Lebedeva, N.V. (1997) Existing health information systems in the Russian Federation and their use in the assessment of environmental health effects. A presentation in the NATO Advanced Research Workshop on Air Pollution in the Ural Mountains, May 26-30, 1997, Magnitogorsk, Russia.
3. Kinney, P.L. and Özkaynak, H. (1991). Associations of daily mortality and air pollution in Los Angeles County, *Environ. Res.*, 54, 99-120.
4. Schwartz, J. (1994) Particulate air pollution and daily mortality in Cincinnati, Ohio *Environ. Health Perspect.* 102, 186-189.
5. Schwartz, J. and Morris, R. (1995) Air pollution and hospital admissions for cardiovascular disease in Detroit, Michigan 142, 23-35.
6. Shumway, R.H. (1988). *Applied Statistical Time Series Analysis,* Prentice-Hall, Englewood Cliffs, NY.
7. Wilson, R. and Spengler, J.D. Eds.(1996). *Particles in Our Air: Concentrations and Health Effects.* Cambridge, Massachusetts: Distributed by Harvard University Press.

EXISTING HEALTH INFORMATION SYSTEMS IN THE RUSSIAN FEDERATION AND THEIR USE IN THE ASSESSMENT OF ENVIRONMENTAL HEALTH EFFECTS

J.J.K. JAAKKOLA, H. OZKAYNAK, M. CHERNIACK, B. WOJTYNIAK, A. EGOROV
Department of Environmental Health, Harvard School of Public Health, Boston, USA

P. RAKITIN, S. KUZMIN, N.V. LEBEDEVA
Environmental Epidemiology Component, Environmental Management Project, Center for Preparation and Implementation for Foreign Projects, Moscow, RUSSIA

1. Introduction

In 1995, the Government of the Russian Federation initiated an Environmental Management Project with a general aim to improve environmental conditions and associated human health [1]. An Environmental Epidemiology Component of the Russian Environmental Management Project was initiated to perform policy-oriented environmental epidemiology and risk assessment, and to establish systems for information transfer to decision-makers in environmental policy. Harvard School of Public Health, representing a broad group of international investigators, was chosen as a consultant to the Environmental Epidemiology Component.

A key element in the management of environmental problems is the availability of valid and relevant environmental and health information. There has been an intensive effort in the Russian Federation to compile and use information on the environment and human health. In 1996-97, we assessed this information as it is collected and analyzed on the local, regional, and federal levels [2]. This paper focuses on the evaluation of health information collection and utilization through illustrative examples from each of these administrative levels. The existing environmental data in the Russian Federation are reviewed in another presentation of this Workshop by Ozkaynak and colleagues [3].

I. Linkov and R. Wilson (eds.), Air Pollution in the Ural Mountains, 219–229.

2. Different Use of Environmental and Health Information for the Assessment of Environmental Health Effects

Environmental and health information can be used in several different formats for the assessment of environmental health effects. Information on concentrations of environmental pollutants (air, water, soil) is necessary for the enforcement of environmental standards on all levels of society. Information on concentrations over time is also needed for the assessment of population exposure; in addition information on population time-activity patterns is often necessary to improve the accuracy of the assessment. Information on population exposure distributions over time can be used to assess the public health impact of known hazardous compounds, when scientific knowledge can provide estimates of exposure-response or dose-response relations. Consequently, beneficial effects of reducing levels of pollution can also be assessed.

Information on disease occurrence in populations over time or space can be used to create hypotheses on health effects of environmental factors. An increase in disease occurrence in a given population over time may inferentially indicate an influence of environmental factors and call for further evaluation. Also geographical differences in morbidity or mortality rates may indicate environmental etiology, but there are always a number of alternative explanations that have to be considered such as differences in the characteristics of the compared populations, and procedures for detection and recording of health events. Geographical differences in patterns of exposure and in patterns of presumably environmentally influenced diseases are the most evident and useful features of large data systems, but also the sources of the most serious errors.

Finally, both environmental and health information as well as additional information on other determinants of health is needed for the assessment of the effects of environmental exposures on health. Sometimes routine monitoring of environmental and health information, as part of a broadly conceived surveillance program, can be used for the assessment of health effects from environmental factors. However, it often proves to be necessary to assess a relation between a given environmental factor and its health effect in specific and well monitored settings where environmental factors, health effects and other relevant factors can be measured validly and efficiently. Targeted and detailed observations in a limited number of selected localities can generate results that can be subsequently generalized to other settings or even nationwide.

3. Methods

We started the review at the grass-root level and followed the information flows towards the central offices in Moscow. Principal generic health information comprises births, mortality, and morbidity which is routinely collected by Russian health care personnel in standardized forms. Completion of these forms is required by law, and it is assimilated into the public health culture. We also evaluated special sources of health information including individual disease registries and epidemiological studies. The local level observations were made in Cherepovets in Vologda Oblast, and in Nizhni Tagil and Yekaterinburg in Sverdlovsk Oblast, and the regional level observations came from Sverdlovsk Oblast. Our review of federal level information focused on institutions receiving primary information from the oblasts, and several important research institutes that made secondary use of this information.

4. Evaluation and Recommendations

4.1. CITY OF CHEREPOVETS - LOCAL LEVEL

The City of Cherepovets, in Vologda Oblast, has a population of 320,000 and is a major industrial center in European Russia. The modern city was developed in the 1950s around the Northern Steel production facility, the largest iron smelter in Russia and the main employer of about 60,000 people. In 1960s two chemical factories producing fertilizers were constructed. Currently there are 20 industrial facilities in Cherepovets. The City is located on the Sheksna River and the Rybinsk Upper Volga Reservoir.

Cherepovets has a computerized recording of maternal and childrenís health from 1992. There are three major types of health information: information on pregnancies and newborns from birth clinics; information on morbidity from polyclinics generated by each physician visit; and information on causes of death maintained by the Registry Offices (ZAGS).

4.1.1 Information on Pregnancies and Newborns from Birth Clinics

Information on each woman admitted to a birth clinic and her newborn is recorded systematically and entered into a DBASE computer file in the SanEpid office. This database includes relevant information such as maternal age, date of admission, residency, address, motherís occupation and employer, number of previous pregnancies, complications during pregnancy and delivery in ICD codes, malformations, birth weight and other indices of size. The computerized database for information on newborn children provides a good population-based framework for studying effects of environmental factors during pregnancy. The most useful health outcomes from the existing data are birth weight, gestational age, and mortality. Malformations form a set of potentially useful outcomes, but at this time, the structure that supports the classification of malformations is unclear. As for malformations, a written description of the major diagnostic groups is needed. There is also the possibility of using perinatal diseases as an outcome.

When studying health effects of environmental factors on birth outcome and perinatal health, there is a need for additional information on factors influencing health. This includes factors in the home environment such as size and nature of living unit, members of household, sources of indoor air pollution, occupational exposures, and factors such as parental smoking and alcohol consumption. Also relevant are social variables, and detailed behavioral information on maternal health during pregnancy. Accordingly we have recommended that the following additional information should be collected routinely for each pregnancy: maternal smoking, smoking of other family members in the home, number of people in the household and size of the dwelling unit, maternal and paternal education, and some indication of economic status.

4.1.2. Information on Morbidity from Polyclinics

Because the private sector is small and access to health care is good, the coverage of polyclinic visit information is expected to be very high,. Each visit is recorded in a standardized form using ICD-9 codes. However, the categories used are crude, e.g. only one code for respiratory diseases which comprise the majority of childrenís visits. A 40 fold difference in the rates of pneumonia within the city indicates probable variation in diagnostic practice. Unexpected low rates of chronic otitis, asthma, and allergic rhinitis indicated a possibility of underreporting of some diseases. The computerized individual morbidity data has a good potential to be used for assessing health effects of environmental factors and as a basis for providing information for decision-makers. The following recommendations are made to improve the usefulness of the information system: diagnostic accuracy should be improved and more effective case definitions need to be introduced for key health outcomes; additional information on health behavior (smoking, alcohol), home environment (type and size of home, presence of pets, type of stove etc.); and socio-economic status should be included in the routine data collection to control confounding. Routine quality control for accuracy of diagnostic practice, coding, and data entry should be established at the raion level.

4.1.3. Information on Causes of Death

The City SanEpid enters data from the death certificates in ZAGS into a computerized database. The database contains all deaths in children from 0 to 14 years of age (approximately from 100 to 150 events per year).

In general, childrenís mortality is a rather insensitive measure of the health effects of environmental factors such as air, water, or soil pollution, because other factors such as congenital or early chronic diseases and accidents are responsible for the majority of these deaths. The relatively small number of deaths within Cherepovets limits the use of these data for testing hypotheses. Adult mortality would be more useful for assessing the impact of environmental pollution, in particular short-term effects of air pollution. Thus, computerization of individual death records is advisable.

4.2. SVERDLOVSK OBLAST - REGIONAL LEVEL

Sverdlovsk Oblast, with a population approximately 5 million people, is located on the east slope of the Central Ural Mountains. It is a historical center of mining and metallurgy, a status which was further reinforced by industrial relocation of major machine producing facilities in World War II. Accordingly, the population is heavily concentrated in industrial cities, with large proportions of the population employed in a small number of industrial complexes. Limited agriculture further contributes to the overall urbanized nature of the Oblast with limited points of population outside of city limits

The Sverdlovsk Oblast SanEpid Center maintains an information system, which covers raion-level data based on annual reports of polyclinic visits, hospital discharges, health of pregnant women, health of infants, and sanitary-hygienic conditions.

4.2.1 General Morbidity Information

All polyclinics and hospitals in the 56 raions (administrative regions) of Sverdlovsk Oblast record the diagnosis at each visit or admission on a talon, which contains name, address, employment (yes/no), place of work, age group (0-14, 15-17, 18+), diagnosis, and first occurrence (yes/no). The annual number of visits and morbidity rates (per 1,000) are reported annually to the Raion Central Polyclinic. The Raion office sends aggregate data to the Oblast Department of Health in annual reports. The Yekaterinburg and Nizni Tagil raions send their reports to the City Department of Health, which then forwards the transmission to the Department of Health in the Oblast. The Oblast level information is then forwarded to the Ministry of Health, Russian Federation, and to the Oblast SanEpid office.

The existing information system seems most useful for administrative purposes such as allocation of health resources. The broad age categories (0-14, 15-17, 18+) is a limitation in the evaluation of the health status of populations. The causal inference based on the available information has serious limitations typical for ecological studies where the unit of observation is a group of individuals rather than individual [4]. The exposure within the units of observation (raions) is not expected to be homogenous, and differences in exposure between units of observation are possibly smaller than the individual differences within each unit of observation. There is very little information on potential confounding factors.

4.2.2. Information on Causes of Death

The Oblast also has computerized mortality information from 1992, and cancer registries in six cities. The mortality data could be used to assess temporal and to some extent spatial variation in relation to environmental factors. The usefulness would be improved through the modest additions of entering all contributory causes of death and place of residence into the file. Useful but somewhat more involved additions could include specifying socio-economic classification, and developing coding of occupations to reflect possible occupational exposures.

The existing mortality data can be used for 1) time-series analyses of the impact of short-term variation of environmental conditions on mortality and 2) analyses of spatial differences in mortality in relation to environmental factors. In the latter type of studies, further information on potential confounders is necessary for valid assessment of the role of the environment.

4.2.3. Automated Information System for Infectious Diseases

All polyclinics and hospitals are required to immediately report all cases of infectious diseases. The reports are made by telephone to the Raion SanEpid office. Information is recorded in the office either directly to computer file or on paper. All diseases are reported and recorded on individual basis, except acute respiratory infections and influenza, which are reported and recorded in aggregated form on a daily basis for the following age categories: 0-2, 3-6, 7-14, and 15+. These individual databases are maintained in the Raion offices, and aggregated data are reported on monthly basis to the Oblast SanEpid Office and further to the Ministry of Health.

The nearly real-time information system is very sophisticated and in principle this type of immediate provision of key data can be highly useful for conducting environmental health studies. A separate study would be required in order to evaluate the validity of the infectious disease diagnoses.

The existing data are suitable for studying the relations between infectious disease distributions in space and time and their potential influence from environmental factors. Studies of disease related to microbiological water contamination come to mind as an appropriate use of this type of database.

4.3. FEDERAL LEVEL

The Ministry of Health (Russian Federation) is responsible for collecting and assessing morbidity data. Health information is recorded at polyclinics and hospitals on special forms (Talons) using ICD-9 codes. The information is reported annually through local (Raion and City Hospital Centers), and regional (oblast) departments of health to the General Computing Center in the Ministry of Health. Besides, infectious disease reporting takes place through SanEpidís local network, and both monthly and annual recording is performed.

4.3.1. General Computing Center in the Ministry of Health

The data management of aggregated federal level health information is very well organized in the MedStat information system. There is systematic, partly automated quality control of information flows with a feedback system linked to producers of information at the regional level.

The information system can produce fast and efficiently time trends and spatial distributions of disease occurrence, indices of health care system and other relevant information. This type of information is useful for administrative purposes when allocating resources for health care. The quality control between oblast and federal levels is well organized. However, there is no systematic quality control of information originating from the health care providers that flows through raion and city levels to the oblast level. Further, the data are aggregated and transformed two or three times between the grass-root level and federal level resulting loss of important information. Consequently, the oblast level averages are not useful for assessing health effects of environmental factors.

4.3.2. Federal Center of SanEpid (Rossriads)

Federal Center of SanEpid (FedSanEpid) has three activities dealing with collection and analysis of health information. First, Federal Center of SanEpid (FedSanEpid) receives the primary information on infectious diseases from the Oblasts. Second, FedSanEpid carries out special studies. Third, FedSanEpid staff are responsible for the development of Socio-Hygienic Monitoring in the Russian Federation which covers also environmental and health information of interest. FedSanEpid also maintains an individually based registry of occupational diseases.

Federal Center of SanEpid (FedSanEpid) receives monthly reports of the frequency of 65 infectious diseases from all the 89 Oblasts on paper and a computer file. Annual reports of 78 infectious diseases are also received on a form and in a file. Furthermore, FedSanEpid receives annual reports of sanitary conditions in the territories.

All data were computerized in 1992 and are pooled into a general information system. Paradox software is used in the interface that allows both numerical and graphical presentation and analysis of data. All the routine tables for the annual reports are produced by existing programs.

A general feature of the data described above is that it represents counts, means or proportions prepared at the level of the oblasts. The oblast level information is aggregated from the raions of each oblast. However, FedSanEpid has access to all the infectious disease data collected by the SanEpid network. The raion level data are in computerized form in about 60-70% of the oblasts.

The management of information from the oblasts is of high quality. The information systems suits well for the monitoring of infectious. Local health records are used for studying infectious disease epidemics. These records are usually on paper, and a team of investigators will cooperate with the regional and local SanEpid staff. The computerized information on monthly and annual counts and rates is not useful for assessing effects of environmental factors for reasons discussed above, although the occurrence of infectious diseases per se can be influenced by air, water, and soil pollution.

4.3.3. Recommendations for Federal Level Activities

Federal level health information based on oblast level data is not useful for assessing the health effects of environmental factors, because of similar but more pronounced limitations discussed in the raion level. At the same time, The Russian Federation maintains a strong network and routines for collecting information on health phenomena which forms a good nationwide framework.

Based on our assessment the following recommendations are given for the federal level agencies to develop health information for the purposes of environmental health:

1. *Develop access to raion level and individual health data.* The federal agencies have a legitimate access to any local or regional level information. Although it may not be feasible or even desirable at this stage to try to concentrate all the individual information in the federal agencies, it is useful to develop capacity to

access this information. Current standardized statistical forms provide an excellent basis. An additional approach could be to develop standardized software to be used in handling routine individual health data.

2. *Focus on the development for selected information in order to maintain high quality.* A key in producing valid and relevant health information with optimal costs is the focusing of data collection. The World Health Organization and European Union are both developing suitable indices for monitoring population health and environmental conditions. It is advisable to follow the international experience in this field, although each country should developed the most suitable indices based on their special needs.
3. *Support and coordinate development of individual-based disease registries such as cancer, malformation, and occupational disease registries.* Information on health and exposure of individuals is needed to assess the relations between environmental exposures and risk of disease. Often studies carried out in selected, optimal settings (cities, small areas) are the most efficient way to learn about exposure-response relations, and this knowledge can then be used nationwide and internationally. This is especially true with relations between common diseases and well defined exposures. However, when studying environmental and other effects on rare diseases such as cancers and malformations, a possibility of bringing together information from large populations crucially improves the possibilities of studying these effects. Thus, the Federal support and coordination of the currently emerging cancer and malformation registries would essentially increase the possibilities of studying the role of environmental factors in the causation of disease. Studies of occupational disease would add to our knowledge of possible effects of high level environmental exposures. Coordination of the registries of common diseases would also be likely to improve the quality of information, although pooling of the information would not be necessary.
4. *Develop national health surveys which collect information on relevant health outcomes and environmental factors in the home, work, and other environments.* Besides routine health information flows and special epidemiological studies, use of national health surveys could provide important information on the relations between environmental conditions and health. An appropriate sampling frame would also allow the assessment of public health impact from some of the most important factors.
5. *Found and train federal task forces to maintain a center of excellence in environmental health and to provide consultation and to disseminate information in the regions.* There is a common belief that development of sophisticated information systems and automated analytical approaches are central to improving the use of empirical data in decision making. However, we think that the primary issue is to improve the knowledge of Russian specialists in solving the diverse problems related to environmental health and rational environmental management. At the first stage, there is an urgent need for knowledgeable federal task forces consisting of experts with multidisciplinary background who can

follow current international development, address complicated ad hoc environmental problems, and develop long-term strategies for improving the quality of environmental and health information and its appropriate use in the decision-making processes. These task forces could form centers of excellence capable of advising regional groups and disseminating information to the regions. Second, in order to ensure future development, the universities should be urgently involved in teaching and research in the fields of environmental health and epidemiology.

5. Discussion

The strengths and weaknesses of Russian health information are discussed in the context of assessing environmental health effects. The availability of existing environmental information or new environmental information naturally influences this assessment.

In general, the Russian Federation maintains an extensive standardized nationwide organization routinely collecting health data in polyclinics and hospitals. The information is collected in standardized forms and reported through two or three stages to the national offices. The system forms a strong basis for acquiring descriptive health data which can be used for administrative purposes such as allocation of resources and assessment of disease trends with certain limitations due to dramatic changes in the society during the 1990s. While extensive in its allocations of manpower and breadth of surveyed health outcomes, health assessment in the Russian Federation is prone to the twin threats of overly ambitious expectations and erratic control of information quality. On the other hand, the hierarchical system of data collection has large advantages over more decentralized or commercial health systems. Some of these advantages are underused. In particular, the current standardized data collection system with potential for completeness is a potentially excellent springboard for surveillance techniques. Registries for cancer and other diseases evolve out of national reporting guidelines, where emphasis for completeness can be introduced starting from strategically important oblasts. Basic cancer registration appears to be feasible within a 5-year interval, provided that registry goals are well-defined and completeness takes precedence over detail. National or oblast level health surveys, directed towards particular measures of health status and specific disease patterns are another underutilized strategy.

A major weakness in the current reporting is the aggregation and transformation of data which may not disturb the production of health statistics, but seriously limits the use of oblast and federal level data. The exposure parameters are not likely to represent a homogenous exposure over the unit of observation (raion or oblast). The measures of health outcomes are likely to be influenced by factors influencing the detection and diagnosis of cases as well as other than studied determinants of disease, i.e. confounders.

The individual visits to polyclinics and hospitals are recorded systematically on paper and ICD coding is generally used in disease registration. In some areas part of the individual information is entered into computerized form. This forms a basis for estimating disease occurrence over time and place. However, there is no general practice of quality control of data collection, entry, and handling and little information to what extent the diagnostic practices vary over space and time. There is reason to believe that diagnostic practice in some diseases may vary considerably by region, and thus spatial comparison of disease distributions may not be meaningful. Also limited information is available on potential confounders either on individual or group level.

In spite of above mentioned limitations, some approaches to the analysis of existing data may be both feasible and fruitful.

1. *Analysis of the effects of short-term exposure to environmental conditions on mortality and morbidity.* Daily monitoring of air pollution is performed in most industrial cities in the Russian Federation. The routine monitoring usually comprises 3 or 4 daily measurements of 20 minutes. Follow-up of daily number of episodes or events in a defined dynamic population eliminates the confounding due to personal factors as well as a number of environmental factors. In modern time-series analysis, the latency period of the effects, seasonal trends of disease occurrence and effects of other pollutants can be taken into account. The accuracy and precision of the disease measures varies and some measures are sufficiently valid. The possible health outcomes for time-series analysis include general and disease-specific mortality, occurrence of respiratory and gastrointestinal infections, episodes of asthma and acute cardiovascular problems.

2. *Analysis of the effects of long-term exposure to environmental conditions.* Use of the place of residence as the basis of exposure assessment could be feasible when the population is expected to stable and there is information on past exposures that can be allocated to residential areas. In such situation associations between disease occurrence and cumulated or time-specific levels of exposure can be estimated in a retrospective cohort study using either fixed or dynamic cohorts. However, there are not sufficient routine data on potential confounders. Age and sex standardization can be carried out in most situations, and some disease registrations include crude information on occupation.

Combination of information from routine data and new data collection is likely to be the most effective way to assess the relations between environmental exposures and diseases. The limited resources allocated for health care may require, in the future, diminution of routine health data collection. This will emphasize careful planning of data collection and performance of well-designed and accurately focused epidemiological studies applying cohort or case-control design. In general small homogenous countries, such as the Nordic countries, have been able to make effective use of standardized disease incidences in assessing potential effects of environmental factors. It is very unlikely that these approaches will have an easy transfer to Russian conditions. While there is a strong and justifiable desire to rapidly translate information of environmental health effects into policy alternatives, a step backwards emphasizing data quality, completeness, and plans for utilization of data is currently more timely.

6. References

1. Konygin E. The largest IBRD-financed project is being implemented in Russia. (1996) In: Environment Management Newsbulletin No 1. Center for Preparation and Implementation of International Projects on Technical Assistance. Russian Ecological Federal Information Agency Publishers, Moscow, pp. 6-14.
2. Ozkaynak H., Jaakkola J.J.K., Spengler J.D., Cherniac M., Ford T., Wojtyniak B., Egorov A., Dumyahn T. (1997) Review and evaluation of existing environmental and health information in the Russian Federation. Task 2 Report prepared for CPPI, Environmental Epidemiology Component. Harvard School of Public Health, Department of Environmental Health, March 31, 1997.
3. Ozkaynak H., Spengler J.D., Jaakkola J.J.K., Kuzmin S., Rakitin P. (1997) Evaluation of existing environmental information systems in Russia applicable for human health effects assessment. A Presentation in the NATO Advanced Research Workshop on Air Pollution in the Ural Mountains, May 26-30, 1997, Magnitogorsk, Russia.
4. Rothman K.J. (1986) Modern epidemiology. Little, Brown and Company, Boston/Toronto, pp. 52, 74, 304-305.

ASSESSING CHRONIC HEALTH RISKS FROM STATIONARY SOURCE AIR EMISSIONS IN VOLGOGRAD, RUSSIA

B. FILATOV and T. KOLODI
SRI of Hygiene, Toxicology and Occupational Pathology, RUSSIA

S. WOLFF and H. ÖZKAYNAK
Harvard School of Public Health, USA

B. LARSON and D. SHAPOSHNIKOV
Harvard Institute for International Development, USA

S. AVALIANI and L. VISHNAVETSKAYA
Agropromproject Institute, RUSSIA

The existing air quality monitoring networks in Russia are typically designed to characterise short-term high concentration impacts rather than measuring long-term average concentrations necessary to study chronic health effects resulting from exposures to low concentrations of pollutants. The problem of estimating impacts from chronic pollutant exposures can be addressed by the application of risk assessment methodology to long-term exposures to toxic chemical substances. The methodology developed by the U.S. Environmental Protection Agency to estimate the individual's probability of cancer risk from lifetime exposures to various carcinogenic pollutants is used in this analysis. The following basic terminology is used in this assessment.

Risk - is defined as the frequency of occurrence of a certain type hazard. Thus, cancer risk may be interpreted as the frequency or probability of tumor formation in a given population or by a single individual.

Individual risk - is the probability of disease occurrence due to exposure to certain chemical(s) (*i.e.* carcinogens) at the location where an individual resides.

Population cancer risk - is considered as the occurrence frequency of specific diseases (malignant tumors) among a certain group of individuals exposed to carcinogens. In other words, the population risk characterizes damage prevalence for different population groups by location and susceptibility.

Common to all of health risk assessment methods is the fact that each step of the risk assessment process requires making multiple assumptions and judgments. This process is based on available scientific information to the fullest extent possible. But scientific understanding is not always adequate to eliminate all uncertainties from risk assessment. Thus, it should be recognized that risk assessment will seldom, if ever, provide final indisputable results.

I. Linkov and R. Wilson (eds.), Air Pollution in the Ural Mountains, 231–238.

Details of the risk assessment methodology developed by the U.S. EPA [1-8] are described more completely in various reports and guidance documents published by the EPA and other researchers. The main purpose of this research was to show how current methods of risk assessment, as recommended by the U.S. EPA, combined with existing Russian environmental data and technical air dispersion modeling methods, could be used to evaluate human health risks from existing industrial emissions in Volgogard. The risk assessment methodology applied was divided into four general, sequential steps:

- Hazard identification involved an assessment of the specific factors with potential risk for human health effects from air pollution. Thus, this initial step focuses on the risk assessment of the potentially most toxic substances first, and also includes selecting the subset of facilities based on significant level of emissions of toxic compounds.
- Assessment of ambient concentration levels and the nature of human exposures. This step characterized human exposures. In this particular, in this analysis air dispersion models were used to estimate population exposures. Based on estimates of concentrations of the emitted chemicals in ambient air (called receptor points), it was then possible to calculate chemical intake levels in the potentially exposed populations of Volgograd.
- Dose-response assessment provided quantitative information on the carcinogenic potency of each of the substances studied in this risk assessment.
- Risk characterization combined dose-response information with exposure point concentrations (EPCs) and chronic daily intakes (CDIs) for carcinogens in order to estimate additional cancer risks, and mortality risks from respirable particulate matter (*i.e.* particulate matter with a diameter of 10 microns or less, PM_{10}).

Considering a number of pollutants generate multiple and sometimes not comparable risks, it was decided to focus on two types of health risks affecting the general population of Volgograd:

1. excess risk of cancer;
2. excess mortality risk from inhalable particulates.

When evaluating the exposure to carcinogens, these chemicals were considered to be toxic without a threshold. The analysis of the particulate matter mortality risks was based on historic data on the relationships between daily particle pollution and mortality in western countries. The assumption was made that chronic health risks can be approximated by the results from acute mortality effects studies. In the Volgograd air pollution health risk assessment only the emissions data from stationary sources were taken into consideration.

The investigation on the Volgograd cancer and particulate risk assessment included the study of the following parameters: facilities inventory to obtain quantitative and qualitative data on quantities of harmful emissions; mathematical modeling of carcinogen and particulate emissions; and, calculation of individual and population risks of cancer incidence and mortality-related particulate risks. Based on analysis of industrial emissions data, it was found that 29 facilities account for about 80% of total reported stationary source air emissions in Volgograd. Fourteen carcinogens contributed to the emissions identified.

For chronic risks, it was necessary to estimate annual average exposure-point concentrations in the study region. Emission estimation was based on existing data and air quality modeling was performed using the Russian Ecolog air dispersion model. This standard air contaminant transport and dispersion model (Ecolog) was developed during the

Soviet period and is mainly used for setting emission limits in population zones near different types of sources (OND-86). The program Ecolog is thus designed to model 20-minute air concentrations assuming worst-case scenario maximum emission rates and weather conditions that lead to the highest concentrations.

Several important assumptions were needed to complete the modeling stage. The 29 facilities selected for this study included several thousand individual stacks and fugitive sources. To save time and financial resources, during the analysis each facility was modeled as if it had one aggregate stack. The characteristics of the aggregate stack (*i.e.* height and location within facility) were defined as a rough average across all sources in the facility. Using this aggregate-stack assumption, a base-modeling run was performed in order to estimate exposure concentrations at the 20 receptor points of the city territory. A separate evaluation study of this approach demonstrated that this modeling assumption worked quite well.

The emissions data obtained for each of the modeled facilities represent estimates based on mass balance or chemical engineering estimates. These represent the maximum short-term estimates of chemical emissions expected to occur throughout an entire calendar year. To convert Ecolog 20-minute maximum emissions to annual average emissions, a method was developed that required two weighting factors. The first factor is defined as the fraction of time the facility operates throughout an entire year. The second factor is defined as the ratio of annual average emissions of the facility for all pollutants converted to an average value (r/c). In order to estimate annual average concentrations (Ci) for each pollutant from each facility, it was then necessary to calculate:

$$Ci = a*wf1*wf2*d \quad (1)$$

where:

a - normalized 20-minute exposure point concentration from Ecolog;
wf1 - weighting factor 1 representing fraction of facility operating time, in unit shares;
wf2 - weighting factor 2 representing steadiness of emissions;
d - maximum emission rate of chemical (i).

The next step in the exposure assessment was to define the levels of pollutants. In this study, the population is defined as the approximately one million individuals who live within the city boundaries and are at risk of being exposed to industrial air emissions. Population densities vary throughout the city, and residential and industrial areas are located in close proximity to one another in several parts of the city. Based on a city map of population densities, 20 receptor locations throughout the entire city were chosen. Each receptor location represents 5% of the city's population (or 50,000 individuals). Each receptor point is located at the nearest node of the air dispersion modeling grid and represents the center of the specific population area. This method of air dispersion modeling allowed the estimation of contribution of each facility emissions to air quality of the territory being studied.

The annual average cancer risk at each receptor point (RP) was calculated with formula (2), in which the sum of average daily concentrations of all registered potential carcinogens is considered to be constant over the entire lifetime of an individual. The Unit Risk Factor $(mg/kg\text{-}day)^{-1}$ is used as the basis for extrapolation of carcinogen potency of the selected chemicals.

$$Kj = \Sigma(Cj^*D^*V)URFi/m^*G, \quad (2)$$

where:

K_j - individual lifetime cancer risk in RP;
C_j - concentration of carcinogens in RP;
D - daily inhalation rate (m^3);
V - average number of life days (365 days x 20 years);
m - average body weight (kg);
URF – unit risk factor
G - average number of life years.

With this algorithm the individual annual average cancer risk for each representative at each population group of the 20 receptor points were calculated. The population cancer risk for all residents within the neighborhood of the given receptor point, the calculated Kj values by 50,000. The estimated risks at all receptor points were then added up to predict the average annual population cancer risk for the population of the entire city.

This analysis estimated the average annual population cancer risk from stationary source air emissions in Volgograd to be approximately 13E-6. That is, if we assume that air pollution levels from stationary sources are constant over 70-years of lifetime, it should be expected that this would result in 13 additional cancer cases per year in the entire city of Volgograd.

According to medical statistics data collected during 1993-1995, there were more than six thousand cancer cases registered in Volgograd. Estimated 13 additional cancer cases posed by carcinogenic emissions characterize comparatively much lower risks. However, most of the reported cancer cases can be attributed to lifestyle factors or aggregate doses to which the city residents were exposed to over previous years of life, mainly from contaminated food and water intake as well as inhalation of ambient or workplace pollution. It should also be noted that industrial production in 1995 in Volgograd, especially at large facilities, was quite lower than in previous years. Therefore, the predicted cancer risk, based on emissions of 1995, is by far understated. Recognizing these assumptions and limitations stated above, it should be mentioned the methodology used for the first time in Russia identified regions with maximal cancer risk (Fig. 1) as well as priority sources of risk (Fig. 2). Volgograd cancer risk analysis showed that rather than the well established carcinogens, such as benzo(a)pyrene, nickel, and chromium, but exposures to vinyl chloride and carbon tetrachloride emissions into air by two large chemical facilities seemed to be contributing most to the total predicted cancer risks.

Particulate matter, and specifically the particulates with a diameter of less than 10 microns (PM_{10}), were also considered in this research. In absence of Russian and U.S. EPA literature on specific dose-response coefficients for particulate pollution, we relied upon the results of recent epidemiological studies conducted in USA. These studies consistently indicated that an approximate one percent increase in daily mortality rate is associated with each 10 $\mu g/m^3$ of PM_{10} pollution in ambient air.

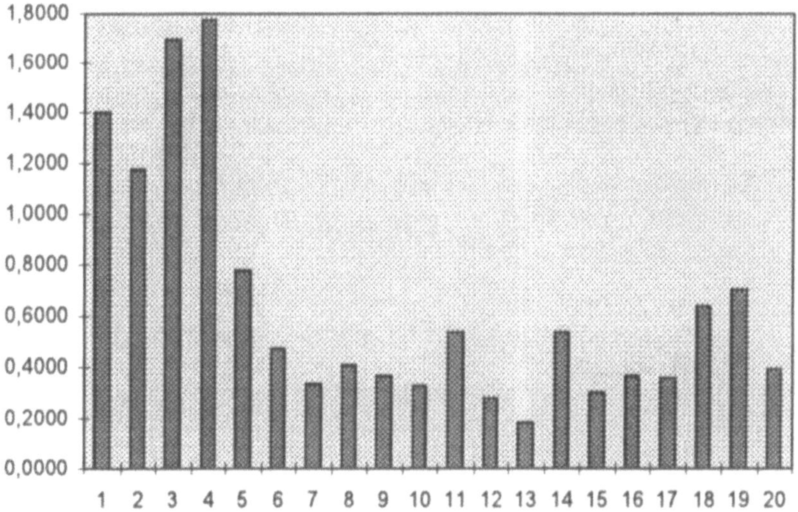

FIGURE 1. Distribution of annual predicted population cancer risks in Volgograd. 1-20 represent different receptors in Volgograd (see [9]).

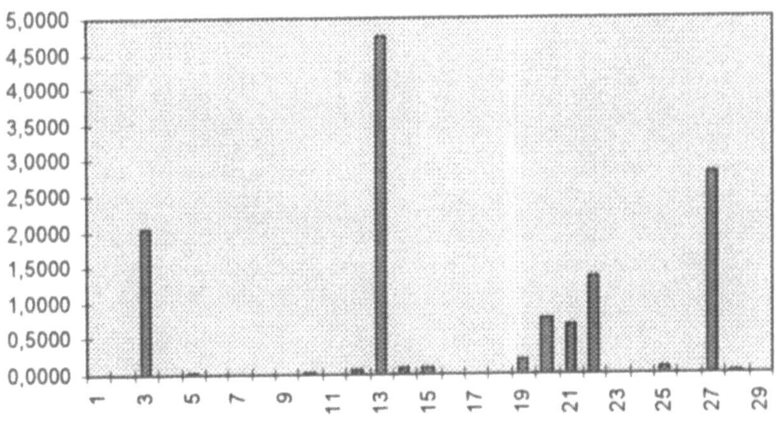

FIGURE 2. Annual predicted population cancer risk from stationary sources of air pollution in Volgograd: 3 - Drilling Machines Plant, 13 - Chemical Industrial Plant, 27 - Aluminum Plant

Based on this information we expect that the rate of daily deaths attributable to air pollution in the Volgograd population can be estimated as:

$$S_{10} = S*0.01/N*d = 2100000*0.01/250000000/365 = 2.3E-7 \quad (3)$$

where:

S - total annual mortality in USA, number of cases;

N - USA population size;

d – 1 year exposure period, 365 days.

The individual annual mortality risk coefficient corresponding to each 10 $\mu g/m^3$ of ambient PM_{10} pollution is then estimated as:

$$S_{PM10} = S_{10}*d = 8.4\text{-}5 \quad (4)$$

Correspondingly, approximately 4.2 deaths/year would be expected in a point population of 50,000 exposed to an average PM_{10} air concentration of 10 $\mu g/m^3$ (assuming linearity for the PM_{10} coefficient).

Our analysis showed two industrial facilities, Red October Plant and Aluminum Plant were responsible for 85% of the predicted mortality risks from chemical air emissions in Volgograd. While particulate air emissions in 1995 from the Chemical Plant (source #13 in Fig. 3) exceeded those from the Aluminum Plant, the mortality risks associated with the Aluminum Plant's emissions were considerably higher (Fig. 3). This was due to the relative location of these plants with respect to the high population areas. The Aluminum Plant (source #27) was estimated to contribute the greatest mortality risk from particulate pollution in Volgograd.

Comparative estimation of air pollution by carcinogenic agents and particulate matter in Volgograd verified an expected domination of particulate pollution to total predicted mortality risks.

FIGURE 3. Annual mortality in Volgograd's population in relation to air pollution by particulates and attendant gases: 27 - Aluminum Plant, 28 - Red October Plant.

The methodology presented in this paper was found to be useful for evaluating the impacts from facilities studied in the project and developing alternative operation and emission procedures as well as improvement of environmental regulation of harmful chemical substances in Russia.

The main objective of this study was to show how existing quantitative health risk assessment methods can be implemented using existing data and air dispersion modeling capabilities that are commonly available in Russia and other parts of the former Soviet Union. The Volgograd study did fulfill this objective. Health risk estimates were derived for both carcinogenic chemicals and noncarcinogenic particulate matter (PM_{10}) via the inhalation exposure route. The mortality risks posed by PM_{10} emissions are estimated at 2,700 additional deaths per year in the city (population of about 1 million), while health risks posed by carcinogenic emissions are estimated at 13 additional cancer cases per year. Even with the substantial uncertainties associated with the risk assessment methods and underlying data, particulate mortality risks appear to be unacceptably high in Volgograd and do warrant closer consideration by environmental and health authorities.

References

1. Daniels J.T. and Holden P.L. Quantification of risk.- IChemE., 1982.
2. U.S. EPA, 1988. Superfund Exposure Assessment Manual. U.S. Environmental Protection Agency, Office of Emergency and Remedial Response. Washington, D. C. EPA/540/1-88/001.
3. U.S. EPA, 1989. Risk Assessment Guidance for Superfund, Volume 1, Human Health Evaluation Manual (Part A). U.S. Environmental Protection Agency, Office of Emergency and Remedial Response. Washington, D. C. EPA/540/1-89/002.
4. U.S. EPA, 1990. Interim methods for development of inhalation reference doses. EPA/600/8-88/066F, U.S. Environmental Protection Agency, Washington, D.C.
5. Masters, G.M., Risk Assessment: Introduction to Environmental Engineering and Science, 1991, 1, 5. -p. 191-210.
6. U.S. EPA, 1992. Guidelines for exposure assessment. Federal Register, U.S. Environmental Protection Agency, May 29, 1992.
7. U.S. EPA, 1996. *Exposure Factor Handbook.* U.S. Environmental Protection Agency, Office of Health and Environmental Assessment. Washington, D. C. EPA/600D-95/002.
8. U.S. EPA, 1997. Integrated Risk Information System (IRIS). U.S. Environmental Protection Agency. Washington, D. C.
9. Wolff S., Ozkaynak H., Filatov B., *et al.* Final Report, May 1997 Assessing Chronic Health Risks from Stationary Source Air Emissions in Volgograd, Russia. Cooperative Agreement Number 118-0003-A-00-5311-00.

PUBLIC HEALTH TRENDS IN A TRANSITIONAL RUSSIAN CITY (1959-1994)

G.I. CHECHENIN
Novokuznetsk Health Information Center and Doctors Advanced Training Institute, RUSSIA

R. DAY
Department of Biostatistics, Graduate School of Public Health University of Pittsburgh, Pittsburgh, PA 15261, USA

Yu. GREGORIEV
Institute for Complex Problems of Hygiene and Occupational Medicine, Novokuznetsk, RUSSIA

Yu. RUSAEV
Novokuznetsk Sanitary Epidemiological Service, RUSSIA

V.D. SURZHIKOV
Institute for Complex Problems of Hygiene and Occupational Medicine, Novokuznetsk, RUSSIA

Novokuznetsk, Kemerovo Oblast, is a large industrial city of 600,000 residents located in South Central Siberia. Major enterprises in Novokuznetsk include the Kuznetsky and the Western-Siberian steel plants (employing a total of 50,000 workers), as well as aluminum, ferro-alloy, chemical, cement and mining machine plants. The city is located adjacent to large coal fields that employ upwards of 200,000 miners in 3 open pit coal mines and 12 underground mines which produce coking and steam coal for commercial uses. The collapse of the Soviet economy and political system have had serious negative consequences for the metallurgical and mining industries that support the majority of working families in the city. This presentation reviewed public health trends in population mortality between 1959 and 1994 using life table data collected by the Novokuznetsk authorities.

Key factors affecting population health in Novokuznetsk over this period were identified as environmental pollution; deteriorating health services; and demoralization. Data were presented to demonstrate that overall levels of environmental pollution appear to have declined over this period due, at first, to pollution remediation efforts and, later, to the decline of industrial production. The decline in the quality of health care services is a more recent phenomena that dates

I. Linkov and R. Wilson (eds.), Air Pollution in the Ural Mountains, 239–240.

from the late 1980's and the decline in the Soviet economy. Demoralization is a psychosocial concept encompassing a number of individual responses -- including risk taking behavior, substance abuse, symptoms of psychological distress -- to chronically stressful situations. It was argued in the presentation that the trends observed in the mortality data from Novokuznetsk cannot be explained on the basis of environmental pollution alone, but must include a recognition of the continuing effects of demoralization in the industrial working population and the more recent effects of the deterioration of health care services.

Age-specific mortality rates for all causes of death and estimated years of life remaining for various age groups were reviewed for the period 1959-1994 using life table data. Two particular trends in the Novokuznetsk data were noted: first, infant mortality rates dropped markedly prior to 1990, then rose again during the period of recent economic and political difficulties; and, second, male and female mortality rates for the age group 20-65 years old showed a continuing increase across the 35 year period covered by these data.

In order to provide greater specification regarding the second trend, the Novokuznetsk mortality data was stratified by sex and by the following combined causes of death: cardiovascular disease, all malignancies, infectious disease, nonmalignant respiratory disease, and trauma and poisoning. The life table data were again reviewed and it became clear that the most striking increase in mortality rates for the 20-65 year-old age group was due to trauma and poisoning. The trend was most apparent in males, although it could also be identified in females. Secondary increases were observed in cardiovascular disease and were likely due to declining health services.

Special attention was given to the combined mortality category of trauma and poisoning. This is a peculiarly Russian category that includes all industrial and other types of accidents, interpersonal and intrafamilial violence, suicide and deaths due to substance abuse. It was argued that it was impossible to explain this increased mortality from trauma and poisoning without taking into account the effects of chronic demoralization as an aspect of the Soviet industrial system for the workers and their families. Western literature on demoralization was reviewed and was related to the functioning of the Soviet social system. The general argument was made that relatively high levels of demoralization existed among workers even in our earliest data. Recent developments, including the collapse of the Soviet political system and the economic depression that has overtaken the Russian steel industry, simply intensified these trends.

The implications of these trends are important. The 20-65 year-old age group is the core of Russian society. These are the most productive years economically and the years during which families are created and children socialized. The mortality data suggest an increasing frequency of additional problems for this age group such as alcoholism, intrafamilial violence, accidents and physical and mental disability. In summary, the data from Novokuznetsk suggest important negative changes occurring in the Russian social structure, the effects of which on the family and the economy may continue for decades into the future.

CONTEMPORARY PROBLEMS OF ECOPATHOLOGY AND EVALUATION OF HEALTH RISK IN MANAGEMENT OF ENVIRONMENTAL QUALITY IN THE URAL REGION

V.S. KOSHKINA
Institute of Biomedical and Ecological Problems, Regional Division, 72. Lenina Prospect, RUSSIA

Management of environmental quality is a complex but extremely pressing question for the Ural Region where metallurgical and chemical enterprises are concentrated. Regional link is a necessary basic modulus of developing improvement of the territory. Regional data on hygenical information, environmental factors, biological effects and health indices of the population are necessary in order to effectively assess health risks and implement management of the environmental quality.

In this regard, we present the socio-ecological situation in Magnitogorsk, an area with highly concentrated sources of ecological risk. The city is in a state of ecological crisis and is also undergoing a period of deep crisis in the social sphere (health care, education, science, culture). This is a typical situation for the Ural Region.

Both the atmospheric environment and the city population have been influenced technogenically during 64 years by solid and gas releases from the metallurgical enterprises. Annual releases during the above mentioned period were up to 1 million tons of toxic chemicals of the 1st and 2nd grades (Table 1) which are either carcinogens, co-carcinogens, mutagens, heavy allergens or general toxicants. The total number of people who are capable of working is more than 100,000 (1/3 of the total city population). Most of them are employed as general workers in industrial plants and suffer from both occupational and environmental exposures. Benzo(a)pyrenes, lead, chromium, iron, arsenic, boron, antimony, sulphur and nitric acid combinations have been the most harmful toxic chemicals polluting the city environment. The continuous influence of toxic chemicals on several generations living in Magnitogorsk creates problems of ecological pathology in medical practice and science. In particular, there is an increased incidence of lung cancer among workers engaged in the dangerous steelworks enterprises (agglomerate, coke-chemical, furnace, founding and rolling shops) that emit carcinogens.

Based on the results of many years of epidemiological cancer research, the above mentioned enterprises are included in an official list of potential carcinogens by the International Agency for Research on Cancer. Working conditions in, and air

I. Linkov and R. Wilson (eds.), Air Pollution in the Ural Mountains, 241–250.

TABLE 1. Chemicals present in Magnitogorsk environment in 1989-1996. Levels above the expected background are marked by +

Chemicals	Air	Water	Soil
Si	+		+
Fe	+	+	+
Ni	+		+
Cr	+	+	
Pb	+	+	
Cd	+	+	+
Be	+		
Br	+		+
V	+		
B	+	+	
Co			
Sr	+		
Aq			
As	+		
Zn	+		+
Al	+		
P			
Na			
K			
Sb	+	+	
Zr			

pollution by, these enterprises creates problems of cancer for both the workers and the residents of the city.

Biological testing of benzo(a)pyrene markers in the urine was conducted for two groups of Magnitogorsk residents, both non-smokers. The first group (exposed) were man and women who worked in coke-chemical enterprises for 5 or more years; the second group (comparison) were men and women not engaged in the steelworks. Both groups showed similar quantities of the carcinogenic metabolites of BP-7,-8, dehydro - BP, 3 hydrooxides BP in the excreted urine (Table 2). However, the average quantity of carcinogenic metabolites in the urine of Magnitogorsk residents was 2 times higher than that found in the urine of St. Petersburg residents by Dr. Linkachov. These results indicate a higher individual cancer risk to Magnitogorsk residents from iron and steel plant emission.

TABLE 2. Carcinogenic metabolites measured in the urine of Magnitogorsk residents in 1994-1995

Group	Concentration (μmol/mol)		Average Total Quantity in 1 liter (μg)	
Exposed (30 men and 10 women)	29.9	3.8	58.2	7.8
	21.4	4.3	50.4	8.4
Comparison (28 men and 11 women)	29.6	3.7	57.7	6.2
	21.1	4.2	49.7	7.6

This research is done in collaboration with Dr. A. Likhachov, St. Petersburg

TABLE 3. Spectral Analysis of Lung Tissues in Magnitogorsk, 1997

Chemical	Concentration		
	Background level	Magnitogorsk	
		Healthy Tissue	Malignant Tissue
Ag	0.0007	6.77	0.72
Zr	2	9.13	8.05
Ca	-	-	222
Co	0.02	0.69	1.46
Ni	0.047	1.79	0.44
Mo	0.03	no	12.23
Sr	-	no	0.22
Si	-	53.33	793.6
Sb	0.06	203.3	153.2
Au	0.063	no	1.63
Cr	0.092	2.8	4.95
Be	0.007	0.16	0.22
Zn	11	61.39	47.53
Cd	0.35	3.11	7.25
P	780	5091	3649
Pb	0.39	6.53	1.14
Cu	1.2	3.06	1.32
Fe	3600	no	125.5
K	1900	558.11	466
Na	1800	1855	1572

Spectral analysis of lung tissue samples from lung cancer patients (Table 3) showed high levels of non-organic chemicals (Co, Ni, Cr, Cd, Pl, Cu, Zn, Sb, Si, Zr and Ag) in malignant and healthy tissues. In addition, Mo, Sr, and Au were found only in malignant tissues. Results of toxicological testing and investigations in Western Europe suggest that chemicals that include Si and Cd in their structure can increase the probability of lung cancer development. It is important to regulate these chemicals in the environment.

Biomarker research in the urine and blood of the children in Magnitogorsk shows a high toxicological burden from such elements as Si, Sb, Cd, Cr, Ni, Be, B, Pl, Zn, Al. Children that were studied were found to excrete high quantities of Si, Sb, B. In addition, 29.7% of them excrete Zn, Cr, Ni and 28,3% excrete Pl and Al. In a few cases Cd and Be were also found. Increased concentrations of Cd and Pl and the presence of Ni and Cr tracers in the blood of children are of concern. About 30% of the investigated children were taken from a cancer risk group. It is necessary to take into consideration the individual sensitivity of the organism and genetic predisposition. However, these findings are not encouraging for the health of the future generations in current Magnitogorsk conditions. Clinical manifestation of prolonged exposure to toxic chemicals, such as diseases of the respiratory organs (mostly chronic bronchitis and lung cancer) and different allergopathologies, are observed among both children and adults.

Earlier epidemiology research showed that the residents of the Left bank district, that is the most polluted have the highest standardized indices of lung, stomach and skin cancers (Figure 1). The cancer rates were calculated per 100,000 people. Figure 2 presents lung cancer rates for Magnitogorsk residents and workers in several steelwork shops. The city morbidity indices for the population were based on data excluding those for the steelworkers. Increased lung cancer morbidity in men compared to women is most likely explained by the differences in smoking. Women engaged in heavy labor conditions at the steelworks were non-smokers. In contrast, 72.8 % of men were smokers. Consequently the high indices of lung cancer morbidity of non-smoking women in such enterprises as hot rolling, agglomerate, furnace, founding, and coke- chemical must be explained by occupational factors (multicomponent dust containing iron, chromium, cadmium, lead, nickel, BP, high temperature, radiant heat). The same chemicals were found in the city environment, but in much lower concentrations. Therefore, the exposure of steelworkers was higher.

Important factors at the steelwork shops are high temperature and radiant heat. These factors probably determine the increased carcinogenic effects in younger workers compared to the rest of the city residents. Rather high levels of lung cancer morbidity were observed for men 20-29 or 30-39 years old working in the coke-chemical enterprises (Fig. 3).

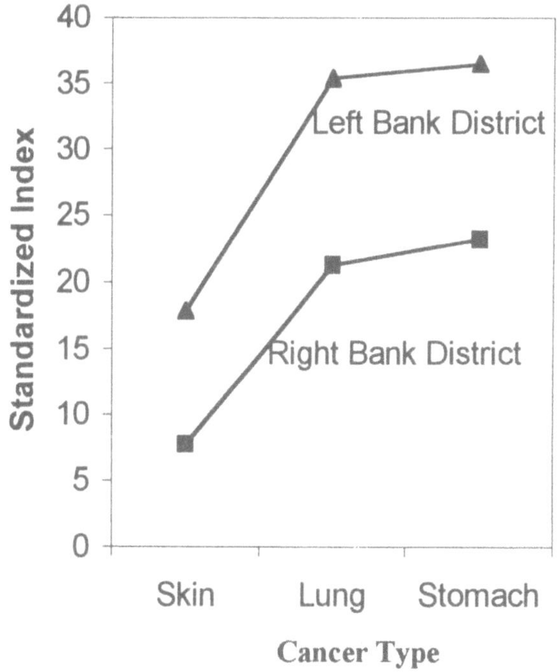

FIGURE 1. Cancer indexes for two Magnitogorsk disricts

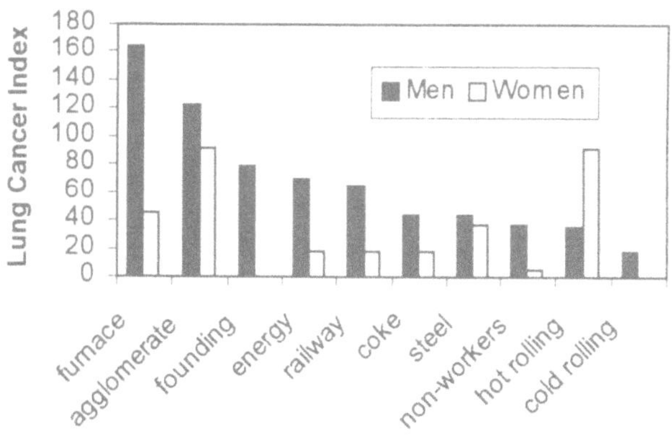

FIGURE 2. Index of lung cancer morbidity among workers of different steelwork shops and Magnitogorsk residents in 1974-1983.

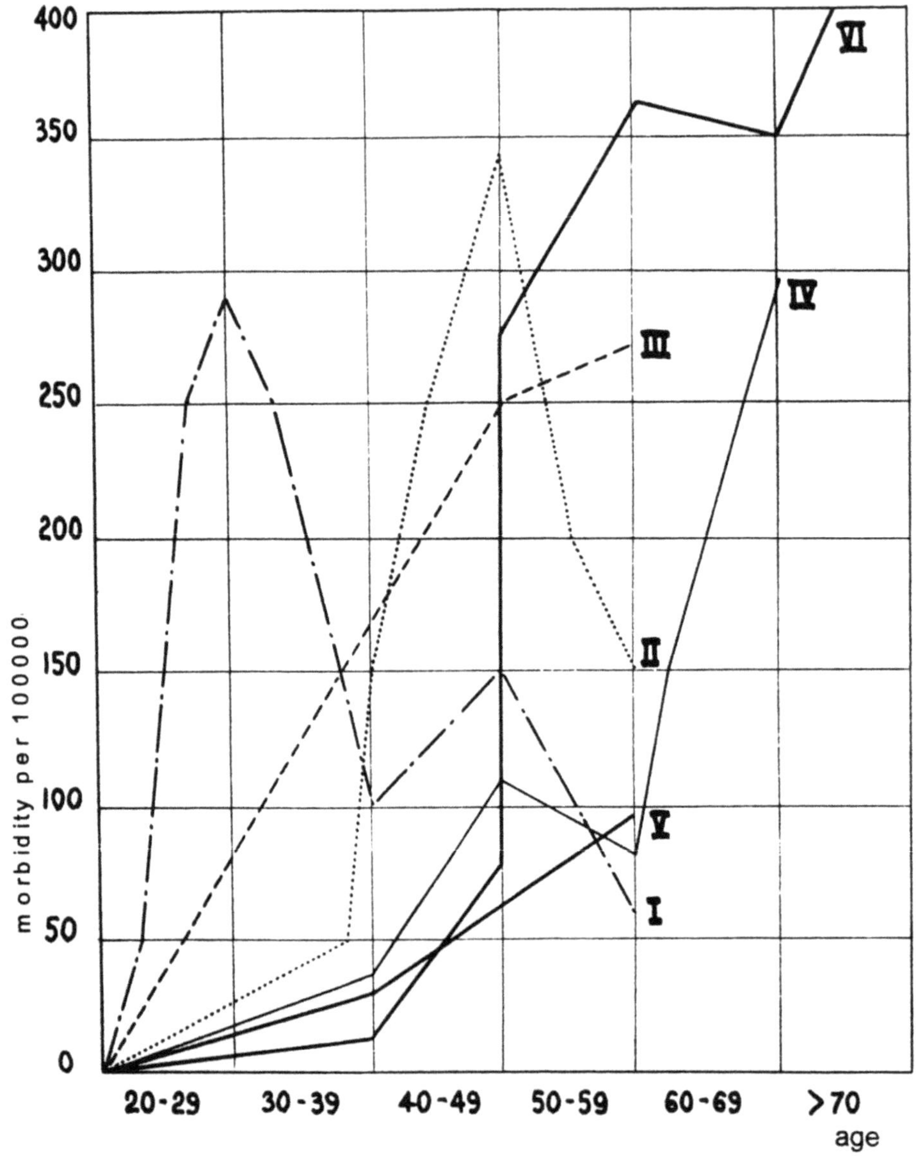

FIGURE 3. Age-specific indexes of lung cancer morbidity among male residents of Magnitogorsk and workers of steelworks according to job category (I coke chemical, II furnace, III agglomerate, IV hot rolling, V cold rolling, VI city residents)

A high prevalence of chronic non-specific diseases was established for workers at the steelworks (both smokers and non-smokers) and is presented in Fig. 4 per 1000 workers. Smoking influences the development of chronic lung diseases among workers and is especially damaging for young age groups (20-29, 30-39). The smoking factor is seen to level off with age. Occupational factors are increased by smoking when a worker suffers from lung cancer.

Background chronic bronchitis increases the occupational risk of lung cancer. Some 82.8% of steelworkers dying of lung cancer had obvious morphological features of non-specific chronic bronchitis shown by autopsy.

Occupational factors are reduced when malignancies are detected and consequently, the exposure time is limited, especially for women. The average working experience of women and men having lung cancer is 15.3 and 22.4 years; the average age of patients is 40 $\pm$ 4.8 years and 47 $\pm$ 2.3 years, respectively. Women's organisms are more sensitive to the occupational carcinogens.

In conclusion, chronic non-specific bronchitis and lung cancer found among the Magnitogorsk residents based on this research can be attributed to environmental factors. According to the epidemiological research and toxicological testing, data can be related to ecologically determined diseases. In children different allergic diseases were found above background values in terms of higher concentration in blood of immunooglobulin E (2000 med/ml and higher); allergopathology to specific and non-specific allergens; and 30 % of children examined have allergy to toxic chemicals (chrome, nickel, lead, sometimes beryllium). Additionally, increased growth of congenital defects of development and congenital anomalies were found. About 1800 newly diagnosed cases were registered annually among children (Figure 5 and 6). These data require additional scientific investigation.

Data presented show that conditions of Magnitogorsk can be used as a model of the Ural Region where a prolonged effect on the population of different exposure factors adversely affect health while the environment is also damaged. A knowledge of environmental and health risk factors can help us to establish ecologically caused changes in functional organism state. These should become an additional tool in management of environmental quality and effective medico-ecological rehabilitation of the population. This problem, undoubtedly, is complex and important for the Ural Region and requires joint efforts by Russian and foreign specialists in order to solve it.

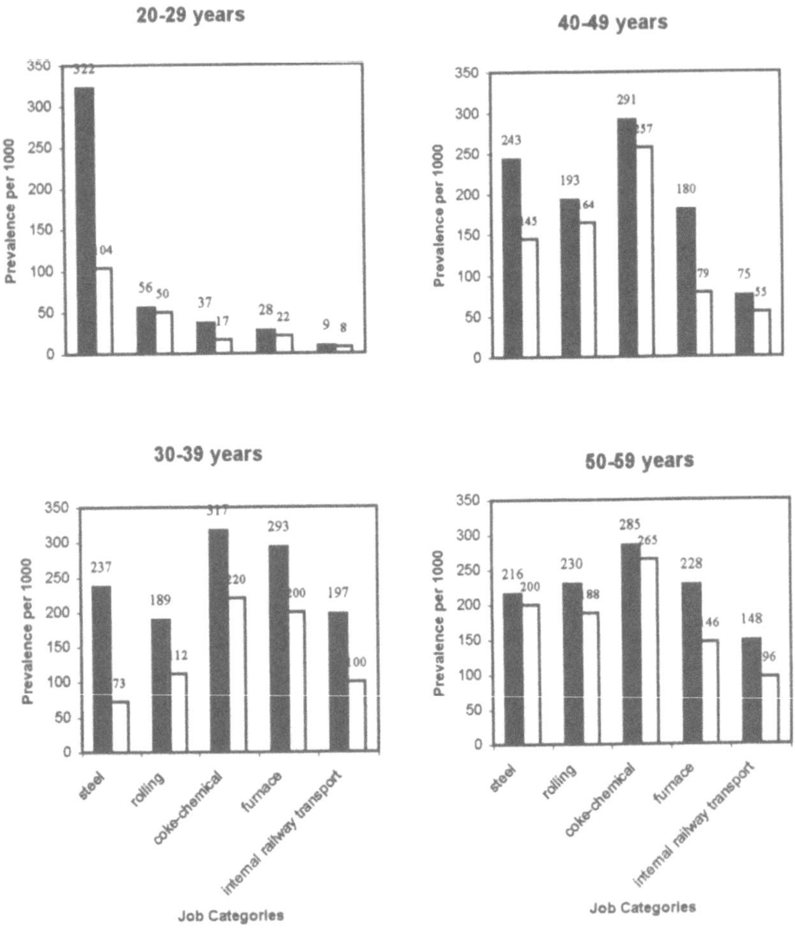

FIGURE 4. Prevalence of chronic malignant diseases among experienced male workers of the main steelworks shops depending on age and smoking habits (open bars-non-smokers, black bars-smokers).

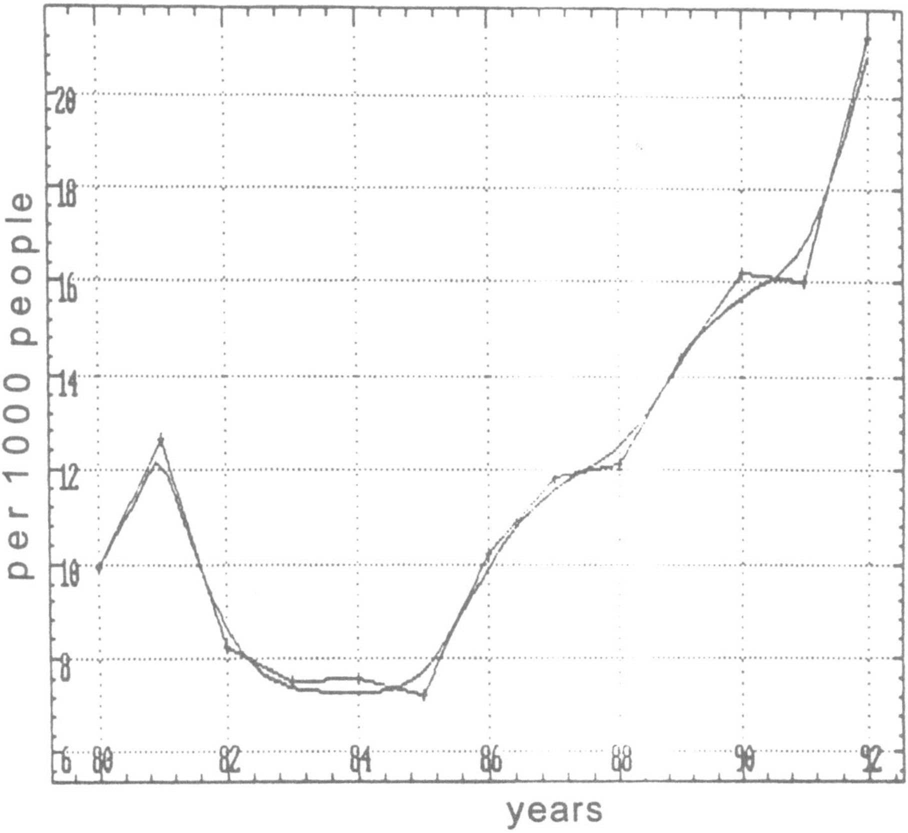

FIGURE 5. Dynamics of congenital anomaly in Magnitogorsk in 1980-1992.

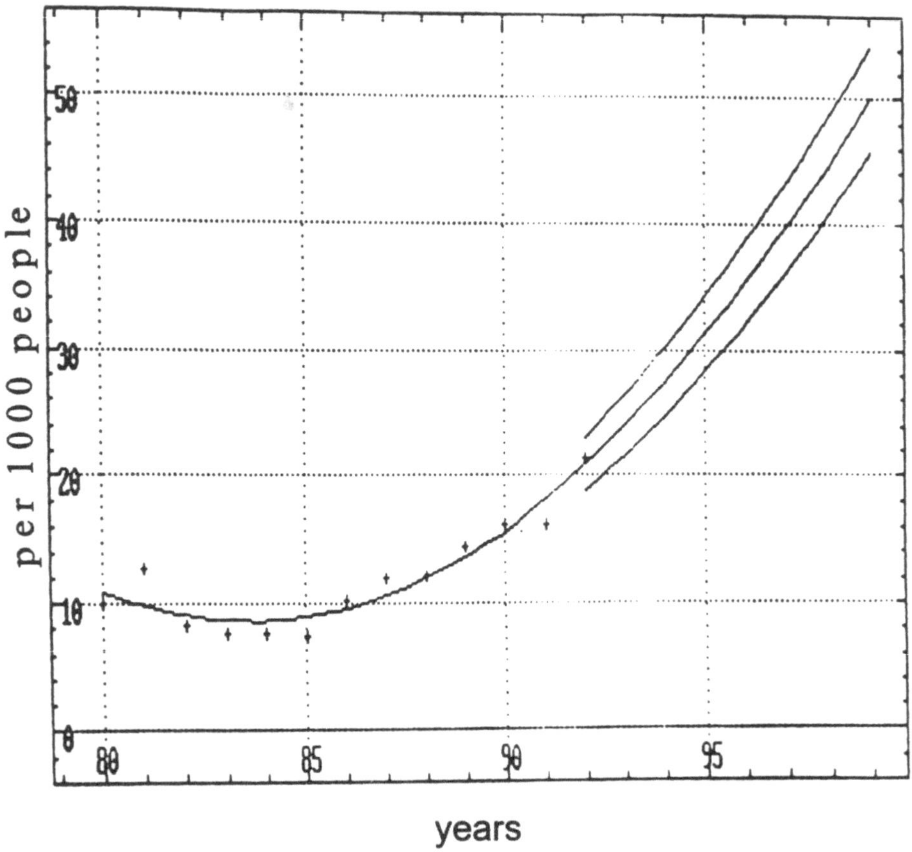

FIGURE 6. Prognosis of congenital anomaly in Magnitogorsk

EVALUATION OF ECOLOGICAL PATHOLOGY FOR THE TASKS OF MONITORING AND NATURE PROTECTION MANAGEMENT

N.V. ZAITZEVA, I.V. MAY, P.Z. SCHUR
Perm Scientific-Research Clinical Institute of Children's Ecological Pathology, 82 Ordgonikidze Street, Perm 614062, RUSSIA

The priority of man's life and health protection is declared in the Federal Law of Russia as one of the basic principles of the protection of nature. Emission and discharge reduction, reconstruction, architectural arrangement of industrial zones, urban housing, recreation territories etc., should be planned first of all for the efficient improvement of the environmental conditions. Thus each administrative level requires a certain data generalization for decision making, correct predictions and scientific criteria substantiation to protect nature efficiently and sufficiently.

The registration of pathologies resulting from the environmental changes for the purposes of environment monitoring and nature protection requires activities in various levels of the organizations. The basic principle is a step by step selection of priorities from the regional to the local level with gradual problem detailing.

At the regional level, the evaluation of ecological pathologies is made based on the population-level statistical data-base of population health indexes (from the state statistics and records from medical organizations), factors of environmental contamination (from reports from industrial companies) and mathematical modeling. Such approach allows the most efficient use of the information already accumulated. It enables us to determine the problem boundaries (considering all ecological, medical and demographic factors) and allows optimal selection of the health indexes. Figure 1 presents general evaluation methodology of the morbidity resulting from environmental changes for the environmental monitoring and management.

Predictive mathematical models for population health are also made on the regional level. The influence of environmental changes on the following factors is calculated: target rates for ecological pathologies, relative contribution of the specific chemicals, time delays in the ecological pathology development, maximum permissible loads (MPL), costs of damages). We used a multiple regression model to calculate the background pathology rates.

After selection of regional priorities for the chosen territories, a detailed site-specific analysis, and if necessary additional research is conducted. Therefore mathematical modeling and statistical analysis is supported by the site-specific data. on contamination sources, meteorological conditions and sampling data.

I. Linkov and R. Wilson (eds.), Air Pollution in the Ural Mountains, 251–260.

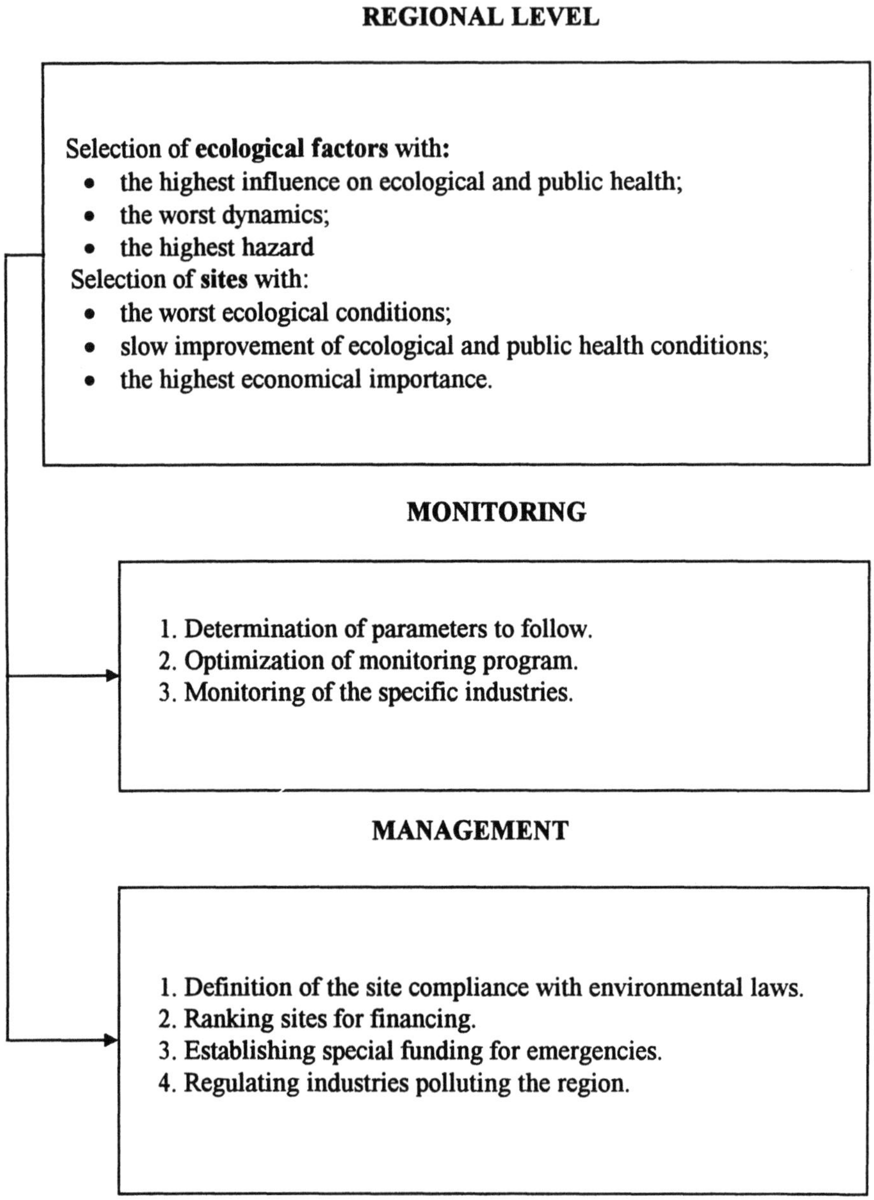

FIGURE 1. Ecological pathology assessment for the management tasks at the regional level.

TABLE 1. Proportional Input (%) of ecological factors on disease incidence for Children in several towns of the Perm Region

Town	Diseases of organs and systems									
	malignant tumors	hormonal system	blood	blood vascular system	respiratory system	reproductive system	skin	musculo-skeletal system	inherited anomalies	average ecopathology %
Alexanrovsk	44.0	41.0	62.0	19.0		16.0	50.0		34.0	38.0
Beresniki		36.0				40.0			6.0	20.0
Gremiatchinsk			66.0	49.0	25.0	35.0			40.0	43.0
Gubakha				32.0	27.0	65.0	52.0	32.0	5.0	31.0
Kisel		63.0	62.0	61.0		62.0	40.0	58.0	24.0	52.8
Krasnokamsk	19.0					21.0				19.2
Kungur			62.0	49.0	56.0	69.0	23.0			51.8
Lysva		42.0				50.0	36.0	58.0	19.0	41.0
Perm					38.0		54.0			46.0
Solikamsk						47.0				47.0
Tchaikovski		77.0	33.0	51.0		34.0	58.0	62.0		48.7
Tchusovoi						42.0			30.0	36.0
Average value	31.5	51.8	57.0	43.5	36.5	43.7	44.7	52.5	22.7	40.5

The site-specific analysis is supplemented by the data on chemical concentration in the environment combined with the information on morbidity registered in local clinics. It provides opportunities to generate priorities remediation of the environmental hazards and to develop preventive and rehabilitation measures for population.

The methodology of selection and evaluation of the antropogenic hazards is an important instrument for the regional and local management. The risk assessment methodology is used to define environmental concentration of hazardous chemicals, their pathway, internal doses absorbed by different population groups, comparative hazard of different chemicals. It is also used to establish the threshold levels of environmental contamination. The results of such analysis allow to direct the environmental protection programs to the primary source of environmental pollution responsible for the highest population risk. There are no doubt, that the most important pathways to be considered are ingestion (primary with water) and inhalation. Chemical readily absorbed by humans must be considered.

The methodology described above was implemented in Perm. Perm's region occupies large area and is divided in 49 units. Industrial facilities, utility companies, municipal services and transport emit more than 300 polluting substances into the atmosphere, including almost all heavy metals, organic chemicals (including aromatic amines, fluorine and halogen-containing substances). The total pollutant emission in 1996 was 672.11 thousand tons. Local plants discharged more than 490.0 million m^3 of waste waters, containing phenols, petroleum, synthetic surface active substances, etc.

Extremely non-homogeneous distribution of the industrial facilities in the area leads different antropogenic loads of the hazardous pollutants. A number of sites, have the greatest air emissions: Perm-Krasnokamsk and Beresniki-Solikamsk industrial centers, towns of Alexandrovsk, Gubakha, Gremjachinsk, Kizel, Kungur, Lysva, Tchaikovsky, Tchusovoi. In these sites, the annual load of polluting is about 327.0 kgs per capita. The loads for the Kizel coal mining, Beresnuaki, Solikamsk, Perm, Krasnokamsk are uncertain. All sites differs on the emission composition and contamination structure, especially for substances of first and second categories of danger. The pathways of pollutants are also different for each site. The worst water quality was registered in Kizel, Krasnokamsk, Alexsandrovsk, Gremjachinsk. Public health consequences differ also from site to site.

The essential part of the environmental protection is financing. To this end, it is necessary to define major directions of financial investment and resource flows maximizing social benefits and minimizing the costs. Evaluation of the public health consequences of the environmental contamination was made based on the developed system of parameters that reflect source terms, concentrations of harmful agents in water and air and their potencies.

Types of developing diseases dependent on the environmental conditions. Table 1 presents relative contribution of the environmental conditions in several towns of the Perm region to the disease frequency found by mathematical simulation.

Mathematical modeling was conducted to evaluate site-specific morbidity.

Diseases with statistically significant increase (index 10% and more, Student's criteria not less than 2.0) during 4-5 years were chosen. The database reveals continuos increase in the incidence of many diseases over the last 10 years (Figure 2).

All administrative units in the Perm region were classified in four groups according to the severity of environmental conditions and public health consequences. This classification was done using methodologies of geostatistical analysis. The first group includes cities and towns with a high level of air and water pollution and a high incidence of pathologies related to environmental contamination. Some examples are the city of Perm and the towns of Beresniki, Solikamsk, Krasnokamsk, Alexandrovsk and Kizel. These sites have severe problems on a global scale. The second group consists of seven sites with severe local ecological problems mainly related to a primary local polluter (Perm Power Station, cement factory "Gornozavodskzement", Nytva metallurgical facility, etc.) or to a group of smaller polluters. The third group consists of 17 sites with local problems, the fourth group consists of 7 relatively clean sites. For these four groups, the maximum permissible loads for contamination were calculated for different hazardous chemicals (Table 2).

TABLE 2. Maximum permissible loads for emissions in Krasnokamsk

Hazard Factor	Maximum permissible load (tons per year)	Average value of discharge (tons per year)
Sulfur dioxide	1,304.8	8,850.1
Hydrocarbon	1,578.0	1,789.2
Nitrogen dioxide	631.8	1,741.1
Carbon oxide	305.1	1,531.8
Hydrogen fluoride	0.006	0.024
Manganese	0.33	0.58
Ammonium	0.186	2.94
Acetone	1.17	2.42
Hydrogen sulfide	4.36	4.4
Formaldehyde	0.36	0.63
Hydrogen chloride	0.001	0.019
Methyl spirit	0.43	0.69
Chlorine	1.01	1.77

Extensive research on the history of contamination is required. Public health consequences were evaluated based on clinical records (clinical examinations, disease incidence, number of pregnancies, etc.) and state registry (birth rates) for industrial zones and administrative regions. Ecological consequences were assessed using data on sources of pollutants and their pathways as well as using information on water and food quality. This methodology provides an opportunity to classify pollution sources according to their impact on the site. An example of such a classification is presented in Table 3.

The largest polluters are enterprises: "Neftechimic" and "LUKoil-Permnefteorgsintez." Nevertheless, the contamination of the surrounding areas by

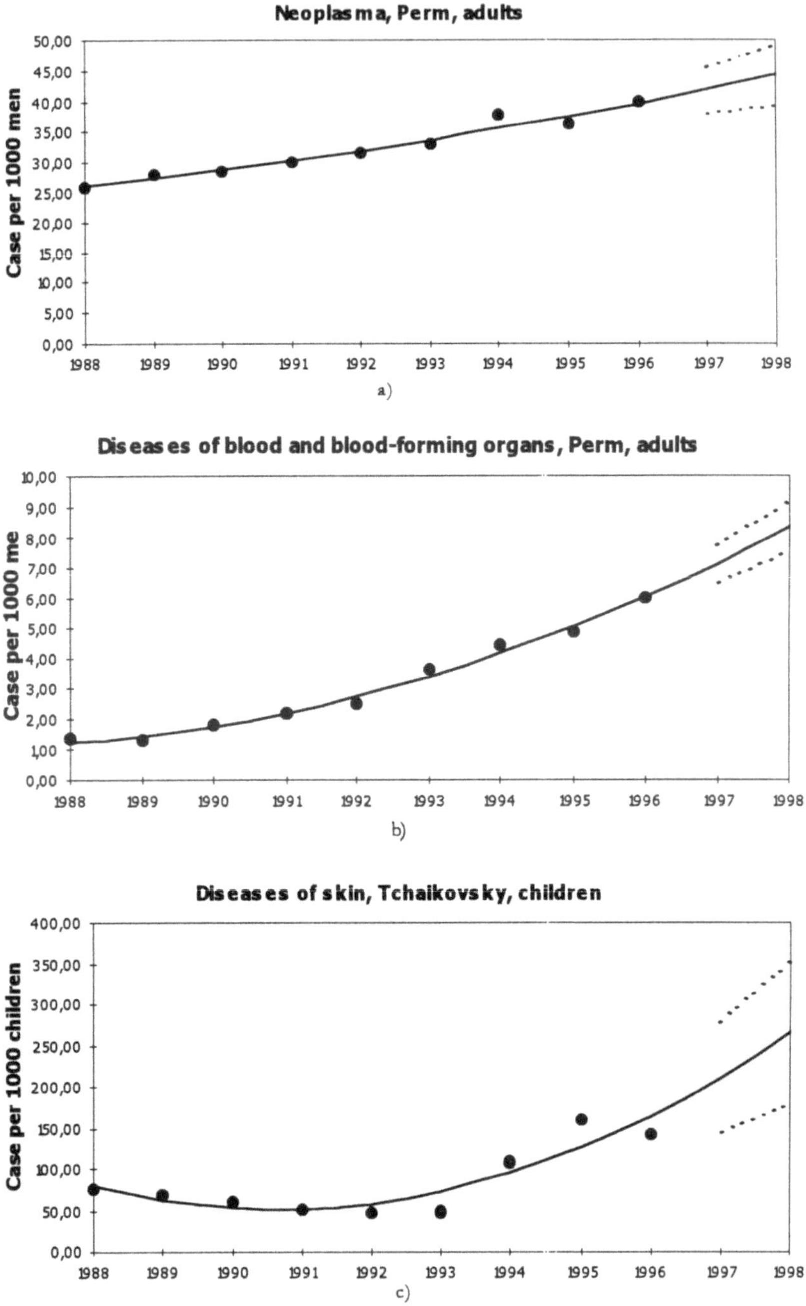

FIGURE 2. Forecast of morbidity for the Perm region

these enterprises is lower than the local contamination generated by the power station. Therefore, a power station is a significant hazard for the local population and requires similar attention to that for oil reprocessing plants.

The methodology of cluster analysis applied to air quality assessment in Perm city has provided an opportunity to select the most dangerous zones according to the factors contributing to the morbidity due to environmental contamination. For example, the highest manganese concentration in air was registered in the Motovilikha district of Perm.

TABLE 3. Comparison of environmental impact by different industrial facilities in Osentsy.

Name	Total Environmental Contamination (arbitrary units)	Site Contamination (arbitrary units)	Facility Contamination (arbitrary units)
"Neftechimic"	655.4992	30.8816	80.7056
"Nefteorgsyntes"	633.7344	41.8496	146.3152
Heat El.Station-9	490.8032	45.9424	30.1136
Fertil facility	140.3744	7.2064	17.3296
Gas-refining plan	101.4176	4.6688	21.8448
Airport	58.1840	10.9200	5.4928
Asphalt plant	5.3344	0.2320	2.1536
Lacquer-pain plant	3.4368	0.2592	0.6848
Tank-washer	2.6832	0.0304	0.8128
Metal-abrasive	1.3200	0.0848	0.3792
"Termosteps"	0.4912	0.0672	0.0496

The model calculations were confirmed by laboratory analysis of the air quality. Manganese was found to increase morbidity. Monitoring revealed significant inhomogeneity among the doses received by children, living in different zones (Table 4). Monitoring of environmental contamination combined with population bioassays (urine, bloodwork, etc.) can provide information for pollutant characterization and serve as an additional stimulus for environmental protection. It should be mentioned, that children are a critical population group for environmental decision-making and litigation. The hazard identification methodology provides an opportunity to define permissible loads of each chemical with complex pathways in heavily contaminated industrial zones.

Table 5 presents internal doses and risks (mg/kg per day) received by children in the "Industrial" district, city of Perm. Based on the analysis of available information, we conclude the following:

- there exists a high risk of exposure to zinc (up to 3.21), nitrates (36.22), cooper (5.97); a moderate hazard for phenol (1.33) and lead (1.24). Food chains are the most important pathways for nitrates, zinc and cooper. Air pathways are most important for phenol. Food chain and air pathways are relevant for the lead exposure.

TABLE 4. Concentration of manganese contamination in children for different districts of Perm city.

District	Concentration in Urine, mkg/l	Concentration in Hair mkg/g
Sverdlovsky	0.0547 ± 0.0185	4.15 ± 1.65
Dzerjinsky	0.0942 ± 0.0226	0.73 ± 0.22
Leninsky	0.0348 ± 0.0117	1.38 ± 1.88
Motovilikhinsky	0.1189 ± 0.0272	5.11 ± 1.65
Industrial	0.0390 ± 0.0171	2.12 ± 1.45
Ordjonikidzevsky	0.0580 ± 0.0320	3.22 ± 1.66
Kirovsky	0.0235 ± 0.0146	4.14 ± 2.57
Average for city	0.0560 ± 0.0365	2.68 ± 2.14

TABLE 5. Internal doses and risks (mg/kg per day) received by children in the "Industrial" district, city of Perm

	Chemical	Age groups, years			
		0-3		7-14	
		Dose	Risk	Dose	Risk
Inhalation	Lead	0.0008	0.1533	0.0004	0.0559
	Chromium	0.0028	0.2394	0.0015	0.2395
	Zinc	0.0014	0.0002	0.0008	0.0000
	Cooper	0.0011	0.0037	0.0006	0.0009
	Phenol	0.0383	1.3330	0.0212	1.3333
Water	Lead	0.0000	0.0000	0.0000	0.0000
	Chromium	0.0000	0.0000	0.0000	0.0000
	Zinc	0.0014	0.0000	0.0006	0.0000
	Cooper	0.0003	0.0002	0.0001	0.0000
	Phenol	0.0000	0.0000	0.0000	0.0000
	Nitrates	0.0095	0.0020	0.0044	0.0000
Alimenteral	Lead	0.0009	0.1719	0.0021	2.9833
	Chromium	0.0000	0.0000	0.0000	0.0000
	Zinc	0.2400	1.4476	0.5600	16.4576
	Cooper	0.0396	1.0557	0.0924	12.1975
	Phenol	0.0000	0.0000	0.0000	0.0000
	Nitrates	3.9060	4.0379	9.1140	45.5819
Total	Lead	0.0017	0.3028	0.0025	1.2439
	Chromium	0.0028	0.1229	0.0015	0.1336
	Zinc	0.2428	0.5280	0.5614	3.2132
	Cooper	0.0409	0.7826	0.0931	5.9700
	Phenol	0.0383	1.3324	0.0212	1.3326
	Nitrates	3.9155	3.6338	9.1184	36.2163

- because air contamination is most important for phenol exposure, it is necessary to monitor phenol concentration in the atmosphere and to control phenol emission;

- 7-14 years old children are the most susceptible for lead, zinc, cooper, and nitrates, while chromium, cadmium and nickel are of significant danger for children less than 1 year old. Risks associated with phenol are equal for all age groups;

- the maximum permissible concentration of phenol in atmosphere is recommended to be 0.001125 mg/m^3. This value was obtained by mathematical modeling for received doses and risks. It is unnecessary to regulate phenol in water.

In summary, assessment of public health is useful for policy decisions on taxation of industries, insurance policies, environmental litigation, etc. The proposed methodology that includes evaluation of the public health consequences of environmental pollution at different levels of details (region, city, city district, industrial zones), geostatistical analysis, predictive modeling and quantification of changes in public health can be used to facilitate monitoring and remedial tasks such as:

- optimal allocation of monitoring station
- initiation of monitoring programs
- selection of environmental protection policies
- monitoring of the doses received by population
- location of the potentially dangerous sites
- zone classification and control of their usage
- initiation of remedial and abatement programs in contaminated zones
- improvement of threshold values for emission of hazardous pollutants
- improvement of environmental audits, taxation and other environmental protection activities by the local and federal governments.

ECOPATHOLOGY OF MERCURY POLLUTION

L.M. KARAMOVA and R.A. ALAKAYEVA
Ufa Research Institute of Occupational Health and Human Ecology, Ufa, RUSSIA

Environmental pollution by heavy metals is crucial for the Republic of Bashkortostan, South Urals, because of the numerous mining and processing industries. In particular the factories of the gold-mining industry contaminate the environment with mercury that is a first class toxicity agent. Mercury was found to lead to neurotoxic effects, it disrupts endocrine system and depresses the immune system. Mercury reduces the non-specific resistance of an organism which results in an increased disease level. Mercury was found to have carcinogenic properties. Environmental contamination by mercury can be dangerous because of its bioaccumulation in the environmental compartments.

We have studied residents of a settlement in the immediate vicinity of a gold-extraction factory. Mercury vapors are produced by chemical treatment of the ore and are released in the air. Three generations of workers have received occupational exposures in the factory. High levels of mercury (12-23 times the maximum permissible concentrations) were detected in the water supplies of the settlement. Therefore, environmental exposure is also significant.

The population of the settlement was screened by a team of physicians that included toxicologists, pediatricians, internal medicine specialists, ophthalmologists and dentists. As the result, 116 people were take to a clinic for detailed examination. Mercury concentration was measured by atomic-absorption methods in blood, nails, hair, teeth and urine. Blood samples were also taken and tested for cholesterol, β-lipoprotein, blood transaminase, LDG, GGT, POL and SH groups. Liver and kidney functions were studied by isotope studies and X-rays. Ultrasound was used to study internal organs. Encephalography as well as cerebral and peripheral hemodynamics were also studied.

The main complaints included headaches, fatigue, erethism, general weakness, insomnia, vertigo, hand trembling, palpitation, anxiety sense, stomatorrhagia, hypersalivation. As viewed from the nervous system the changes manifested a picture similar to a neurosis state, more often as asthenic, astheno-vegetative syndrome, neurocirculatory distonia coming to a manifestation of toxic encephalopathy. In the course of examination special attention was paid to fine symmetrical tremor of eyelids, finger-tips, more seldom of head, symmetrical revivication of knee, Achilles reflexes, widening their reflexogenic zone. Vegeto-

I. Linkov and R. Wilson (eds.), Air Pollution in the Ural Mountains, 261–262.

vascular disturbances were exhibited as hyperhidrosis, "vasomotor play" of face, acrohypotermia often in conjunction with distal hypesthesia of polyneuritic type, red steady deffuse dermographism. Practically all the settlers revealed a parodontium lesion, and mucous membrane diseases in oral cavity.

On the side of gastrointestinal tract and hepatobiliary system the changes revealed as gastritis, erosive gastroduodenitis, dyskinesia of biliferous tracts. Increased transferases activity was noticed in evaluating liver functional stste. Ananlyses for mercury content in biomedia were made by the atomic-absorption method at the Institute test centre. All the examined people showed a high mercury content in blood, urine, 32 people hair, nails. Permissible level, recommended by WHO (1990) - 50 μg/l, was exceeded in 57% of adults and 31% of children. Of particular attention was the fact that mercury content in the children'blood was 1.9 times higher than that in adults.

70% of the people examined revealed mercury content in urine, in 12% it was within the limits of the critical level. 20 factory workers showed chronic intoxication by a complex of toxic substances such as mercury, cyanides, sulfur compounds. Clinically it was revealed as nervous and hepatobilair system disorder and parodontium. All this provided support for the fact of a mercury load on organism of settlers living in the zone of technogenic pollution by factory wastes.

In conclusion, gold-extracting factories seriously contaminate soil, air, water, and food products. They pose a high risk for population health. Physical development was impaired, there was an increased frequency of acute and chronic diseases, and a disturbance of nervous system function in the population living in the vicinity of the gold-extracting factory.

PERSPECTIVES ON UNITED STATES METHODS OF HUMAN AND ENVIRONMENTAL RISK ASSESSMENT

J.C. WILLSON
Kleinfelder, Inc.
5953 South Iola WayEnglewood, CO 80111, USA

1. Introduction

The intention of this paper is to highlight the parts of United States risk assessment methods that are valuable worldwide, and to point out how the methods can be misused at great and unnecessary cost. Industries in many countries are essential to the economy, but they operate on profit margins that make it difficult to spend money on protecting the environment. Blind use of US risk assessment methods can result in wasting money, but careful use can reduce the cost of identifying the real sources of health effects.

The *question* addressed by risk assessment can be summed up very simply: "*We know the environment is dirty, but is it too dirty? How clean is clean?*" The answers, unfortunately, are proving hard to come by. Twenty years ago, we Americans thought we could just clean up every bit of every chemical that had been released by agriculture and industry. Ten years ago chemistry had gotten very good but technology was still bad at removing the chemicals that the laboratories could detect. Cleanups were found to be far, far more expensive than expected. Five years ago, real people who were not sick found that because of environmental remediation their homes were torn down, their yards were torn up, and their lives were torn apart when the factories closed and their jobs were gone - all in the name of cleaning up pollution, in this specific case lead. During this time, methods for assessing risk and guessing how clean is clean were developed but never validated.

There are two main approaches to evaluating human health and environmental risk: the hypothetical approach and the epidemiological approach. (A third approach, experimentation, is not appropriate with humans, nor with valuable ecological resources such as endangered species, but information from animal experiments can be useful.) Ideally, each would be used to provide a check on the other. What is called "risk assessment" is the unvalidated, hypothetical approach. In the US, for the most part we use the hypothetical approach to recommend cleanup of hazardous waste sites. This results in a lot of wasted money, only because the hypothetical approach used now is a very conservative, protective approach. For example, it results in claiming that levels of arsenic, antimony, beryllium and manganese that are below some typical natural background levels in soil or water can cause a health risk and must be cleaned up. No nation in the world can afford to "clean" almost all of its soil. No nation in the world is

I. Linkov and R. Wilson (eds.), Air Pollution in the Ural Mountains, 263–275.

rich enough to "clean" soil, water, or air that does not pose a real health risk. It is true that even 20 or 30 years ago; the US had some serious chemical problems. It is true that one of our rivers (the Calumet in Ohio) had so many organic chemicals in it that it was able to burn. Cleaning up the river did not require this hypothetical approach to risk assessment. Many of the problems in Russia don't require a hypothetical approach; it is obvious what should be cleaned up. In other situations, it is not clear whether the presence of chemicals is the cause of observed health effects.

The US is now struggling to find a balance between getting the environment too clean and not protecting enough. Risk assessment as it is currently practiced violates many rules of good science. It is a start, but not enough. It is only one part of a three-part "triangulation" approach to building solid evidence. It is really only a model that helps develop hypotheses about chemical risk:

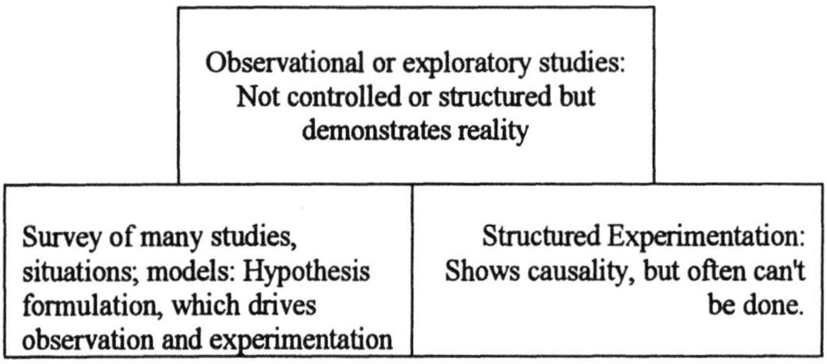

"Triangulation" is a process of making measurements from many different angles or vantage points. With many studies, you only see part of a picture, like the blue or yellow or red printer's separation of a full-color printed picture. Triangulation brings together evidence so that you see the full picture. In environmental assessment, triangulation should be used instead of just doing risk calculations <u>or</u> studies of health effects. The different types of studies complement each other; the study of a chemical or exposure situation should include as many different types of studies as possible: Laboratory experiments, relevant environmental measurements, epidemiology studies of many kinds, surveys of other exposures, hypotheses based on model predictions. Without triangulation, it is difficult or impossible to show a strong link between a chemical cause and a health effect.

In May, 1997, the Internet press reported on a terrible situation in Krasnouralsk, home of the Krasnouralsk Copper Works. Residents and doctors all observed that everyone was sick with many types of diseases, especially the children; 75% of the children showed signs of being mentally retarded. Exposure to lead was claimed by some people to be the cause of the illness, especially the mental retardation. However, the children had an average blood lead of 13 μg/dl; 86% of the levels were below 20 μg/dl. Lead at those levels could not possibly cause measurable retardation in individual children. These levels of lead were typical nationwide in the United States just 15 to 20 years ago, and certainly 75% of children in the US did not show signs of mental retardation. Levels of 40 to 100 μg/dl would be required to see effects on intelligence due to lead. In this case, other chemicals, poor nutrition, and other factors are responsible for observable effects. Soil lead levels at Krasnouralsk were not specified, but they were said to be "high." We have studied both soil levels and lead levels at Leadville, Colorado, where the soil lead levels are also "high" (up to 11,000 ppm) but the blood lead levels are mostly low. In other words, lead in soil at mining sites has only a small and weak relationship to lead in childrens' blood. It would probably be far more effective to reduce air emissions in Krasnouralsk rather than to clean up soil. But first, the situation should be studied to clearly identify the real sources of health effects so that money can be spent most efficiently.

Without additional information about this town other than the newspaper article and general information about copper smelters, a two-part approach appears appropriate:

1. Conduct a brief risk assessment according to US EPA guidelines, which will provide hypotheses about which chemicals or other sources are potentially causing harm, and then
2. Conduct an epidemiological study to see if there is a causal link between potential identified sources and illness in the children (but see comments below about drawbacks of epidemiology).

The steps for the risk assessment are as follows:

1. Prepare a conceptual site model to identify potential chemicals, sources, pathways, and receptors. A copper smelter is likely to produce sulfuric acid, copper, lead, arsenic, and other metals, depending on the source ore.
2. Sample media (air, water, soil, food, others) on those pathways to estimate chemical concentrations that could reach the children. Sample reasonable variation, such as "the winds that blow the blue smoke to the school."
3. Identify known screening toxicity values for chemicals that are found
4. Estimate which chemicals could be contacting the children in quantities that COULD hypothetically pose a risk using estimated exposure and toxicity values. Only chemicals that may be providing a dose that is 10 to 100 or more times the EPA's estimated safe dose should be considered to pose a potential observable risk.
5. Note that cigarette smoking gives Americans a far higher chemical exposure than most hazardous waste sites and causes far more health effects than environmental chemicals do. (What is the situation in Russia?)

The focus for the epidemiology study would be as follows:

1. Prepare by making local inquiries about potential sources of pollution, and NON-environmental-chemical causes of health effects, including mycotoxins from mold, allergens from plants or insects, and genetics.
2. Identify potential confounders (such as cigarette smoking or other chemical sources) and eliminate them if possible. Measuring and controlling for them is a distant second choice.
3. Conduct a health evaluation for each child, including biological sampling and medical effects evaluation with a focus on effects that are relevant to identified chemicals
4. Evaluate the non-chemical aspects of the environment, including nutrition and parental and medical care, parental occupation, reliance on home-grown food, hygiene (hand-washing), and particularly smoking habits of the parents or caregivers.

5. Conduct environmental sampling relevant to the child, including home, school, and day care water, soil, air, and typical food sources. This information should be matched to each child, so that the specific concentrations of specific chemicals in THAT child's environment are known.
6. Avoid using surrogate measures of exposure, such as "distance of house from smelter," which introduce large measurement errors.
7. Conduct a statistical analysis of the data.
8. Now put the entire story together
   - Does actual exposure occur?
   - Does the response in children increase as the dose increases?
   - Are other conditions needed to prove causation met? (Hill's other attributes, or correlates to Koch's postulates- strength, consistency, and specificity of the association, timing, biological plausibility, coherence of the data)

Epidemiology has several great drawbacks, however. It is not sensitive to very low incidences of health effects or very small changes in health. (However, in some societies, if a risk is not observable, it may not be considered significant. The significance of very small risks is a social judgement, not a scientific one.) It can be very expensive. It requires that exposure has already occurred, and it cannot predict the future. It is hoped that a few good epidemiology studies may provide enough information to make decisions about remediating many different cities and sites, thus justifying the cost and permitting some reasonable predictions into the future.

2. Explanation of the Risk Assessment Process

The risk assessment process used in the US has five main parts that form a logical progression based on what is known about the relationships between toxicity and exposure. The process is designed to use specific conditions at a site to evaluate risk. This can also be thought of as a framework to evaluate specific conditions in different countries to risk. The parts of the process are:

1. Data analysis and problem formulation using a preliminary conceptual site model
2. Evaluate exposure using a refined conceptual site model
3. Evaluate toxicity
4. Calculate potential risk
5. Evaluate uncertainties

Fortunately for all of us, the mere presence of a chemical anywhere is not enough to cause a risk. Enough of it must (1) move to and (2) contact someone (a receptor) before there is a risk. Actually we can be more specific than that about the requirements for significant exposure that might indicate complete exposure pathways from a source to a receptor, for example via air:

1. A source must exist: an incinerator or ventilation stack, an evaporation pond, fugitive (non-point) emissions from an industrial facility, or any other significant source of chemical that is open to the air. A secondary source might be water in a home that releases aerosols when used for showering, cooking, flushing toilets, watering a vegetable garden, and so forth.

2. A release mechanism is required. For air, look for (1) volatilization, (2) wind release of particulates from contaminated soils, (3) emission through ventilation of stacks, etc., and (4) negative pressure that develops inside basements that sucks in volatile chemicals, radon, etc. from soils or groundwater-derived vapors surrounding building foundations.
3. A transport mechanism may be required if potentially exposed people are located away from the release point. For air, "transport mechanism" means "wind."
4. The obvious exposure medium is air, but consider also: (1) Deposition of particulate matter on outdoor soils that may be eaten directly, tracked inside and eaten as "incidentally ingested" house dust, or absorbed by including garden vegetables, which may be eaten; (2) Deposition in indoor house dust; and (3) Attachment to dust particles, which are then readily deposited in the lungs; this is the mechanism for radon exposure. Other potential routes to exposure media are possible.
5. An exposure point is required. The amount of chemical that actually reaches a person or an ecological receptor is the amount that is significant, not the amount that is emitted from the source. Transport models or monitoring stations should take this into account. Air measurements should be made in the breathing zone (3 to 6 feet off the ground), not at the ground from a flux chamber or on a telephone pole. Also note that direct measurements made away from a source are likely to measure other sources as well. We have found that measured cadmium and other metals may have come from domestic wood burning and not from mining sources, for example.
6. Receptors must be present, now or in the future. Is that downwind cabin a year-round residence or just a summer home? Did the transport model predict concentrations at the housing development or in the middle of a fallow field? If the only possible receptors are maintenance people or occasional visitors, what is their expected exposure frequency? Risk assessors generally add here that there must also be a route of exposure — oral, inhalation, or dermal. Since everyone has to breathe, but not everyone has to pump and drink the groundwater, this is not much of an issue for air exposure. There is an additional route of exposure that may be important for airborne chemicals, and that is eye exposure that may result in significant irritation and tearing.

3. Indirect Exposure Pathways

In the early 1980's, the infancy of environmental risk assessment, it was deemed sufficient to make simple assumptions about daily intake of water, soil, and air by adults living full time with chemicals in those media; simple calculations of chemical intake were then made, and risk was calculated. As risk assessment grew up, it became apparent that children were not tiny adults, that exposure rates could vary widely, and that chemicals moved from one medium to another; these things needed to be accounted for. In the field of air toxics, additional pathways became known as "Indirect Pathways." Guidance for evaluating indirect pathways has been formalized under some programs; see for example guidance for hazardous waste and other combustion facilities [1-3]. This and similar federal and state guidance show how a full and complex set of exposure pathways can be evaluated. Such pathways include deposition on plant leaves; deposition on soil, root uptake and translocation; uptake by cattle and accumulation in beef and milk; and so forth. The approach is conservative and protective, but the resulting risk models have not been validated. For example, measurements may show that the modeled concentrations may not actually exist. There is a real danger of multiplying many conservative decisions regarding pathways and exposure assumptions together and producing an unrealistically high estimate of risk. Some people argue, however, that such estimates are appropriate to protect people fully in the absence of data. The models provide a good starting point for understanding the fate of air toxics. In many cases, it will be very worth while to gather supplementary information to refine the risk estimates. It may also be worth using quantitative uncertainty analysis methods (some examples include probabilistic or Monte Carlo analysis, or fuzzy logic analysis).

4. Developing data quality objectives for an investigation and risk assessment

The conceptual site model is really a collection of hypotheses about what could be happening to chemicals at a site or in a situation. An investigation leading to a risk assessment is an evaluation of these hypotheses. The best (some would say only) way to evaluate hypotheses is to use the scientific method; the first steps are these: Ask your questions, design your investigation to answer the questions, and check to see that your investigation will really answer the questions you originally asked. Some version of this process has been formalized in many fields (e.g. economics, psychology) under different names. In environmental investigations, it is called the Data Quality Objectives (DQO) process. EPA has created some good guidance describing the process. Often, however, people think that the DQO process is nothing more than getting a high enough count of soil or air samples and a low enough detection limit ("Gee, 5 nanograms per microgram sounds low enough to me!"). It is much more than that: it is a project manager's *most powerful tool* to demonstrate to *senior management, agency personnel, and the public* that the environmental project is doing what it should. The process also documents decisions that are made, so that if project personnel change, or the project is so long that at the end no one can remember the beginning, it is less likely that previously settled matters will be reversed or challenged.

EPA has proposed a three-step, and more recently a seven-step, process for developing data quality objectives. The original three steps can be restated as questions:

1. Exactly what question are you trying to answer; what decision are you trying to make?
2. What do you need know or learn to answer that question or make that decision?
3. What data collection and study design will provide the needed information?

The seven-step process is laid out in massive detail in the EPA guidance already mentioned. We'll outline the process here with the objective of demonstrating its value and scope.

1. State the problem. Summarize the contamination problem that will require new environmental data, and identify the resources available to resolve the problem. Example: An old industrial site is found to have specific chemicals in its soil; the weather is dry and the area is dusty.
2. Identify (define) the decision. Write down the decisions that need to be made and that require new environmental data to address the problem. Example: Determine whether chemicals released by wind from a bare site pose a risk to nearby residents. Or, determine whether driving trucks on the site will raise dust that could pose a risk.
3. Identify inputs to the decision. Identify the information (new measurements or old, existing information) needed to support the decision. Specify which inputs need new environmental measurements. For the above examples, measurements in the residential area or measurements during typical truck usage may be appropriate.
4. Define the study boundaries. Specify the spatial and temporal aspects of the environmental media or potential exposure that will bear on the decision. Example: weather patterns over a year may result in different emission rates that should be factored in to realistic estimates of health risk.
5. Develop a decision rule. Develop a logical "If_ then_" statement that defines the conditions that would cause the decision-maker to choose among alternative actions. These rules often take the form of "If measured levels do not exceed calculated levels then a "No further action" alternative is appropriate; otherwise, conduct additional evaluation." Smaller decisions that are part of the larger ones are also laid out in this step. Figure 2 provides an example flow chart with potential decisions.
6. Specify limits on decision errors. Specify the decision-maker's acceptable limits on decision errors, which are used to establish performance goals for limiting uncertainty in the data. This is the step where sampling protocols, detection limits, statistical power calculations, laboratory performance requirements, and other performance goals and specific "Data Quality Objectives" come in. (However, effective and appropriate DQOs can only be established in the context of the rest of this process.)
7. Optimize the design for obtaining data. Identify the most resource-effective sampling and analysis design for generating data that are expected to satisfy the DQOs.

Project managers who are reviewing a study planned using the Data Quality Objectives process should consider the following questions. They are harder than they look, because they all ask, "What is missing?"

1. Do all proposed data, tests, analyses, etc. address an identified need or decision? (If not, do not collect the data.)
2. Have all of the significant questions and decisions been identified?
3. Will the selected inputs answer the questions? If not, what will it take to answer them?
4. What uncertainties in the risk assessment make the assessment too conservative and could be reduced with additional data—are they addressed?
5. What are the weaknesses of the study from a logical standpoint? Can they be reduced in the design phase?

Data collected through the use of Data Quality Objectives should be a useful and economical way to evaluate exposure.

5. How toxic are the chemicals? Occupational exposure and risk assessment

Designing a site evaluation using a conceptual site model and data quality objectives is only part of the risk assessment process. The toxicity evaluation portion of the process is critical, and is currently undergoing revision in the US. Current toxicity information provided by the EPA should be looked at as valuable but hypothetical screening information that may be helpful but not completely accurate, especially in other countries. Much of the toxicity information used in risk assessment in the US comes from animals, not humans. Mercury, arsenic and lead are among the few exceptions. The Mad Hatter in *Alice and Wonderland* was known to be exposed to mercury.

The truth is that we have lots of information from actual inhalation human exposure about the toxicity of many significant chemicals. The bad news is, in the US we haven't use much of it for public health risk assessment. This is because the exposed humans were workers exposed on the job, and EPA has assumed that worker exposure is not applicable to the general public. Because they are generally covered medical monitoring programs and guarded by the watchful eye of industrial hygienists and physicians, and because they theoretically work with chemicals willingly, the thinking goes that they can be exposed to more chemicals in larger quantities and thus be at greater risk than the general public. Furthermore, workers are often healthier or less naturally susceptible than the general public. It's assumed that workers can safely be exposed to more. Yet another reason that worker information hasn't been used extensively is that even though hundreds of workers may receive exposures, the exposure may not be high enough or the number exposed may not be large enough to determine whether chemicals cause cancer.

The counter argument is that workers believe that they are working in safe conditions in the US in the 1990s; they don't expect to incur greater risk than their family at home. Workers cannot avoid breathing chemicals that are widespread in workplace air even if they are trained in safety measures. While training may allow people to reduce the amount of chemicals that they ingest accidentally, they can't help breathing whatever is in the air where they work. In any case, information derived from occupational exposures is not often used in evaluating toxicity and risk of chemicals typically found at hazardous waste sites or emitted from industrial facilities. There is still a great deal to be learned about chemical and radiation exposure from ongoing human exposures, both occupational and public.

6. Describing toxicity: Reference concentrations and unit risk

In the past 10 years, the US EPA has developed toxicity values[1], which it generally requires to be used in Superfund risk assessments; these values have been almost universally adopted for other risk assessment uses, including RCRA evaluations and state-led risk assessments. (Any applicable state regulations or guidance should always be checked.) The toxicity values are developed directly from human data in a very few cases; more commonly they are derived from a small number of animal studies. The general process is to identify an experimental dose that causes either no effect, or the smallest degree of effect, and then to divide that dose by one or more safety factors (typically between 10 and 10,000) to derive a dose that is probably safe for human consumption. This dose is called a "Reference Dose" or "RfD." Chemicals that cause cancer in experimental animals are subjected to a different modeling process to estimate the risk of cancer associated with ingestion of a milligram per day[2] of the substance. This is called a "slope factor."

The values are updated frequently, so current sources of toxicity information must always be consulted. Values presented here are examples only. The values are generally not derived from occupational experience, but from controlled animal studies, or accidental exposures to the general public. Specific derivations of these values are described in IRIS, and this information should be consulted if it is suspected that a certain chemical may pose a risk. It's important to know whether the identified toxicity value is very near a level that may cause harm in humans, such as the one for mercury, or if little information is available in humans and information from animals was used along with a safety factor (generally 100 to 10,000) to derive a toxicity value.

However, risks are described a bit differently for exposure to chemicals by the inhalation route (as compared to the oral route). For oral exposure, a specific dose (in milligrams of chemical per kilogram of person) or a cancer slope factor is given instead of a concentration in water or soil. For inhalation exposure, a concentration or unit risk value is often given, instead of or in addition to inhalation reference dose and cancer slope factor, partly because the toxicity information is collected in terms of exposure concentration rather than a measured dose relative to body weight. The concentration provided is expected to present a hazard index of 1.0, and the unit risk is the cancer risk per milligram of chemical per cubic meter of air — both assuming full time exposure.

Effect	*Oral route*	*Inhalation route*
Non-Cancer	Reference Dose	Reference Concentration
Cancer	Slope Factor	Unit Risk

[1] The values are published in IRIS, the EPA's Integrated Risk Information System, available through several data base services, and available on CD through Government Institutes. Other values are published in HEAST, the Health Effects Assessment Summary Tables (9200.6-303; EPA 540-R-94-020; call EPA for most current edition available.)

[2] every day for a lifetime

EPA defines unit risk as "The upper bound excess lifetime cancer risk from continuous exposure to an agent at a concentration of 1 $\mu g/m^3$ in air." However, in the author's opinion, the only biologically active chemical that any human receives a continuous lifetime inhalation exposure to is oxygen. The unit risk concept is by definition overly conservative and automatically overestimates risk both by using upper bound toxicity estimates and upper bound exposure estimates; it should be modified to reflect reduced exposure time, and the conservativeness in the toxicity component must be revealed to risk managers and to the public. Because the unit risk factor and the reference concentration roll in exposure assumptions with toxicity information, the difference between a unit risk factor and a slope factor can be explained in terms of the risk equation as follows:

The risk equation: **RISK = Concentration x Toxicity x (Exposure Frequency)**
For non-inhalation exposures:
RISK = Concentration x Slope Factor x (Exposure Frequency)
For inhalation exposures:
RISK = Concentration x Unit Risk

(Equations for Reference Doses and Reference Concentrations can be written in the same way.)

Doses may also be provided by EPA, although its current policy is to provide only unit risk values. Doses may be converted to exposure concentrations, and vice versa. Note in the examples given of toxicity values that the most carcinogenic chemicals have high slope factors (think of this as highest risk per milligram of exposure) and the most toxic gases or vapors have the lowest references dose or RfD (think of the reference dose as the lowest acceptable exposure dose). So arsine gas is considered to be 1,000 times more toxic than benzene, which we still breathe from gasoline fumes, and TCDD (the most potent of the dioxins, but not clearly a human carcinogen at low doses) is treated as though it is 10,000 times as carcinogenic as arsenic, which is known to be a human carcinogen. The toxicity values therefore are not proportional to the strength of our knowledge about the effects of the chemical or the likelihood of exposure.

To use the unit risk factors or reference concentrations, multiply the appropriate value times the exposure concentration. For example, if there are 10 $\mu g/m^3$ of benzene in the air to which someone is exposed full time, the estimated cancer risk for lifetime exposure using this approach is 8.3×10^{-5} or about 8 in 100,000. To use the inhalation slope factors or reference doses, some exposure assumptions are necessary: An adult is assumed to breathe in — and absorb the chemicals from — 20 cubic meters per day. That is about the volume of air in a 9 x 10 foot, or 3 x 3 meter, bedroom or den. Unless there is other information, they are assumed to absorb all the chemical of interest from that air, even though they only absorb a fraction of the oxygen in it (otherwise we couldn't do artificial respiration!). Adults are assumed to weigh 70 kg and live 70 years, as for other risk calculations. These assumptions are then used in the following equation, or a similar version that may include more information:

TABLE 1. Examples of EPA inhalation toxicity values (E-notation is used: E-2 = 10^{-2}).

	Unit risk (cancer) concentration (per μg/m3)	Inhalation slope factor (kg-d/mg)	Reference dose (mg/kg/day)
Arsenic	4.3E-3	1.5E+1	--
Arsine	--	--	1.4E-5
Benzene[3]	8.3E-6	2.9E-2	1.7E-3
Formaldehyde	1.3E-5	4.6E-2	--
Carbon Tetrachloride	1.5E-5	5.3E-2	5.71E-4
TCDD ("Dioxin")	--	1.2E+5	--

General Risk Equation

Intake = (C x IR x EF x E) / (BW x AT)
Cancer Risk = Intake x CSF
Hazard Quotient = Intake / RfD

Where:

C = Concentration (in air or other medium)
IR = Inhalation Rate, or intake rate for other medium
EF = Exposure frequency, in days/year
ED = Exposure duration, in years
BW = Body weight, in kg
AT = Averaging time, in days (lifetime for cancer risk, equal to ED for non-cancer hazard quotients)
CSF = Cancer slope factor
RfD = Reference dose.

So, for example, the calculation for risk from chemicals in water would be the same except that the inhalation rate of 20 cu.m/day would be replaced by the water ingestion rate of 2 liters per day. Although many people argue that most people do not drink that much water, in fact that is the minimum widely recommended for good health in the United States.

That's all there is to doing a single pathway, single chemical risk calculation. The model is simple, linear, and easy to compute. It may or may not be accurate, but it

[3] For many chemicals that are used in large quantities in the US, our experience or practice with occupational or consumer exposure is at odds with the conservative risk assessment approach used by EPA. Benzene is an example. According to the unit risk value, continuous exposure to just over 10 μg/m^3 would pose an estimated upper bound risk of 10^{-4}, the high end of the acceptable risk range under Superfund. However, the occupational limit is currently 10 ppm, or about 30 mg/m^3—3,000 times higher. Even adjusting for the length of the work week versus a full, 168 hour 7 day week, and even acknowledging that the relationship between exposure and cancer risk is not linear at high doses, a discrepancy remains. This has provoked many new studies and reevaluations in recent years. New findings are provoking major evolution in how cancer risk is evaluated, the results of which are partially reflected in the EPA's 1996 Cancer Risk Guidelines.

is accepted, conservative, and widely used. If there are multiple pathways or multiple chemicals, the risks from each should be summed unless there is reason not to do so.

However, the results of such risk calculations are conservative. Possible solutions include these:

1. Use better exposure, toxicity information
2. Triangulate, using epidemiological studies
3. Use quantitative uncertainty analysis

7. Ecological Risk Assessment

Ecological risk assessment takes a different approach from human health risk assessment in that a variety of approaches may be used to identify actual or potential risk, but the emphasis may be on actual risk, depending on the risk managers' goals. In fact, ecological risk assessment follows the approach recommended above for Krasnouralsk. It is true that many people have conducted what they call ecological risk assessments using the "Quotient Method" in which the exposure dose is divided by the safe dose, and if the quotient is less than one, then the chemical exposure is considered safe. There is nothing wrong with this approach if the answer is "Safe." However, if the quotient indicates a problem, then the other parts of the ecological approach should be conducted to determine whether the predicted risk is actual (even if it is a future risk) or only hypothetical. The current approach follows the scientific methods, including the triangulation approach where appropriate.

1. Interact with the risk managers (learn objectives, needs, available resources)
2. Problem formulation (collect information, prepare a conceptual site model, and state your problem)
3. Analysis: Characterize Exposure, Characterize Ecological Effects (Design studies that will answer your questions; collect information; conduct experiments if necessary)
4. Risk Characterization (Evaluate the data, look for causation, identify the sources of the effects, and recommend safe exposure levels. Describe uncertainty in the assessment.)
5. Interact with the risk managers (again).

The ecological risk assessment process in the US has focussed much more on rational decision making than the human health process has. The EPA and Risk Assessment Forum documentation about the process is well worth studying and applying to human problems as appropriate. A limitation, of course, is that we can never intentionally experiment with humans and expose them to potentially dangerous chemicals, as we can in some cases with ecological situations. However, the unfortunate natural experiments in which people have already been exposed can be carefully observed and used to prevent further exposure if exposure and effects are scientifically documented.

8. Conclusions

The goal of risk assessment is to help those who must decide on what to do about remediating a site, controlling emissions, or other environmental releases in the past, present, or future. It is these decision makers, and the society in which they live, who will decide what health risk is acceptable; the risk assessment only shows what the risk might be. If the risk assessment is conservative and protective, that should be carefully explained. A risk assessment does not determine whether a situation is "safe" or "dangerous"; the risk managers do that. A good risk assessment will present enough information so that the decision makers, or risk managers, will understand the meaning of the risk assessment and will make reasonable decisions that are appropriate for the situation and its the economic and social context.

However, that is often not the case, especially when the risk information needs to be provided to a country other than the United States. The United States has been accused of "environmental imperialism" - that is, spreading our environmental agenda like an empire throughout the world, and forcing our risk management policies on other countries who can't afford them. When this happens (and it does) it is sometimes because fearful people with good intentions blindly believe distorted science, but also often because policy makers impose their values as well as their science on other people. In the United States, enormous progress has been made in cleaning up environmental releases of chemicals, and in preventing future releases. The progress has been made at great emotional and financial cost. The emotional cost is incurred when people become unnecessarily fearful of safe levels of chemicals. It is the job of environmental scientists to find a balance between the tendency to alarmist false science and environmental imperialism on one side, and release of truly dangerous and destructive chemicals on the other. It is also our job to provide an appropriate basis for making decisions about how to spend a limited amount of money most effectively.

9. References

1. US EPA. 1994. Exposure Assessment Guidance for RCRA Hazardous Waste Combustion Facilities. EPA 530-R-94-021, April 1991.
2. US EPA. 1990. Methodology for Assessing Health Risks Associated with Indirect Exposure to Combustor Emissions. EPA/600/6-90/003. Interim Final. January 1990.
3. US EPA. 1993. Addendum to the Methodology for Assessing Health Risks Associated with Indirect Exposure to Combustor Emissions. EPA/600/AP-93/003. Review Draft. (Update may be available).
4. US EPA. 1993. Data Quality Objectives Process for Superfund. EPA 5400R-93-071. Interim final guidance.

PRIORITIZING ENVIRONMENTAL ACTIVITIES IN NIZHNII TAGIL BASED ON THE ASSESSMENT OF HUMAN HEALTH RISKS DUE TO AIR CONTAMINATION

P.S. ZHILIN and I.D. KOPYLOV
Nizhnii Tagil Interraion Environmental Protection Committee,
Nizhnii Tagil, Lomonosova 7, 622001, RUSSIA

1. Introduction

The Nizhnii Tagil Environmental Management project is a multi-component project designed to demonstrate new methods of environmental management and decision-making for local governments in the Russian Federation. The project is overseen by a joint Russian-American Steering Committee. In accordance with the Russian-American cooperation program, the Priority Setting Committee (a local public committee enacted directs environmental planning and priority-setting activities by a decree of the city Major). Local scientists as well as the general public (95% of polls) believe that ambient air pollution is the main factor affecting the health of Nizhnii Tagil residents. Therefore, both scientific and public opinion support the Project activities for environmental improvement. The priority-setting activity consists of the following phases:

1. Collect data on pollution sources
2. Rank specific contaminants based on their potencies
3. Identify cancer and non-cancer effects of the top chemicals of concern
4. Rank the pollution sources according to their impact on human health
5. Develop and implement low cost environmental activities that produce the maximal environmental health benefit.

Three years ago, the concept of risk analysis was introduced by the American counterparts as a method to help establish environmental priorities. We have learned both the methodology and capabilities of risk analysis and are using it to make a quantitative assessment of the health risks posed by environmental air quality. What is the main difference from the common practice of setting norms for hazardous emissions (permitted emission limits), and for environmental protection activities which target the reduction of the pollution rates to a value not over the permitted concentration for a residential area and 0,8 of the permitted concentration for resorts and mass recreation areas?

We have the right to be proud of the fact that in 1951 Russia was the first country in the world to set permits on 10 of the most widely spread ambient air pollutants. The first air quality norms enacted were based on the works of V.A.

I. Linkov and R. Wilson (eds.), Air Pollution in the Ural Mountains, 277–282.

Ryazanov, K.A. Bushueyeva, and M.A. Pinigin, and were characterized by an average daily permit limit that prevents a chronic unfavorable effect. In addition to the average daily permit limit, single peak permit limits to assess drastic concentration spikes were introduced for pollutants with odor or irritating properties. Since the establishment of the daily permit limits, we have not always succeeded in obtaining these permitted rates for all pollutants. In 1970, within the framework of the Clean Air Act, the USA introduced national air quality standards. These standards have promoted considerable success and approached the limits of economic efficiency of environmental protection activities by the early 1980s. The concept of risk-based decision-making was formed in 1983. The main goal is to set priorities for risk reduction based on an identification of human health risk sources (risk assessment) and the most efficient ways to reduce risk (risk management). Health risk is a calculated value of the potential for human mortality or morbidity due to exposure to pollutants found in food, water, the work place, home and particularly, in the ambient air.

Barbara Felitti, Eurasia Program Director for the Institute for Sustainable Communities (ISC), provides methodological guidance of the Project in Nizhnii Tagil by involving experts from the American side, and I as a representative of the Interraion Environment Committee, am assigned to manage the implementation of the Project from the Russian side. Project progress is reported to the public, and also to the Russian-American Steering Committee, with representatives of the Russian Nature Ministry, Oblast Environmental Protection Committee, Nizhnii Tagil City Administration and Ecoprom from the Russian side; and USAID, USEPA and ISC from the American side.

2. Approach

The scope and new nature of some of the tasks for this Project required additional valid and objective information on the ambient air pollution level from data normally used, as well as the setup of a unified, integrated database of air pollution sources and meteorological information for air modeling. It was decided to use Russian norms and standards to identify the ambient air pollution rate. This ensures that the data generated conform to Russian standards, and the information will be beneficial to the city no matter what the Project results are.

Optimization of the work was our task at all the stages. Emission reports provided by industries, 2-TP (air) analyses, revealed 4705 stationary pollution sources and 141 pollutants. In addition to industrial air pollution, the impacts of the following emissions were also considered: vehicles, household gas burners, firewood and coal burning in individual houses, burning waste dumps and sludge landfills. Data on many of these sources were not available, and it was necessary to find ways to calculate or estimate the data for dispersion modeling. To calculate health risks it was necessary to identify pollutant exposure concentrations and the number of humans exposed to this pollution. For the latter, we divided the city into districts and determined the coordinates of their centers. However, the task of identifying the associated population proved to be difficult. An equally difficult task was the development of a unified coordinate system for the electronic city map. Pollution sources from several industrial coordinate systems had to be combined into the unified coordinate system (e.g. there were 3 such systems just at the NTMK steel mill). The coordinates of ambient air observation posts and city waste sites were also added, and the borders of 22 regions of the city were put on the map.

At different stages of the Project implementation, various creative groups were involved including those of the Polytechnic University, Uralgidromet, Sanitary Epidemiological Center, road police, architectural institute, Tagilgrazhdanproyekt and others. Our critics questioned and opposed our data minimizing methods (in a database of 4705 pollution sources we analyzed in detail 571 that are responsible for 75% of all air emissions), and the selection of chemical pollutants based on a tons/year emission rate, as a gram/sec rate is the accepted basis for the norm setting. We took into account their suggestions and included 11 additional low fugitive emission sources of specific chemicals.

At the first stage a list of pollutants of concern was developed based on the required existing air concentrations (reference dose (RD) 52.04.186-89). Carcinogenic chemicals were added to the list based on Russian Nature Ministry Committee's conclusion about the "Nizhnii Tagil Complex Program of Environmental and Human Health Rehabilitation for the Period up to the Year 2000". Subsequently, 2 chemicals were excluded from the list based on the assessment of hazardous impacts on human health. As a result of the initial assessment, a total of 21 contaminants were identified for further analysis based on the results of the ground level concentrations obtained, and a review of a number of medical aspects such as the degree of health problems and carcinogenic qualities. American methods were used to calculate health risks, with the initial health risk data taken from IRIS (Integrated Risk Information System). Validation of harmful properties of chemicals was also found in references "Hazardous chemicals" and "Carcinogenic substances".

The location of the center of each of the city's 22 regions were defined as the exposure point. Concentrations of the 21 pollutants of concern were determined for the 22 regions and four ambient air monitoring observation posts. Eventually, 13 contaminants were preserved and ranked on the basis of individual cancer risks and a non-cancer risks. The 22 districts were then ranked based on the human health risk rate for each district. For 3 districts (Uralvagonzavod, Tekhposyolok, Krasny Kamen) a comparison of the expected disease rates with medical statistics of actual incidence of disease was also made. The results between expected and observed disease rates were consistent.

Ground level concentration calculations for Gidromet posts (ambient air monitoring stations) and the city districts were modeled using "Ecolog-1.32" software. The work was examined at the regional Oblast Environmental Protection Committee norm setting department, and the primary activities of the Project and the Interraion Environmental Protection Committee's database were approved. The work done was also found useful for assisting the city ambient air monitoring system network.

As a result of the health risk assessment, 8 of the 22 regions of the city were identified for further evaluation. The top 20 pollution sources of the 13 pollutants of concern in these districts were then identified. Next an algorithm was developed to determine the contribution and effect of each pollution source on the city in general.

To develop the next stage of source control strategies, the city industries were provided with lists of pollution sources at their plants indicating the impact of the source on Tagil resident's health deterioration. For instance, there is no need to first spend funds to obtain permitted levels of naphthalene as its hazardous irritant properties are not comparable to carcinogenic properties of benzopyrene, hexavalent chromium, formaldehyde, or benzene. To be fair, the high concentrations of the mentioned naphthalene cause polyneuritis, and stomach secretion impairments.

However, we have not yet determined the primary environmental protection control strategies for risk management. A number of challenges emerged, and we have not yet solved all of them.

To evaluate health risks posed by air contamination, the annual average concentrations are required using the following factors as its basis: the average human inhales 20 m^3 of air a day, weighs 70 kg and lives 70 years. Calculating concentrations with the computer program OND-86, single peak concentrations are obtained that can be converted into average daily and annual average concentrations using a conversion factor. According to references the conversion factor ranges from 1/10 to 1/40 and depends on many variables. American studies indicate the concentrations for 20 min exposure are approximately 16 times more than annual concentrations (Turner, 1969). Professor Evgeny Lvovich Genikhovich, Voyeykov Observatory in St. Petersburg, recommended that we use a conversion factor of 1/10. At the same time, Professor Genikhovich suggested we wait for a final recommendation on the conversion factor as he was completing and validating an annual average concentration calculation model for the city of Volgograd. But such models are already available. "Atmosphere" software by "Logos" in Novosibirsk calculates C single peak concentrations using OND-86, the single peak based on Gauss flare, and also annual average based on Gauss flare (developed by Anatoly A. Bykov). Assisted by Uralgidromet, we received a meteorological file processed in Novosibirsk and based on the local weather station observations. As we expected there was nothing consistent found with the conversion factor K=1/10 -- if there is no wind from the source towards the monitoring point, there can be any variation. Also, estimated single peaks are not comparable to anything -- they by virtue of their definition, differ from background and underfire concentrations, as well as from maximum measured concentrations. It is tempting to compare annual average concentrations with those at the monitoring posts, and we did so. The correlation between multi-ton chemical emissions and ambient air data showed that estimates exceeded the observed concentrations for manganese, iron and suspended solids. The measurements considerably exceeded the estimated for some other specific chemicals. We tried to enter correction factors but we chose to leave them off, as the valid estimate of each source is required for risk management. The meteorological file was adjusted under assistance of David Sullivan, an American specialist, and the American ISCLT3 modeling program was used to calculate annual average concentrations which take into account the actual operation time of emission sources. Vehicle emissions were also considered in the program as area sources (residential sites).

After these re-calculations were conducted, calculated annual average concentrations were compared to those measured at the air pollution monitoring posts, with a noted improvement in some correlations. Health risks were then evaluated based on annual average concentrations, and, subsequently, the ranking of pollution sources based on annual average concentrations will be adjusted. The results will be used to evaluate the decisions to identify environmental activities to reduce pollutant concentrations which have the greatest health impact. The economic efficiency of possible environmental control strategies must also be evaluated. Health risk and its relationship to average annual and average daily concentrations differs for non-cancer and cancer chemicals. Therefore, for benzopyrene, at the concentration equal to the permitted average daily concentration, the health risk 2 per million is close to the American standard 1 per million, and for formaldehyde is 39 per million. By our calculations, existing benzene and chromium permit limits will allow an additional 830 and 18000 cancer diseases per million, respectively, during human life.

These results have sometimes created a paradox, when the pollution rate is under the permitted limit, yet, as for the above listed chemicals, cause major health risks. This implies that medical specialists will have to move chromium and benzene permits closer to international standards.

3. Errors and Uncertainities

1. On-road emissions data was insufficient and distribution was estimated proportional to the number of residents in the districts, traffic intensity, and the length of roadways and streets.
2. Modeled formaldehyde concentrations are considerably lower than actual measured ambient concentrations and we will have to further evaluate known pollution sources as well as the alternative ones such as motor vehicles, firewood burning, etc.
3. Measured and estimated concentrations for benzene, benzopyrene, manganese and phenol data do not correlate well. (observed vs. predicted emissions 3,1; 3,9; 3,5: 2,4 times, respectively)

4. Conclusions

1. The average annual concentrations obtained are good for evaluating health risks and making decisions on reducing risks. The results to date require the continuation of this work, particularly in the area of PM10, for which we are completing a health risk assessment.
2. No conversion factors to estimate single peaks are capable of providing valid annual average concentrations required for health risk evaluations.
3. The experience gained can be beneficial for other areas suffering from air pollution problems. We have managed to surmount multiple challenges of adaptation of American approaches to Russian conditions.

The health risk method authors proved to be right stating that risk management is not as much an environmental but an economic, political and social problem. The fact that similar work has been started in Krasnouralsk and planned for other cities of the region speaks for its prospective usefulness. The difficulties and problems involved in implementing these approaches are surmountable.

5. Acknowledgements

The work of the Institute for Sustainable Communities to support the priority setting process described in this paper was funded by the United States Agency for International Development via its interagency agreement with the United States Environmental Protection Agency.

HEALTH EFFECTS ASSOCIATED WITH AIR POLLUTION IN THE URAL MOUNTAINS

Report of the Working Group on Health Effects[1]

J.O. NRIAGU
Department of Environmental & Industrial Health, School of Public Health, University of Michigan, Ann Arbor, MI 48109 USA

J.G. WILKINSON
Carnegie Mellon University, Department of Engineering and Public Policy, 5000 Forbes Avenue, Pittsburgh, PA 15213. USA

1. Introduction

Magnitogorsk is a city of 465,000 people that is choking under the cumulative load of pollution from 70 years of sustained production of iron and steel using less-than-adequate technologies. The main enterprise and principal employer in the city is the joint-stock company, "Magnitogorsk Iron and Steel Works", which boasts of an history of decades of continuous production growth of all kinds of metal products. The sustained intensification of production processes peaked in 1989 when 7.4 million tons of coke, 11.7 million tons of cast-iron, 16.1 million tons of steel and 11.9 million tons of rolled stock poured out the various plants in the city (Starikov, 1997). Most of the growth was coaxed out plants that were built in the 1930s and were operated with little or no modifications to deal with the accompanying increase in emissions of toxic substances to the environment.

Magnitogorsk Iron and Steel Works is an integrated system with sintering plants, coke ovens, blast furnaces, basic oxygen furnaces, rolling mills, galvanizing shops, pickling baths and finishing shops. The polluting nature of these processes is legendary and steel towns in many countries (such Pittsburgh and Gary of the United States and Hamilton Canada) often acquire a notoriety related to the red or brown atmosphere symptomatic of the foul nature of the air (EPA, 1984). Magnitogorsk is no exception. In 1989, emissions of toxic wastes averaged about 1,800 kg for every inhabitant of Magnitogorsk (Starikov, 1997). The key pollutants emitted by the

[1] Members of Working Group: J. Nriagu (Chairman), J. Wilkinson (Rapporteur) S. Bzezgina, N. Checka, J. Jaakkola, N. Katler, V. Koshkina, H. Ozkaynak, V. Peckorin, O. Ponomareva, L. Privalova, M. Smuts, N. Vavilova, and N. Zaitseva

I. Linkov and R. Wilson (eds.), Air Pollution in the Ural Mountains, 283–288.

various steel making processes include Ni, Cr, Cd, Be, Pb, As, Si, suspended matter, coke oven gases (benzo(a)pyrene, ...) as well as SOx and NOx. Most of these contaminants do not biodegrade easily and when the backward trajectory of emissions is considered, it comes as no surprise that Magnitogorsk has recently been declared an "ecological disaster area" (Anikushin et al., 1997).

The toxic contaminants have exacted a substantial toll on the health of local population, with the children and women being particularly vulnerable. Indeed, morbidity and mortality studies that have been conducted in these areas show that there is serious cause for concern about the state of the public health. From 1992 to 1996, birth rate in Magnitogorsk fell from 10.2 to 8.3 per 1000, infertility increased by 63% (from 1.3 to 3.5 per 1000), preterm delivery increased by about 16% (to 9.5% of all births), and 48% of pregnant women had suffered anemia, renal disturbance, pregnancy onset diabetes, and/or hematological disorders (Kushina, 1997). Marked increases were observed in neonatal and newborn diseases including defects of the central nervous system, sepsis, congenital heart problems, defective kidney and urinary tract development among others. The increase in morbidity figures was accompanied by a 50% increase in infant mortality. Equally noteworthy is the fact that new cancer cases are appearing at a rate of about 360 per 100,000 population per year in Magnitogorsk as of 1996 which is higher than the rate from 1971, 236.2 per 100,000 population per year (Varava and Kostyukov, 1997).

The reported deterioration in maternal-child health occurred at a time of technological rearrangement in the Iron and Steel Works. A new converter shop was installed in 1990, and several installations were closed down including open-hearth furnaces, two-bath steel making unit, ore dressing shop, agglomeration shop, cake battery, and other units (Starikov, 1997). The combination of ecological and economical forces has reduced the steel output from a maximum of 16.1 million tons to 1989 to about 6 million tons in 1996. The apparent growth in health problems in the face of shrinking steel production in the city raises a number of questions about interacting determinants of health in the particular population. Many women work in the plants, which makes it difficult to differentiate between the risks of occupational from environmental exposures. The recent collapse in Soviet economy and political system has undoubtedly played a role in the worsening of the health of people in many industrial cities of the Urals including Magnitogorsk (Day et al., 1997). The exceedence of the carrying capacity of local ecosystems by accumulation of past industrial emissions would imply that human exposures are less dependent on new inputs of toxic contaminants. Relating the prevalent chronic and acute diseases in the city to environmental risk factors clearly represents a challenge for the scientific and public health experts. The NATO Workshop served to bring together foreign and Russian experts to apply their collective wisdom to this problem.

2. Environmental Exposures

The integrated steel mill in Magnitogorsk is a vast industrial complex that consumes large quantities of energy (coal and electricity) and raw material (iron ore, lime, coke, etc) and in turn releases wastes into the air, water and soils. Steel-making processes associated with substantial emissions of pollutants include raw material handling, sintering, coke making, blast furnace, basic oxygen furnace, rolling, processing and finishing, slag and hazardous waste handling (US EPA, 1980). Environmental pollution and associated impacts in Magnitogorsk area are described in greater detail in the other Group Reports. The Health Group dealt primarily with how people get exposed to the contaminants and the associated health effects. The emphasis was on the environmental factors that impact human health with little attention given to such variables as lifestyle factors, occupational factors, or factors risk that predispose an individual to health problems (e.g. genetics). Furthermore, the discussions revolved around quality of available database and future critical measurements needed to link environmental risk factors to health outcomes in the Magnitogorsk area in particular and the Ural region in general.

The atmosphere remains a major route of exposure to many toxic chemicals from the iron and steel plants. Aerial emissions from the steel mills include black and red dusts, toxic metals, sulfur and nitrogen oxides, carbon monoxide and coke oven gases (polychlorinated aromatic hydrocarbons, hydrogen cyanide. ammonia, hydrogen sulfide, and phenols). The fact that this wasteload can be hazardous to human health is well documented in studies done in steel cities in many parts of the world. Only very limited attempt has been made to assess the human health risks associated with air pollution in Magnitogorsk and surrounding region. Various strategies for estimating exposures to air toxics were discussed. It was noted, for example, that the feedstock into the plants can be quantified to a large extent and the technology is well known so that a rough inventory of atmospheric emissions is feasible. The emission data can then be linked to dispersion model to approximate ambient air levels (Peterson, 1997) that can be used in first-cut exposure and risk assessments. Inadequacies in available database and in measurement techniques were noted. The needed to collect data that can be used in exposure assessment and risk management for air pollution in the city was clearly highlighted.

Food was another exposure route that was considered to be of some concern. With the economic dislocation in the country, many residents increasingly rely on foods grown in their gardens to meet their dietary needs. Furthermore, the food stores are forced to stock mainly produce from local "communal" farms. Over the years, some of the soils have become severely contaminated with toxic metals which can be transferred to the human food chain. The extent to which foods grown on such soils are compromised by the toxic metal burden is currently unknown. The Group felt that this problem deserves some attention. Increased health risk from locally grown foods is an issue not only in Magnitogorsk but other urban areas of the Urals.

Iron and steel industry is a prodigious user of water and can adversely affect the water supply in two ways: by polluting the lakes, rivers, streams and encatchments with toxic wastes and through disruption of hydrological cycle on which the renewability and sustained availability of potable water depends. Either effect can influence human health directly or indirectly. Some health risk may be associated with the use of ground water and untreated surface water in the region. There was no clear sense as to whether the quality of municipal water supply in Magnitogorsk may be inimical to health.

The discussions emphasized the need for a multi-media strategy in assessing the risks of environmental pollutants on the health of people in Magnitogorsk.

3. Local Investigations

In response to the suspected high prevalence of environmentally-related disorders, the local health authorities have established a number of programs to address the problems. For instance, intensive therapy centers have been set up in Magnitogorsk for new born children with congenital health problems. There is a center for children with asthma other respiratory diseases, infertility centers, programs for better and earlier diagnosis of birth defects, as well as clinics devoted to the health of the pregnant mother and the unborn child. The health statistics given above were actually derived from these environmental health programs (Koshina, 1997). The clinical studies, however, have not been coordinated with any systematic environmental measurements, so that the results are of limited use in risk assessment.

Although the medical and scientific establishments in and around Magnitogorsk have undertaken a number of small-scale, short duration studies to identify, and in a few cases quantify, some environmental risk factors that impact human health (e.g. Varava and Kostyukov, 1997), the available resources are inadequate to either continue the studies to their logical conclusion, or to implement plans that have been designed to address the environmental problems. Furthermore, insufficient resources are available to undertake large-scale or long term projects to evaluate the relative significance of the various environmental risk factors.

Magnitogorsk is but one of the many urban areas in the Urals (others include Asbest, Kamensk-Uralski, Nizhni Tagil, Ekateringburg, Chelyabinsk and Pervouralsk) that are on the Russian list of 45 towns with the extremely high air pollution (Kyamkin et al., 1997). Most the Uralian cities have also developed around mining industries. Nizhi Tagil, for instance, has many factories (heavy metals primarily) similar to those in Magnitogorsk and hence shares similar health problems - infertility and lower birth rate, high rates of endochrine disorders in girls, high rate of premature delivery, birth defects (including heart and kidney diseases), and maternal diseases (diabetes, anemia, kidney and heart problems) (Zhilin and Kopilov, 1997). Any risk assessment model, risk management program or intervention strategy developed for one city can easily be ported to other towns. It became obvious during the group discussions that little interaction exists between the Uralian cities on how to deal with their common environmental problems and associated health effects.

4. Recommendations

Despite the efforts of the public health sectors in the Magnitogorsk and Ural areas, the populations in these regions continue to experience high levels of suspected, environmentally related health problems. A number of low cost programs have been implemented by the local health authorities in these regions to address the various environmental health risk factors. Significantly more work in a mutli-media context needs to be done to fully assess the environmental conditions and their impacts on public health in the areas so that proper intervention strategy can be designed. Because of the perceived seriousness of existing conditions, the recommendations took on a prescriptive tone. It goes without saying that the environmental problems are unlikely to be resolved unless the local economy is viable and sustainable.

1. In order to maximize the use for the limited resources available, some effort should be made to coordinate the fragmented studies (clinical and scientific) and programs dealing with environmental and ecological health in the city of Magnitogorsk. A coordinating body may be established to serve the role of (a) opening the lines of communication among the various groups and competing interests, (b) facilitating exchange of information, (c) minimizing the duplication of efforts and (d) ensuring that the best available data are using in program planning.
2. Related to the first recommendation is the need to establish closer ties with other towns in the Urals that are experiencing similar environmental health problems. For instance, experts from Magnitogorsk should be encouraged to become observers of the two-year epidemiological study, which starts July 1997, that is being conducted jointly between researchers from Sverdlosk obelisk and the Harvard School of Public Health; a goal of this study is the transfer of technical know-how to public health researchers in the Ural Mountains region. Many towns in the region are also involved in engineering low-cost or no-cost intervention strategies which Magnitogorsk can benefit from.
3. There is a definite need to develop local capacity on a number of critical research methods especially environmental epidemiology, risk assessment and risk management. This can be facilitated through development of joint research projects, exchange of experts and/or training of staff locally (using foreign specialists) or in institutions overseas.
4. The need was also identified for developing proposal writing skills for competitive grants to enable local experts tap into outside potential funding sources to continue or expand existing programs on biomedical monitoring of children, cancer prevention, and health education. NATO can play an important role by providing seed money for this activity.
5. Concern was expressed, based on the Chernobyl experience, about the adequacy of emergency plans to deal with chemical accidents in the plants. While such plans reportedly exist, it is not clear that an effective structure is in place for their implementation in Magnitogorsk. Any existing emergency response plan should

be re-examined, a team put in place for its implementation and the local population educated on what to do in case of chemical accidents. The team may wish to seek for advice and assistance from the emergency planning agencies of the European Union and the United States.

5. Conclusion

People in the regions around Magnitogorsk and, in general, the Ural Mountains continue to be exposed to environmental risk factors that impact their health. Despite ongoing efforts by local health agencies, it appears that public health in the Magnitogorsk and Ural Mountains areas is still deteriorating due, in part, to the recent collapse of the Soviet economy and political fabric. The NATO Advanced Research Workshop provided an opportunity for local and foreign experts to apply their collective wisdom to the environmental crisis in Magnitogorsk and other mining towns of the Urals. A number of recommendations have been made with the understanding that the over-arching economic problems must also be addressed to guarantee improvements in human and ecosystem health.

6. References

1. Anikushin, V., A. Alyaev, L. Varava, V. Veligon, A. Kostyukov, N. Checha, 1997. "Draft Federal Program For 1996-2006 On The Leading Out The Territory Of Magnitogorsk From The Condition Of An Extreme Ecological Situation And On The Population Sanitation." In: Preceedings of the NATO Advanced Research Workshop on Air Pollution in the Ural Mountains. 25-30 May (in Russian).
2. Day, R., V. D. Surzhikov, G. I. Chechenin, Y. Ruseav, and Y. Gregoriev, 1997. "Public Health Trends In A Transitional Russian City." In: Preceedings of the NATO Advanced Research Workshop on Air Pollution in the Ural Mountains. 25-30 May (this volume).
3. Kyamkin, A., A. Uspin, V. Vasilenko, 1997. "Characterization Of Environmental Pollution In The Ural Region" In: Preceedings of the NATO Advanced Research Workshop on Air Pollution in the Ural Mountains. 25-30 May (in Russian).
4. Starikov, A., 1997. "The Strategy Of The Joint-Stock Company 'Magnitogorsk Iron And Steel Works' In Solving Ecological Problems." In: Preceedings of the NATO Advanced Research Workshop on Air Pollution in the Ural Mountains. 25-30 May (in Russian).
5. Varava, N., A. Kostyukov, 1997. "Survey of Health Indices of Pregnant Woman and Newly-born in Magnitogorsk over Five Years." Report of the Head of the Department of Ecology of Magnitogorsk, Magnitogorsk, Russia.

Part 3

Air Pollution and Environmental Policies

The steel town of Magnitogorsk from the east.
Photp: R. Wilson

THE ROLE OF ECONOMIC VALUATION IN DECISION MAKING ABOUT AIR POLLUTION

A. MARKANDYA and I. MILBORROW
Department of Economics and International Development
University of Bath, Bath, UK

1. Introduction

This paper begins by presenting the case for quantifying the damages caused by air pollution on the environment. The development of work in this area is reviewed. The current state-of-the art known as the *impact pathway approach* is then presented and critiqued. Chapter 2 then discusses the theoretical issues that arise when estimates of damages are transferred. A wide range of recent studies are then considered and the best range of estimates for SO_2 and particulates for the Russian situation are presented.

Chapter 3 reflects on the discussion and discusses how damage estimation can be relevant to policymakers who are taking decisions about the regulation of air pollution. It is suggested that due to the particular problems of transition economies a broad range of issues need to be considered, particularly the social costs of unemployment arising from the closure of polluting activities. A broad cost-benefit framework is proposed as a way forward.

An important feature of this paper is that it is only partially completed. We present valuation data, methodologies and dose-response functions along with some of our ideas on their transferability to a transition economy such as the Russian Federation. However, we invite input from Russian experts on the availability of the required data and the feasibility of our assumptions.

1.1 WHY VALUING AIR POLLUTION IS IMPORTANT

Many activities are involved in the generation of electric power, ranging from fuel extraction to decommissioning of the power plant. These activities comprise a *fuel cycle*. Some discharges damage ecosystems, others affect human health or productive activities (agriculture, buildings etc.) and others damage recreational services. The emissions result in *impacts,* which in turn have economic **costs** associated with them. In addition to the impacts that are part of the environmental chain, there are others not associated with the environment that are important from a social point of view. Among these are issues of employment, energy security, and administrative cost of regulation.

I. Linkov and R. Wilson (eds.), Air Pollution in the Ural Mountains, 291–305.

Where the marginal impacts described above are not properly reflected in the prices charged to the final consumers of the energy sources, they are said to result in *externalities*. We leave open at this stage what is a 'proper reflection' of these impacts in the price. Monetary valuation of the external costs of energy generation, in this case the costs attributable to air pollution is useful because it can facilitate policy-makers in making more rational choices about the use of energy. In addition, where it is not possible to value environmental damages the impact pathway approach can enable useful physical indicators to be developed.

1.2 DEVELOPMENT OF WORK IN THIS AREA

It is only really since the end of the 1980s that serious studies were undertaken to attempt to quantify externalities arising from electric power generation. The seminal piece was that of Hohmeyer [6]. He used a 'top-down' approach. First, he identified other studies' estimates of the total damages (e.g. health costs) attributed to air pollution. Next he estimated the fraction of the total emissions that are from electric power generation with fossil fuels (e.g. 28%). The two figures are then multiplied to obtain the health damages from fossil fuels. The methodology assessed the 'big picture', but relied heavily on approximations and previous estimates of total damages. The total fuel cycle is not considered and it cannot provide a tool for assessing site-specific effects, which may be very important.

Another well-known study is by Ottinger *et al.* at Pace University [14]. They used a 'bottom-up' approach to: (i) estimate the emissions and their amounts, (ii) estimate the dispersal of the pollutants, (iii) determine the population, flora, and fauna exposed to the pollutants, (iv) estimate the impacts on the population from exposure to the pollutants, and (v) compute the monetary cost of that impact. Pace relied on numerical estimates from previous studies, and used those values to compute damages. Pearce *et al.* [15] made a study in similar spirit to Pace, drawing quite heavily on the latter. However, by taking a fuel cycle approach they addressed a more complete picture and estimated some externality 'adders' (costs per tonne of emissions) for electricity production in the UK. Such studies were not site specific therefore ignoring differences due to topography or population concentrations.

A problem with these early studies was that much of the work was derivative, with each 'new' study taking its basic information from the earlier studies. In 1991, a joint study by the US Department of Energy and the Commission of the European Communities was launched to "develop a comparative analytical methodology and to develop the best range of estimates of [damages] costs from secondary sources". It attempted to develop methods to measure all the externalities (environmental and non-environmental) associated with electric power. The study is known as the ExternE Project[1]. This paper draws on the work of that project and tries to develop guidelines for its extension to the Russian situation. ExternE is currently in its third

---

[1] Professor Markandya is the senior economist for the ExternE Project and therefore a core member of the team that developed the economic valuations.

phase, with around 40 teams from 15 countries participating in the dissemination, completion and application of an accounting framework suitable for implementing at the policy level. The methodology aims to be consistent, transparent and comprehensive. It covers eight fuel cycles as well as for four conservation options for electricity generation. ExternE goes beyond the earlier approaches in terms of:

a) a more thorough characterization of the energy technologies and their discharges into the environment on a site-specific basis,
b) considering all major stages of a fuel cycle rather than just electric power generation,
c) modeling the dispersion and transformation of pollutants rather than relying on previous estimates,
d) engaging in a more extensive, critical review and use of the ecology, health sciences, and economics literatures than previous studies,
e) estimating externalities by accounting for existing market, regulatory, insurance, and other conditions that internalize some damages so that they are not externalities,
f) and using models and analysis, rather than direct use of numbers from the literature.

The aim is to derive estimates for the total level of externalities associated with each fuel cycle. However, the impact pathway approach only considers environmental costs and benefits. As mentioned previously, there may be various non-environmental costs and benefits. These also need to be incorporated into the framework. Only then can the true level of externalities and hence the social costs/benefits of the fuel cycle be determined (see Figure 1 overleaf).

1.3. THE IMPACT-PATHWAY APPROACH

The *impact pathway approach* spells out the different stages of fuel cycles. For instance, the coal fuel cycle consists of: coal mining, transportation, generation, and transmission. The different gèneration of waste is included in the analysis, as is the construction of the additional facilities (to the extent they are needed) for mining, generation, and transportation. Also identified are the main burdens generated by such activities in terms of water, solid waste, and air. The impact pathway is shown in the left hand box in Figure 1. It traces the burdens through the different environmental media into ambient concentrations. These concentrations will have impacts that depend on the receptors' responses—i.e., the ability to absorb wastes which affect human and ecological functions. The impacts are classified in terms of health, ecological, and 'other' categories. This constitutes the impact assessment stage. The second stage is economic valuation, which will utilize one of three methods (direct, indirect or hypothetical) depending on the receptor being considered.

FIGURE 1. Quantification of Social Costs - The Impact Pathway Approach

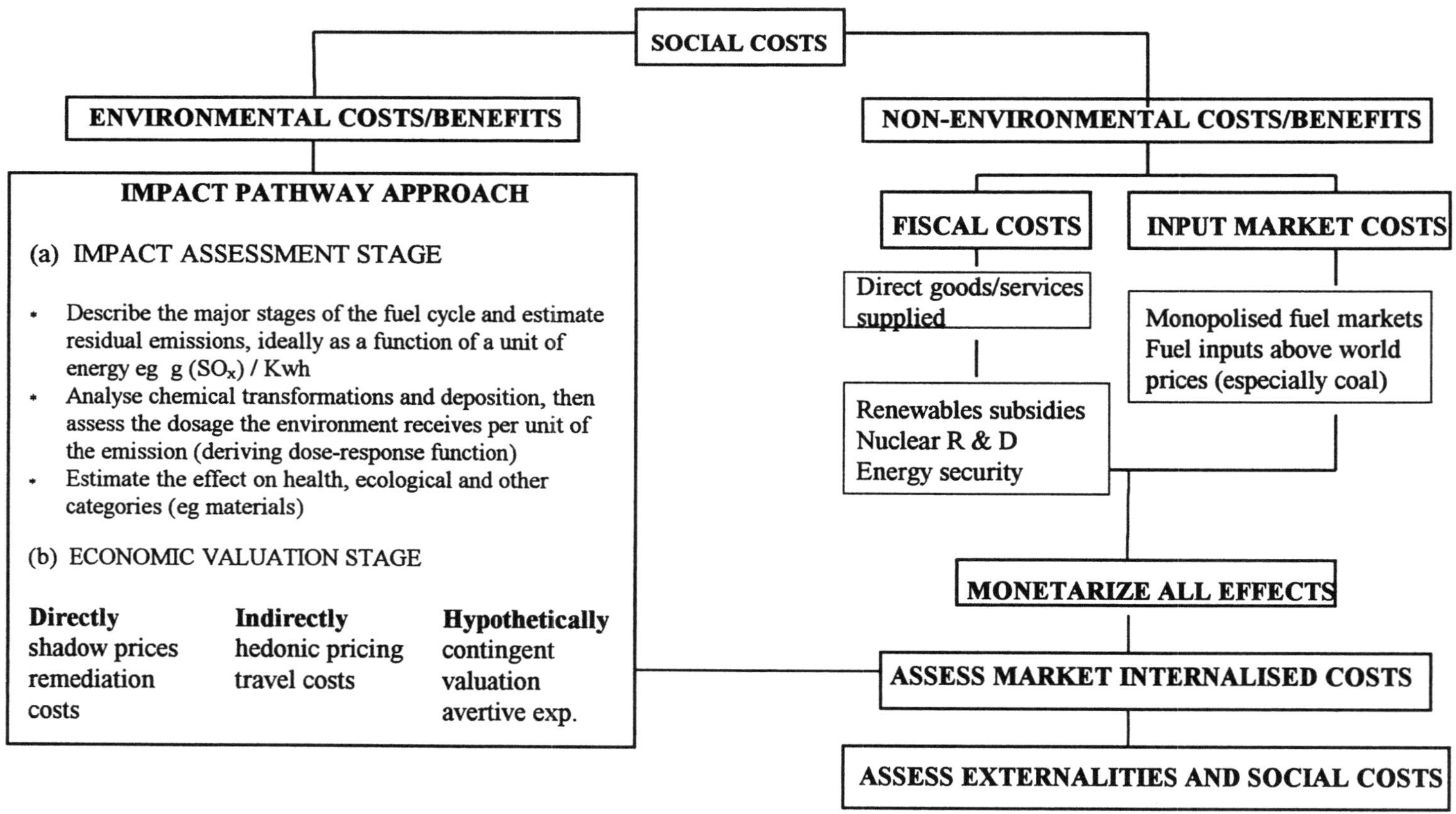

The impact assessment stage relies almost entirely on the use of *dose-response functions* to evaluate physical damages. The basis for economic valuation are the concepts of willingness-to-pay (WTP) or willingness-to-accept (WTA). WTP measures how much individuals would pay to avoid environmental damage or secure an improvement in the quality of life. Correspondingly, WTA measures the compensation required if environmental damage and a lower quality of life were to be tolerated. These include market values (adjusted for market imperfections where appropriate), estimates based on contingent valuation studies, and estimates based on travel cost studies. Hedonic studies, which value damages by estimating the impact on property values and that largely bypass the use of dose-response functions, were only used in the valuation of noise costs.

An important issue that emerges is the choice of discount rate, since pollutants generated from a fuel cycle may exist in the atmosphere for a considerable period of time. For the nuclear fuel cycle, potential effects have to be assessed over thousands and millions of years. The debate on what rate should be used has not been resolved and is unlikely ever to be fully resolved. It was decided that three different rates would be used by the EU/US study that spans a reasonable range: a zero rate, a rate of 3%, which reflects the best guess at the social discount rate, and a rate of 7%, which reflects the kind of rate typically used for appraising large scale investments in the EU and US.

1.4. SOME LESSONS LEARNED FROM THE EXTERNE PROGRAMME

1.4.1. *Health impacts are most important and easiest to transfer.*

The estimation of health impacts using other epidemiological and valuation studies is the easiest, although even here there are problems. Health damages account for the largest share of total damages and have received most interest at the policy level. The mortality valuation studies are based on situations where individuals have voluntarily accepted an increased risk of death or sacrificed something to reduce that risk. Issues such as the way the death occurs, the fact that it is not voluntary, the latency period, and the age groups affected are not allowed for in the Value of a Statistical Life (VOSL), which is used in most of the work carried out so far. To address some of these issues, recent developments have explored the possibility of using a concept called Value of Life Years Lost (VLYL)[2].

1.4.2. *Transferring damages to market related products such as crops is relatively easy*

Dose-response functions that are transferable exist where small changes in ambient concentrations are being valued. Local price data can be applied. However, where the changes in concentrations are large, it will be necessary to model supply and demand for the different products and estimate the changes in consumer and producer

[2] See Ref. [10] for discussion of the VLYL concept.

surpluses. This will require an extensive local model that cannot be transferred from outside.

1.4.3. *Ecosystem damages are the most difficult to quantify and transfer*

Assessing and valuing damages to environmental assets is the most difficult and the most site-specific. The main problem is the establishment of reliable dose-response functions linking pollutants to the various 'endpoints' that we would wish to value (e.g. plants, fauna etc). The approach adopted has been to establish critical values for specific pollutants and apply these to broad groups of natural vegetation. There is very little literature on valuing air pollution impacts on ecosystems. Some studies have attempted to value species loss, amenity loss and the loss recreational benefits (e.g. fishing). The treatment of damages to terrestrial and aquatic ecosystems still remains seriously inadequate.

1.4.4. *Global warming damages are highly uncertain.*

The impacts of global warming are predicted to affect a huge range of receptors. Estimation of damages is complex, being scenario dependent, very uncertain and very long term[3]. The most comprehensive assessments are largely qualitative. Valuation is even more problematic, because of the macro-scale of the impacts and the interactions between them A consensus on valuation is unlikely, yet global warming damages may well be the most serious fuel cycle impacts.

1.5. AREAS OF UNCERTAINTY

It should be emphasized that the assessment procedure is extremely difficult and controversial. There are so many areas where uncertainties arise in the impact pathway that, by the time the final figure has been reported, it contains many cumulated possible errors. It is important to convey the extent of the uncertainty to policy-makers. This could be done by developing an uncertainty message system that offers a structured qualitative appraisal of each study of data entry used to generate damages, or to identify confidence intervals at as many stages of the analysis as possible. Two main areas of contention can be summarized.

1.5.1. *Scientific uncertainties*

In order to estimate and model the impacts of just one emission, a huge amount of primary research, estimation and modeling is necessary. The local, regional and global consequences of air pollution depend on atmospheric, biological, chemical, ecological and physiological relationships across time, space and socio-economic conditions. These various pathways involve feedbacks and interactions with one another over different time scales, with both reversible and irreversible consequences. This is clearly subject to considerable uncertainty.

---

[3] For a comprehensive assessment of all global warming issues, see Ref. [7].

1.5.2. *Valuation uncertainties*

Economic science has progressed considerably in the development of indirect valuation techniques such as contingent valuation surveys applied to environmental project planning. However, innate problems remain such as lack of information, potential biases and a limited perspective within which decisions are taken. Ethical questions also arise over the monetary valuation of human health and the environment.

2. The Scope for Transferring Damage Estimates

The environmental damages associated with a particular pollutant will depend on a great deal on the dose-response functions and damages per unit of pollutant. Clearly, it would be infeasible to estimate all environmental damages for each location and pollutant *ab initio*. Therefore, the transfer of estimates from one study to another is highly important. Benefit transfer is "an application of monetary values from a particular valuation study to an alternative or secondary policy decision setting, often in another geographic area than the one where the original study was performed" [12]. There are three main biases inherent in transferring benefits to other areas:

(a) original data sets vary from those in the place of application, and the problems inherent in non-market valuation methods are magnified if transferring to another area;

(b) monetary estimates are often stated in units other than the impacts. For example, in the case of damage by acidic deposition to freshwater fisheries, dose-response functions may estimate mortality (reduced fish populations) while benefit estimates are based on behavioural changes (reduced angling days). The linkage between these two units must be established to enable damage estimation;

(c) studies most often estimate benefits in average, non-marginal terms and do not use methods designed to be transferable in terms of site, region and population characteristics.

Benefit transfer application can be based on expert opinion or meta analysis. Expert opinion looks at the reasonableness involved in making the transfer and in determining what modifications or proxies are needed to make the transfer more accurate. In many cases expert opinion has been resorted to in making the benefit transfer for this project, as was the case in the ExternE Project. In general the more 'conditional' the original data estimates (e.g. damages per person, per unit of dispersed pollution, for a given age distribution) the better the benefit transfer will be.

2.1 META ANALYSIS

Where several studies, reporting a similar final estimate of environmental damage, exist, and where there are significant differences between them in terms of the background variables, a procedure known as meta-analysis has been developed to transfer the results from one study across to other applications. What such an analysis does is to take the estimated damages from a range of studies of, for example, coal fired plants and see how they vary systematically, according to affected population, building areas, crops, level of income of the population, etc. The analysis is carried out using econometric techniques, which yield estimates of the responsiveness of damages to the various factors that render them more transferable across situations. This can then be used to derive a simple formula relating environmental costs to per capita income, which could then be employed to calculate damages in countries where no relevant studies were available.

2.2 CONCLUSIONS ON BENEFIT TRANSFER

Transferability depends on being able to use a large body of data from different studies and estimating the systematic factors that would result in variations in the estimates. In most cases the range of studies available are few. More can be done to carry out meta-analysis of the type indicated, but it will take time. The best practice in the meantime is to use estimates from sources as close to the one in which they are being applied and adjust them for differences in underlying variables where that is possible. Often the most important obstacle to systematic benefit transfer, however, is a lack of documentation in the existing valuation studies.

From the environmental damage-energy source linkages identified above, one can identify an increasing order of difficulty (in terms of modifications that have to be made) with which estimates can be transferred from the original study to the situation in which they are to be used;

(a) the most easily transferred data is the dose-response function itself, relating environmental impacts adjusted for population. Thus numbers in the form 0.8×10^{-6} excess deaths per µg m^3 would be transferable across studies as long as adjustments to the other variables in the dose-response function were made (e.g., relative humidity). The additional local information that is required to use such data is simply local market conditions, costs and prices;

(b) the next ones in order of difficulty are monetary estimates of damages per unit of pollutant by concentration. Results are reported, e.g., in ECU/µg m^{-3}, or in ECU/km/person of lost visibility. Estimates may vary according to population affected, in which case an analysis of such variations would be desirable. Other socioeconomic variables that would be of relevance are income levels of the affected population, age, background environmental variables such as rainfall etc., and socioeconomic variables such as medical services and how they are paid for. If enough studies are available, a meta-analysis can be performed (see below), in which the mean estimated value is regressed against these variables. Then the relevant adjustment to the estimates are made, given the local values of the explanatory variables. No additional local variables should be required;

(c) similar to (b) above are estimates of monetary damages in terms of emissions or units of energy produced. In such cases one needs all the information listed above, plus details of how the emissions or energy units relate to the concentrations or whatever impacts are responsible for the damages. For example, damages may be quoted as \$x/kWh for coal. The relevance of this estimate to a different situation will depend on how the kWh is related to emissions and how the emissions are converted into concentrations in the area where the impacts were measured, plus the variables with which the relationship between concentrations and damages vary. Thus most work will have to be done in these cases and, in the context of a fuel cycle study it is unlikely that such estimates could be used at all.

2.3 SCOPE FOR PREPARING DAMAGE ESTIMATES FOR RUSSIA

The correct approach would be based on a spatially disaggregated analysis of damages. Such a disaggregation is essential to achieve the kind of detail that is required for an adequate valuation, as well as providing regional information likely to be of great value to policy makers. Regional data are probably the most valuable for a country as administratively vast as the Russian Federation. Given data on the levels of air pollutants, damage estimates could be made using dose-response functions and unit valuations based, as far as possible, on a WTP approach The method is a somewhat restricted version of a full *bottom up* analysis adopted in the ExternE Project. Sources of pollution are not estimated directly, only by attribution after the analysis is complete. Priority pollutants to be analysed are particulates (PM_{10}), SO_2 and NOx. Priority impacts to be quantified from a policy making perspective are health damages; mortality and morbidity effects.

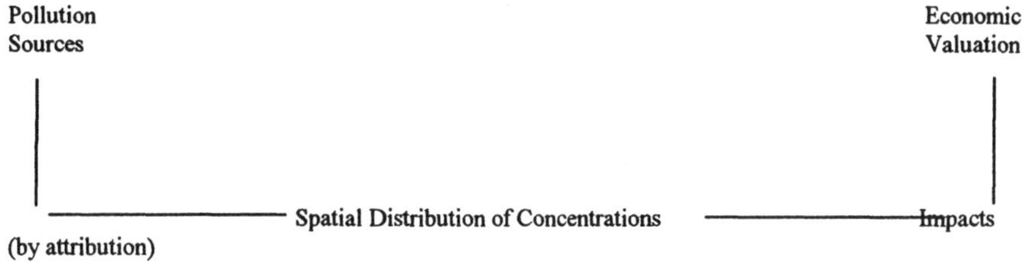

FIGURE2: Modified Impact Pathway Approach

Differences between countries in terms of attitude to risk are not clearly enough identified in the VOSL literature for us to be able to make separate country estimates. One adjustment that has been proposed for lower income countries is to adjust the

VOSL by the ratio of the real per capita GDP in the country concerned, to the GDP in the US or EU [9]. Real GDP implies purchasing power parity (PPP) adjustment. This implicitly assumes an "income elasticity" of willingness to pay to reduce mortality risks of one. A different value for this elasticity is cited in Krupnick *et al.* [8], who, referring to the work of Mitchell and Carson [11], argue that a case can be made for a value of 0.35.

Russia's *per capita* GDP in 1994 was $4610 in after correcting for PPP [18]. That would make it 18% of that of the US in the same year. ExternE currently uses a VOSL of ECU 3.1 mn. (1995 prices). Based on this value the US VOSL would be ECU 4.1mn, or $5mn (assuming an elasticity of one for VOSL with respect to GDP). With the same elasticity the corresponding Russian value would be $0.9mn and using an elasticity of 0.35 it would be $1.7mn. Given the large uncertainties about the VOSL value we have taken a range that goes from $500,000 to $1.7mn.

Although a number of estimates of environmental damage had been made in the pre-transition period, the methodology was not based on WTP/WTA. Furthermore, the changing economic, social and environmental situation would render the earlier estimates invalid. Estimates of health damages from air pollution have been made by Krupnick *et al.*. [8]. They took data on air quality in selected sites in four countries (Bulgaria, Hungary, Poland and Ukraine). Using this they calculated the change in exposure to the human population that would result from a reduction in concentrations to meet EC standards. They valued the changes in exposure by applying benefit transfer from the US. Morbidity impacts were scaled down for the difference in average daily wages ($7.5/day in the CEE region and $90/day in the US, giving a scaling factor of 8.3 percent). Mortality impact were also scaled down on the same basis. In addition a second estimate of mortality impact was made using an income elasticity of demand for reduced risk of death of 0.35 (see above). The corresponding values of a statistical life are $0.87mn and $1.84mn respectively, (1995 prices).

We have taken the estimated health benefits from Krupnick *et al.* and divided them by the changes in emissions. In doing so we have assumed that the fall in total emissions is equal to the mean population-weighted concentration reductions required to meet EC ambient concentrations standards. Emissions data are for 1991 and are taken from the World Resources Institute figures.

The resulting health benefits for TSP and SO_2 are given in Table 1 below, converted to 1995 prices. As our interest is in total benefits (not just health), we have scaled up the health damages by a factor of 9.8 for SO_2. This is based on estimated total damages for the UK [10], as a percentage of health damages. For TSP the damages are almost all health related (except for soiling) and therefore no adjustment has been made. Total damages for the four countries are reported in $1995/tonne of emissions. Krupnick *et al.* estimate very low benefits for sulphur reductions, because they assume no mortality effects. This is questionable; in the ExternE study it was argued that an SO_2-mortality link probably did exist, but that it could not be regarded as additive to the TSP-mortality link, as the two variables tended to pick up a common effect. As a comparison, Table 2 reports damages for PM, NO_x and SO_2 for studies from OECD countries.

TABLE 1: Total damage costs for air pollution in selected transition economies (based on Krupnick *et al.* [8])

Country	Pollutant	Damage Per Tonne
Bulgaria	Particulates	n.a.
Hungary	Particulates	5668
Poland	Particulates	4303
Ukraine	Particulates	n.a.
Bulgaria	SO_2	59
Hungary	SO_2	0
Poland	SO_2	13
Ukraine	SO_2	0

Most of these are 'bottom-up' studies similar to ExternE, combined with some benefit transfer of dose-response function and unit valuation. A comparison of the results in Tables 1 and 2 reveals the following:

a) Total damages for SO_2, in Table 2 show considerable variation. Within one study, using a consistent methodology, the estimates vary be +/- 30 percent. These reflect differences across sites. Across studies there is a much wider range, with the European studies showing damages of $2100 to $3700/tonne, and US studies showing damages of $1068 to $1686/tonne. All these above numbers are largely derived from health impact which remain controversial; hence the big differences with Table 1.
b) For NO_x the damages are even more variable. The Rowe *et al.* [16] study argues that damages are negative for combined cycle gas plants and pulverized coal plants as a result of reducing zone concentrations in the urbanized areas. The figures in Table 2 are for the natural gas combined cycle plant. Other US studies also demonstrate a large variation, indicating that NO_x damages are highly site specific.
c) For particulates, figures within one study, which adopts a common methodology vary by a factor of 15 between the lowest and highest values [16]. Across studies, the lowest US value is around $3,000/tonne and the highest $59,000/tonne, a factor of 20. European studies range from $15,500/tonne to $94,000/tonne. The higher European estimates are largely a result of the inclusion of chronic mortality effects, in addition to the acute ones. Most US studies, as well as CSERGE [1] do not include chronic effects.

The conclusions from these tables are therefore that:

- There is a substantial variation within one country of damage estimates, based on location.
- Depending on what dose-response functions are taken, values can be altered by a factor of ten or more.

For policy purposes, in an economy in transition there are two implications. First, in transferring values from OECD countries, we must allow for the considerable uncertainty that exists in the OECD estimates themselves; particularly on the health impacts. Second, average values such as those reported by Krupnick *et al.* are of little use in decisions involving specific plants, locations etc.[4]

2.4 RECOMMENDED VALUES FOR RUSSIAN STUDIES

On the basis of the above discussion we recommned using a range of values in Russian studies on the health impacts of air pollution. If we scale down damages by 18% from the OECD figures, reflecting the differences in PPP measures of US and Russian GDP, the range for SO_2 damages is $186/tonne to $671/tonne. This is higher than the figures from Krupnick *et al.*, but reflects the possibility of health damages, which should not be ignored. For TSP, the corresponding range of values based on the OECD figures is $540/tonne to $16,920/tonne. This embraces the figures given in Table 1 by Krupnick *et al.*. Further orginal work is needed to test the suitability of these ranges and to investigate other pollutants such as NO_x and O_3.

TABLE 2: Estimates of damage costs for air pollution in OECD countries

Study / date	Study Area	Pollutant	Damage $/Tonne
CSERGE (1993) [1]	UK (Europe)	SO_2	2925
Rowe *et al.* (1995) [2]	New York	SO_2	1068
Thayer *et al.* (1994) [3]	California	SO_2	1686
NY Energy Plan (1995) [4]	New York	SO_2	1035
ExternE (1996) [5]	UK	SO_2	2108
ExternE (1996)	Germany	SO_2	3727
CSERGE (1993)	UK (Europe)	NO_x	1276
Rowe *et al.* (1995)	New York	NO_x	-124
Thayer *et al.* (1994)	California	NO_x	18072
NY Energy Plan (1995)	New York	NO_x	5070
Northern States Power (1992) [6]	Minnesota	NO_x	242
CSERGE (1993)	UK (Europe)	Particulates	15553
Rowe *et al.* (1995)	New York	Particulates	26058
Thayer *et al.* (1994)	California	Particulates	59420
NY Energy Plan (1995)	New York	Particulates	2973
Northern States Power (1992)	Minnesota	Particulates	3052
ExternE (1996)	UK	Particulates	38241
ExternE (1996)	Germany	Particulates	94435

(1) The CSERGE figures were originally in £1992. Inflation factor is 11.1% and x-rate £1=$1.56. Damages are from across Europe from UK sources.

(2) The Rowe *et al.* figures were originally in $1992. Inflation factor is 12.4%. The figure is the

[4] This is not intended as a criticism of Krupnick *et al..*, [8] Their study is about national policy, although the above discussion suggests that a national policy based on uniform standards may not be the best form of regulation.

straight average of the lower (787) and upper (1349) values. The negative figure for NO_X is due to ozone scavenging near urban sites for a combined cycle gas plant.

(3) The Thayer *et al.* figures were originally in \$1992. Inflation factor is 12.4%.

(4) The NY Energy Plan figures were originally in \$1992. Inflation factor is 12.4%.

(5) Original ExternE figures were in ECU1992. Inflation factor is 13.2% and x-rate 1ECU=\$1.269. Damages for accidents are divided between SO_2 and particulates in proportion to emissions.

(6) The Northern States Power figures were originally in \$1992. Inflation factor is 12.4%. Reference to study is Desvouges *et al.*. [2].

3. Using Air Pollution Valuations within a Decision Making Context

The estimates of damages from air pollution have a number of practical applications. The following are some that we have been working on and that should be of potential interest to other researchers. Indeed, we invite Russian experts to comment on our methodolgy and its suitability for applying to the Russian context.

3.1 DECISIONS ON THE UPGRADING OF POLLUTION-INTENSIVE INDUSTRIES IN HIGHLY POPULATED AREAS

The costs/benefits derived here will provide an indication of where the benefits of upgrading energy technology are significant. They can be used to influence decisions on energy planning and fuel choice options as well as retrofitting technologies such as greater sulphur capture through flue gas desulphurisation technologies. Expenditures on environmental protection may not be popular due to more immediate concerns such as economic growth and job creation. A comprehensive cost-benefit analysis including the benefits from air pollution reduction would enable decisions to be placed in the appropriate, rational framework.

3.2 DECISIONS ON THE CLOSURE OF POLLUTION-INTENSIVE INDUSTRIES

In some industries upgrading the energy technology may not be possible or prohibitively expensive and pollution plants may be forced to close. This is likely to be a sensitive issue due to the effects on local unemployment. Ideally, the *social* costs of unemployment should also be considered which include not only the loss of income but also the value of additional time and the health consequences. With regard to health issues, many cohort studies have reported a statistically significant excess mortality rate among unemployed men. The above issues should again be integrated be integrated with the preceding discussion on health impacts from air pollution to formulate a cost-benefit analysis. It would then be possible to ascertain whether immediate or phased closure was the optimal policy.

3.3 INCENTIVES FOR THE ADOPTION OF RENEWABLE TECHNOLOGIES

In areas where there is potential for such technologies it may be possible to estimate the benefits of reducing fossil fuel use. The inclusion of social costs will enable a more level playing field to be established for such technologies which are often criticized on grounds of high capital outlay and long pay back periods.

3.4 EVALUATION OF PROGRAMS TO REDUCE CARBON EMISSIONS

The issue here is known as *secondary benefits*. Separate environmental objectives, particularly the attainment of greenhouse gas emissions targets may call for levels of CO_2 to be reduced. In adopting cleaner technology or fuel switching, levels of SO_2 and particulates will also be reduced, resulting in secondary benefits. Damage estimates can therefore assist in an assessment of the wider benefits of CO_2 emissions reductions.

4. References

1. CSERGE (1993) *Externalities from Landfill and Incineration*. HMSO, London.
2. Desvousges W.H. *et al.*. (1994) *Assessing Environmental Externality Costs for Electricity Generation (Final Report)*, Research Triangle Institute, Chapel Hill.
3. European Commission (1996) *ExternE: Externalities of Energy*, Vols. 1-6, European Commission, Brussels
4. Fallon, P. and D. Verry (1988) *The Economics of Labour Markets* Philip Allan
5. ECO Northwest (1987) *Generic Coal Study: Quantification and Valuation of Environmental Impacts*, prepared for the Bonneville Power Administration, Portland Oregon.
6. Hohmeyer, O. (1988) *Social Costs of Energy Consumption: External Effects of Electricity Generation in the Federal Republic of Germany*, Springer-Verlag, New York.
7. IPCC (1996). *Climate Change 1995: Economic and Social Dimensions of Climate Change*. Contribution of Working Group II to the Second Assessment Report of the IPCC. Cambridge University Press, Cambridge.
8. Krupnick, A. *et al.*. (1996) "The Value of Health Benefits from Ambient Air Quality Improvements in Central and Eastern Europe", *Environmental and Resource Economics*, 7, 4, 307-332.
9. Markandya, A. (1994) "Measuring the External Costs of Fuel Cycles in Developing Countries", in O. Hohmeyer and R.L. Ottinger (eds.) *Social Costs of Energy*, Springer-Verlag, Berlin.
10. Markandya, A *et al.*. *Green Accounting in Europe: The Role of Damage Estimation. Four Case studies.* Commission of the European Communities, draft February 1996.

11. Mitchell, R.C. and R.T. Carson (1986) *Valuing Drinking Water Risk Reduction Using the Contingent Valuation Method: A Methodological Study of Risks from THM and Girardia.* Report for the US Environmental Protection Agency, Washington DC.
12. Navrud, S. (1994). *Economic Valuation of External Costs of Fuel Cycles: Testing the Benefit Transfer Approach.* Forthcoming in A.T. de Almedia (ed.), 1994: Models for Integrated Electricity Resource Planning. Kluwer Academic Publishers.
13. New York State Energy Office (1994) New York State Energy Plan Vols. II-III.
14. Pace University Center for Environmental Legal Studies. (1990). *Environmental Costs of Electricity*, Oceana Publications, New York.
15. Pearce, D., C. Bann, and S. Georgiou. (1992). *The Social Cost of Fuel Cycles*, Centre for Social and Economic Research on the Global Environment, University College London, London, United Kingdom.
16. Rowe R.D. *et al.* (1996) "The New York Environmental Externalities Cost Study: Summary of Approach and Results" in *The External Costs of Energy: Proceedings*, CEC, Brussels.
17. Thayer, M.A. *et al.*. (1994) *The Air Quality Valuation Model*, Regional Economic research Inc. and TRC Environmental Consultants.
18. World Bank. (1996). World Development Report 1996, World Bank. Washington D.C.

OPTIMAL CONTROL APPLIED TO AIR POLLUTION

B. MOREL
Carnegie Mellon University
Department of Engineering and Public Policy
5000 Forbes Avenue
Pittsburgh, PA 15213. USA

> Mathematics has become the dominant language of the sciences, not because it is quantitative -- a common delusion -- but primarily because it permits a clear and rigorous reasoning about phenomena too complex to be handled with words.
> H. Simon, Nobel Laureate in economics, 1976 [1].

1. Introduction

The industrial revolution has been the beginning of an experiment consisting in changing the composition of the atmosphere by projecting gigatons of chemicals. In recent years evidences have accumulated that the impact on the biosphere and human health, is complex and potentially disastrous. The costs associated with the health effects of pollution keep increasing with evidences of new effects [2]. The mitigation of air pollution comes at a very significant cost and may even imply some difficult societal changes. On the other hand, the damages due to pollution are so substantial that it has become clear that failing to limit air pollution is potentially as costly as, or costlier than, engaging in a vigorous mitigation policy. So, if there is a trade-off between industrial activity and pollution level, then the relation between economic prosperity and pollution level has a different nature. The indirect cost of pollution is such that economic prosperity requires that the industrial activity be made environmentally friendly.

In the case of the southern Urals and in particular Magnitogorsk, pollution is associated with the production of steel and cannot be easily mitigated without endangering the economy of the area. On the other hand, pollution has been allowed to reach alarming levels and cannot be permitted to stay at present levels, let alone grow further.

Although pollution and its mitigation have a strong regional dimension, for a variety of reasons they have also a national dimension. The regulation of pollution, definition of norms for air quality, their enforcement and part of the cost of

I. Linkov and R. Wilson (eds.), Air Pollution in the Ural Mountains, 307–314.

implementing them requires the involvement of the central government. The different situations between the regions and the central authority is a potential source of tension. The economic impact of regulations varies drastically between regions. In the case of Russia, the Urals will probably be far more affected than the rest of the country.

The basic trade-off of air pollution is illustrated in Figure 1. It is between investing in mitigation or coping with the cost due to the negative impact of pollution. Although not much is known about the cost of mitigation, it is safe to assume that it increases non-linearly with the level of mitigation. In the same way, the damage cost of pollution increases non-linearly with the level of pollution. The damage depends on the atmospheric concentrations in pollutants. It is a poorly known function, as it varies from one region to the next, it depends on the details of the local biosphere and the health effects that differ from place to place. But its general shape is certainly concave, i.e. as suggested in Figure 1. The mere fact that we can assume such shapes to those curves is already a very significant mathematical piece of information.

The two costs add their effect to build the overall amount associated with pollution. From a strict economic point of view, the optimal point is to minimize the total cost associated with pollution. This point cannot be determined precisely in practice. It is not the same for different regions. The same health effects do not have the same societal cost in different societies. New negative effects on health or the biosphere continue to be discovered. The exact cost due to environmental damage is never precisely known.

What the Figure 1 fails to capture is the dynamical character of pollution. If pollution is allowed to affect the environment, the degradation is progressive and open-ended. Corrective measures have to be taken to redress the situation.

The existence of an economically optimal point implies that failing to invest enough in mitigation can lead to a higher overall cost due to pollution. This is what the lack of resources, available to support a vigorous mitigation policy, threatens to accomplish in the Urals.

The economically optimal point may not be optimal from every point of view. An additional reduction of the pollution level may result in a desirable improvement of the standard of living deemed affordable, at least in affluent societies.

In societies like the US, the communities concerned by the pollution problem have far more resources available for pollution mitigation than communities of the southern Urals, while the problem they face is less acute. This suggests that the methodology used in the US to address the problem of pollution may not be the best ones for Magnitogorsk. They may tend to consist, far too much in detailed assessments of impacts, instead of focusing on finding immediate cost-efficient mitigation measures. The most cost-efficient policy is the one that is closest to the solution with an optimal control approach to the problem.

In the balance of this paper we illustrate how optimal control works in that context and can help in designing an optimal long-term policy for a place like Magnitogorsk. The analysis is superficial, but hopefully suggestive.

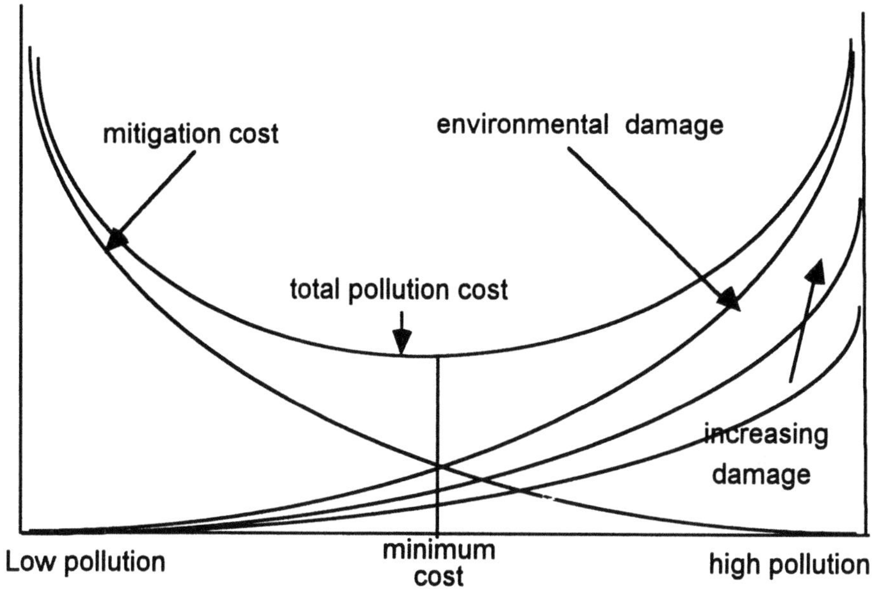

FIGURE 1. The mitigation problem in a diagram: There is a trade off between mitigation cost and environmental damage. Environmental damages are not well understood. With time the cost associated with these damages increases. Not investing enough in mitigation may result in a net cost.

2. Optimal Control and Air-Pollution Policy

In the context of policy optimal control is a normative approach, i.e. it provides a reference point: the path of minimal cost. The optimization of an air pollution policy supposes a good understanding of the physics and chemistry of air pollution and of its interaction with the economy. The former draws its complexity from the large number of factors involved. The latter varies between regions and tends to be poorly understood.

Optimal control provides a unique mathematical framework to put together all the elements of the problem and their uncertainties. It is based on a mathematical translation of what is understood of the problem. It should not involve additional assumptions. The ability to state the problem precisely is a measure of how well we understand it. The existence of an optimum solution is equivalent to the existence of a most cost-efficient solution. Alternatively, if the uncertainties of the problems were such that a well-defined solution to the optimal control problem did not exist, it would mean that the policy problem itself has no good solution or is based on a dilemma.

In optimal control language, the problem is to find the mitigation policy (expressed as the time dependent value of abatement $\mu_i(t)$) which minimizes:

$TCT(T) = \int_0^T e^{-\rho t} Cost(t) dt$, subject to the constraints that over time, the emission levels of the pollutants are to be reduced to a prescribed value $E_i(T)$.

The emissions can be expressed as: $E_i(t) = (1 - \mu_i(t))\sigma_i(t)Q(t)$, where $\mu_i(t)$ are the level of abatement at time t, and also the *control variables* of the optimal control problem, $\sigma_i(t)$ are technological factors of abatement and $Q(t)$ is the level of economic activity related to the emission pollutants. The coupling of the pollution to the economy is made through that quantity. Its form changes with the situation. In the case of Los Angeles for example, it is proportional to the population (number of cars) and the economic prosperity of the area. In the case of Magnitogorsk, it is proportional to the steel production.

The *dynamical variables* of the optimal control problem are the atmospheric loads of pollutants, $M_i(t)$. We assume that they obey equations of the kind [3].

$$\dot{M}_i(t) = -\delta_M \left(M_i(t) - M_0^i\right) + \beta_i E_i(t)$$

To solve the optimal control problem we need to give an explicit form to the cost function. The total cost of air-pollution is the sum of two components (Figure 1): mitigation cost and environmental damage.

Because of the general form of the curve, the cost of achieving an abatement μ in the emission of a given pollutant ($\mu_i(t) = 1 - \frac{E_i(t)}{E_i(0)}$) can be approximated by[4]: $C(\mu) = a\mu^{\nu}$. In the case of Magnitogorsk, the major part of the mitigation effort should be in abatement of emissions of sulfur oxides. This would have the additional effect of also reducing the atmospheric load of particulates [5] which is the other major pollutant in Magnitogorsk.

The damage function can be approximated by: $D(m) = bm^{\lambda}$, where $m = M(t) - M_{ref}$, the difference between the actual atmospheric concentration of pollutants, and the level of concentration deemed safe[6] The total cost function has therefore the general form $bm^{\lambda} + a\mu^{\nu}$.

This is not the function we use in the optimal control problem. There is a huge asymmetry between cost due to environmental damage and mitigation costs. The latter is a direct cost. Whereas the former is more indirect and can only be roughly estimated. Although the damage due to pollution is the main motivation for the mitigation policy, it does not enter into the design of an optimal mitigation policy in the same way as mitigation cost. The target levels of abatement are computed according to criteria which include the environmental damage. International human health norms also enter. Mitigation strategies have to be fine tuned to minimize the cost to achieve those predetermined target levels of abatement. The cost-efficiency of mitigation technologies is the driving factor for the choice of policy.

For the purpose of our computations, this means that we can assume that the level of abatement $\mu(T)=1-\frac{E(T)}{E(0)}$ will be achieved eventually. It has been chosen, taking into account all the damage effects and the only cost relevant for the optimal control problem is the mitigation cost: $C(\mu)=a\mu^{\nu}$.

With those assumptions the optimal control problem can be readily solved [7]. The time dependence of the optimal implementation of abatement policy is exponential:

$$\mu(t)=\mu(T)e^{\frac{(\rho+\delta_M)}{\nu-1}(t-T)}, \text{ with } \mu(T)=\left(1-\frac{E(T)}{\sigma(T)Q(T)}\right) \text{ as target value.}$$

The total cost of mitigation is:

$$TCT(T)=a\mu(T)^{\nu}e^{-\frac{\nu(\rho+\delta_M)T}{\nu-1}}\frac{(\nu-1)}{(\nu\delta_M+\rho)}\left\{e^{\left[\frac{\nu(\rho+\delta_M)}{\nu-1}-\rho\right]T}-1\right\}$$

It depends on the target value of the abatement, on the time at which this value should be reached, on the exponent ν and coefficient of the cost function, on the discount rate ρ, and on the rate at which the pollutant is cleared through natural channels δ_M. These parameters are variably known. Important to notice is the robustness of the form of the result to uncertainties in the values of those parameters.

3. APPLICATION TO THE CASE OF MAGNITOGORSK:

In the previous section, we computed the solution to the optimal control problem without asking the question: at what level of mitigation does the production of steel cease to be profitable? The cost of production of steel may be affected by the use of abatement technology. If we assume that the revenue coming from the production decreases proportionally to the level of abatement, (i.e. $R(Q,\mu)=R-\Gamma\mu$), there may be a value $\tilde{\mu}$ such that the revenue vanishes. This defines an upper limit for the degree of mitigation (Figure 2).

If the optimum level of mitigation is economically sound $\tilde{\mu}$ is lower than the ideal level of mitigation μ_{id}, in order to maintain the target atmospheric load of pollutant, a reduction of production $\Delta\tilde{Q}=Q-\tilde{Q}=\frac{Q(\mu_{id}-\tilde{\mu})}{(1-\tilde{\mu})}$ is necessary.

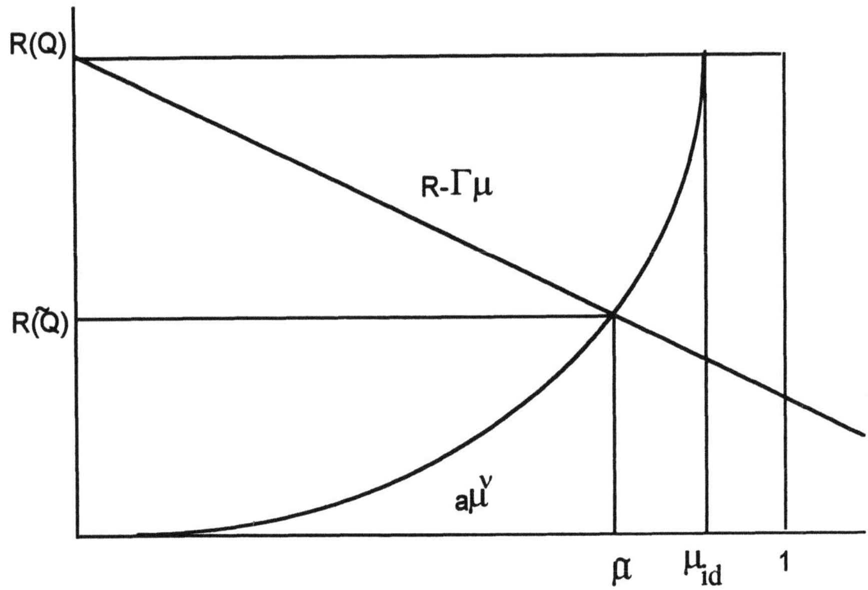

FIGURE 2. The optimization problem in diagram: the "ideal" level of mitigation mid is such that mid does not correspond to the optimum level of mitigation. This is $\widetilde{\mu}$. The optimum policy involves a certain reduction of polluting production.

The revenue coming from the production of steel is therefore bound to be reduced twice: because of the increase in production cost due to mitigation measures and because of the reduction of the production.

This is not the end of the optimization problem. The loss of revenue due to the reduction of production has to be compared with the cost (in taxes and environmental damage) of polluting a bit more. It is probably economically more sound to suffer some pollution than to meet the norms.

This takes place on the backdrop of a somewhat weak steel economy. As an example, the City of Pittsburgh, among others, demonstrates that old-fashioned production of steel is not a solid industrial activity.

The uncertainties relevant for policy are the values of the parameters "a" and "ν" which control mitigation cost, "δ_M" the rate of clearance of the pollutant, "ρ" the discount rate, $E(T)$ the permissible asymptotic emission level, $\sigma(T)$ the abatement level due to technological development, and $Q(T)$ or $\widetilde{Q}(T)$ the optimum asymptotic level of production, and to a lesser extent "b" and "λ" which control environmental costs.

Some of those uncertainties may be reduced with time: for example the level of abatement achieved at a certain cost and the impact on production cost of mitigation measures, and the evaluation of the damage due to pollution. Some uncertainties as the economics of steel production may remain. None of those uncertainties have the potential to change fundamentally the result.

There is a diminishing return in investing endlessly in assessing the health effects, for example. A more valuable information would be in understanding better the economics of mitigation. A better estimate of the cost of mitigation technologies, for different levels of mitigation, is of essence. Depending on the ultimate level of mitigation needed, different combinations of approaches are optimal. These different approaches have to be evaluated in the context of the specifics of the region, the economic prospects of the production and the ability of the plant to modernize. The choice of mitigation technologies involves a non-trivial optimization.

In practice, the policy will have to go through frequent if not permanent re-evaluations in the spirit of dynamical programming. This kind of approach requires the support of an optimal control algorithm.

4. Conclusion

The fundamental policy challenge of pollution is to translate into economic terms interactions between scientific considerations like the physics of pollution and its impact with economic realities like the industrial activity and to define an optimal path toward an environmentally friendly prosperous economy. Optimal control is the only mathematical framework allowing such an integrated discussion. Even if the results are normative in nature, they reveal a lot of structure of the general problem. An important finding is the robustness of the general results to the uncertainties.

The uncertainties associated with the effects and mitigation of air-pollution, do not preclude the usefulness of optimal control. Quite the opposite: a message of the optimal control study is that the design of an optimal mitigation path is not significantly affected by the uncertainties of the value of the different parameters of the problem. This is due to the fact that however large the uncertainties may be, they do not affect the general shape of the mitigation curves and/or damage curve. They only affect the value of the parameters. They affect the details of the implementation of the optimal policy, not its general design.

This means concretely, that many of the uncertainties surrounding the pollution problem of cities like Magnitogorsk do not need to be completely suppressed or even significantly reduced before sound policy decisions can be taken.

The mere fact that the shape of the curves is known, is in and of itself, a very significant piece of information which makes it possible to solve analytically a simplified version of the problem, without having to give explicit values to the parameters. As seen in the previous section, a rough knowledge of the value of the parameters is sufficient to characterize the contours of an optimal policy. Furthermore, the general formulas of the previous sections can be used to estimate the value of additional information.

The lack of resources available to be invested in mitigation means that it may not be possible to implement the mitigation at an optimal rate. Optimal control can be used to estimate the cost of not being able to stay on the "optimal" trajectory. The "solution" of the pollution problem of the Urals involves probably a reduction of the industrial activity responsible for that pollution, i.e. steel production. Such difficult decisions belong to the long term economic plan for a cleaner and prosperous Magnitogorsk. However imperfect optimal control may be to discuss policy, there is no alternative to optimal control arguments to conduct an integrated assessment and provide a compelling basis for such decisions.

Optimal control has its share of complexity and its limitations. The result of an optimization algorithm, depends critically on the constraints. One of the challenges of applying optimal control in the context of policy in general, and air-pollution mitigation in particular, is to understand the constraints. Otherwise the solution obtained is completely unrealistic. The constraints put in the statement of the optimal control problem are therefore very important, and must be carefully examined and revisited.

What could be construed as an additional obstacle to the usefulness of the approach can also be seen as one of its major virtues. The mere exercise of identifying the constraints specific to a region like Magnitogorsk (which we did not do in this paper as it can only be the result of a dedicated research), is in and of itself an important contribution to understand its solution.

5. References

1. Models of Man, J. Wiley and Sons, p. 89, 1957.
2. See for example, R. Day in these proceedings.
3. More complicated equations are also possible. For example tropospheric ozone is the product of a reaction between NOx and VOC. The evolution of the mass load is described by: $M_{ozone}(t)=kM_{voc}(t)M_{NOx}(t)—KM_{ozone}(t)$. The concentration distribution involves transport.
4. This formula can be justified on more theoretical grounds. Cf W.D. Nordhaus, *Managing the Global Commons, the Economics of Climate Change,* MIT Press, 1994. For clarity, we limit the number of pollutants to one.
5. J. Seinfeld: *Atmospheric Chemistry and Physics of Air pollution*, John Wiley and Sons, 1986.
6. If more than one pollutant is present, the damage function would presumably have terms where the product of the concentration enter expressing the synergistic effect of the presence of different pollutants at the same time.
7. B. Morel, Air Pollution Control as an optimal control problem, in preparation.

UNCERTAINTY ASSESSMENT OF BIOGENIC EMISSIONS ESTIMATES AND ITS IMPACT ON OZONE ATTAINMENT CONTROL STRATEGY SELECTION

J.G. WILKINSON
Carnegie Mellon University, Department of Engineering and Public Policy5000 Forbes Avenue, Pittsburgh, PA 15213. USA

Abstract

The regulatory community continues to issue guidance on air quality planning which encourages planners to rely on best, single value emissions estimates for use in the air quality modeling process. Little if any guidance is issued on how to assess and use uncertain emissions estimates to evaluate control strategy selection. In particular, biogenic emissions are interesting in that they are highly uncertain, and they are unaffected by traditional anthropogenic control technologies. Yet, biogenic emissions impact air quality particularly impacting the effectiveness of stringent controls on volatile organic compounds. Therefore, because emissions controls cost billions of dollars annually, it is prudent to investigate the impacts that uncertainty in biogenic emissions estimates (and in general, all emissions estimates) has on emissions control scenarios.

In this study, the consequences are shown of how improved biogenic emissions estimates impact air quality modeling results and the degree to which uncertain biogenic emissions affect emissions control strategy selection in the South Coast Air Basin (SoCAB) of Los Angeles, California. First, the ranges in the biogenic emissions estimates due to uncertainty in data and model parameters are as follows (all values reported at the 95% confidence interval): isoprene ranges from 0.03 to 3.7 times the expected value (1800 tons/day); nitric oxide ranges from 0.22 to 72 times the expected value (99 tons/day); other volatile organic compounds range from 0.54 to 1.7 times the expected value (980 tons/day); and monoterpenes range from 0.54 to 1.8 times the expected value (700 tons/day). Second, the improved biogenic emissions estimates (50th percentile) for the SoCAB increase the modeled ozone peak from 178 ppb (base case simulation with previous biogenic emissions estimates) to 216 ppb for August 29, 1987 which is the last day of the three day episode. Finally, when the uncertain biogenic emissions estimates are propagated through a photochemical model, the results indicate that NO_X control strategies are favored over VOC control strategies. This finding contradicts the recent findings by the South Coast Air Quality Management District (SCAQMD) which show that VOC emissions controls will preferentially help mitigate the ozone problem in the SoCAB.

I. Linkov and R. Wilson (eds.), Air Pollution in the Ural Mountains, 315–340.

1. Introduction

Emissions control programs throughout the United States have resulted in the reduction of peak ozone levels over the last fifteen years, though controls have not been as effective as desired, and many urban areas still experience elevated ozone levels [1]. One of the problems is that ozone is formed through complex, photochemical reactions of reactive organic gases (ROG) and oxides of nitrogen (NO_X) [2, 3], and it is not readily apparent which emissions controls are most effective to reduce high ozone levels. Further, since the ROG and NO_X loadings in an airshed are comprised of both anthropogenic and biogenic sources, accurately estimating both anthropogenic and biogenic emissions and the related uncertainties are necessary to develop effective emissions control programs. Furthermore, in some airsheds, biogenic emissions are the dominant ROG emissions source [4, 5, 6].

Air quality modeling relies on emissions inventories to estimate the impacts of emissions control strategies on future air quality. Emissions inventories are compilations of emissions estimates from numerous sources in an airshed. The effort to prepare emissions estimates is resource intensive and is fraught with uncertainty. In particular, existing national and regional biogenic emissions inventories are typically not well suited for use in air quality planning [1]. Although numerous studies have improved our understanding of biogenic emissions [7, 8, 9, 10, 11, 12], biogenic emissions estimates remain highly uncertain [13, 14]. Also, since regulatory guidance still focuses on the use of *single value* emissions estimates for both anthropogenic and biogenic emissions in developing control strategies, little has been done to quantify the uncertainty in emissions estimates let alone its impact on emissions control programs. The objectives of this study are to:

- Prepare improved biogenic emissions estimates for the South Coast Air Basin (SoCAB, Los Angeles, CA);
- Evaluate the improved emissions estimates which are based on advanced biogenic emissions estimates models;
- Quantify the uncertainty in the biogenic emissions estimates due to the biogenic model parameters and data inputs;
- Quantify the impact of uncertain biogenic emissions on predicted air quality;
- Examine the impact that uncertain biogenic emissions estimates have on the choice of emissions control programs in the SoCAB; and
- Develop an improved understanding of how well the current state-of-the-science biogenic emissions models correspond to observations.

This manuscript briefly addresses the ozone problem in the SoCAB and the emissions control strategy as adopted by the SCAQMD in the 1994 Air Quality Management Plan (AQMP). Next, a discussion of the computational algorithms for the *Bio*genic *M*odel of *E*missions (BIOME) is provided, and the uncertainty in the model parameters and the data is described. This is followed by a brief description of the California/Carnegie Institute of Technology (CIT) air quality model, and the method that was used to elicit the impact on control strategy selection due to uncertainty in the biogenic emissions. Finally, the results are discussed followed by a brief description of possible future research.

2. Methods

This study included two components: an uncertainty analysis and an impact analysis. New biogenic emissions estimates were developed using algorithms that are considered state-of-the-science. Further, the uncertainties in the new estimates were quantified. Photochemical modeling was conducted to elucidate the impacts that uncertain biogenic emissions estimates have on predicted air quality. This section further describes the methods used in this study.

2.1 SOUTH COAST AIR BASIN

The South Coast Air Basin (SoCAB) of Los Angeles, CA (Figure 1) is the only air basin in the United States to be classified in *extreme* non-attainment of the National Ambient Air Quality Standard (NAAQS) for ozone which is 0.12 ppm on a one hour average basis [15]. An air basin is in *extreme* non-attainment of the ozone NAAQS when the fourth highest, one hour average ozone value exceeds 0.28 ppm over three contiguous years. Although the number of days where ozone values above 0.12 ppm continues to decline in the SoCAB (e.g. 179 days in 1992, 147 days in 1993, 99 days in 1995 [16, 17]), the SoCAB continues to record the greatest number of days exceeding the ozone NAAQS of any area in the country [18]. This is despite significant efforts of public and private concerns to control emissions in the SoCAB over more than thirty years.

The SCAQMD proposes to control NO_X emissions by 77% and VOC emissions by 69% to attain the ozone and particulate matter NAAQS [19]. Control measures are to be gradually implemented until attainment is reached in the 2010. However, the proposed control measures appear to have been developed without regard to how biogenic emissions impact air quality in the basin, let alone any regard for the uncertainty in the emissions estimates (i.e. the air quality management plan [AQMP] emissions summary tables report all biogenic emissions estimates as zero [19].

2.2 BIOGENIC MODEL FORMULATION

Numerous models to estimate biogenic VOC emissions exist [5, 9, 20, 21]. The model that is currently accepted as representing the state-of-the-science [22] has been successfully applied by Geron et al. [23] in the eastern United States and by Wilkinson et al. [14] in the Coastal Oxidant Assessment Study for Texas. Unlike biogenic VOC models, very few models exist to estimate biogenic NO_X. Williams et al. [24] present a model to estimate biogenic nitric oxide that has gained wide acceptance.

The *Bio*genic *M*odel of *E*missions (BIOME) is a component of the Emissions Modeling System -- EMS-95 [25]. BIOME computes biogenic emissions estimates of isoprene, monoterpenes, other volatile organic compounds, and nitric oxide from biogenic emissions factors, land use data, and meteorological data. Figure 2 is a simplified processing diagram of BIOME. Details on the operation of BIOME can be found in Wilkinson et al. [25].

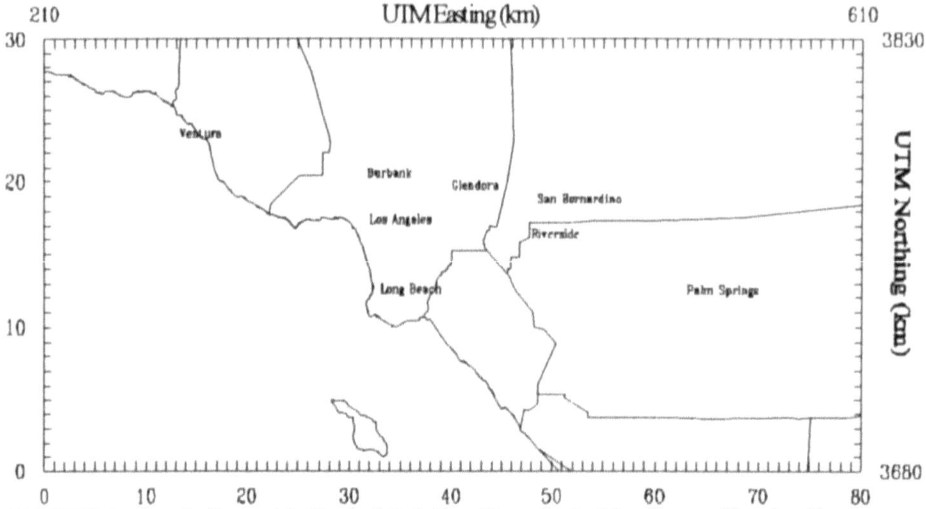

FIGURE 1. South Coast Air Basin Modeling Domain (grid cells are 5km by 5km).

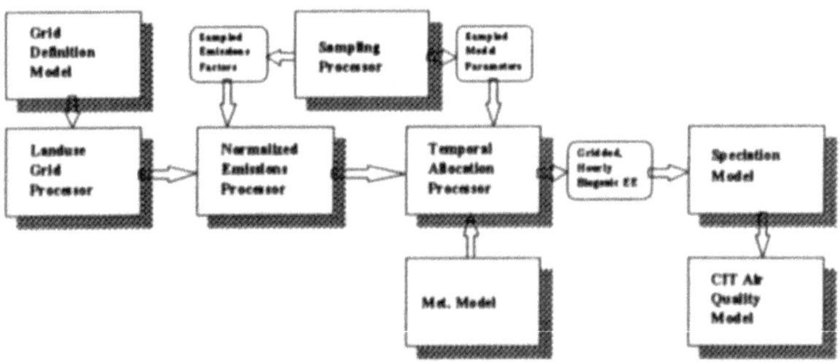

FIGURE 2. Simplified Schematic of BIOME.

2.3 VOC EMISSIONS ESTIMATES

Recent studies suggest that photosynthetically active radiation (PAR) and leaf temperature can account for short term variation in isoprene emissions [9, 26]. However, short term variation in monoterpene emissions appears to be dependent only on leaf temperature [e.g. 21, 26]. Equation 1 is the general form to estimate biogenic VOC emissions with the isoprene environmental correction factor shown in Equations 2 to 4 and the monoterpene and other VOC environmental correction factor shown in Equation 5 [22].

$$E = E_S \cdot C_f \cdot A \tag{1}$$

where: E is the meteorological-specific emission rate ($\mu g \cdot hour^{-1}$);
E_S is the species/pollutant-specific emission flux at 30°C and 1000 $\mu mol \cdot m^{-2} \cdot s^{-1}$ ($\mu g \cdot ha^{-1} \cdot hour^{-1}$);
C_f is the hourly, pollutant-specific environmental correction factor; and
A is area of species coverage (ha).

$$C_f = C_L \cdot C_T \tag{2}$$

where: C_L is the light intensity correction factor; and
C_T is the temperature correction factor.

$$C_L = \frac{\alpha \cdot c_{L1} \cdot L}{\sqrt{1 + \alpha^2 \cdot L^2}} \tag{3}$$

where: L is the PAR flux ($\mu mol \cdot m^{-2} \cdot s^{-1}$);
c_{L1} is an empirical coefficient derived from emission rate measurements; and
α is an empirical coefficient derived from emission rate measurements.

$$C_T = \frac{\exp\left(\dfrac{c_{T1}(T - T_S)}{R \cdot T_S \cdot T}\right)}{1 + \exp\left(\dfrac{c_{T2}(T - T_S)}{R \cdot T_S \cdot T_M}\right)} \tag{4}$$

where: T is the leaf temperature (taken to be ambient temperature -- K);
R is the ideal gas constant (8.314 $J \cdot K^{-1} \cdot mole^{-1}$);
T_S is the normalizing temperature (303 K);
T_M is an empirical coefficient derived by fit to species of several genera;
c_{T1} is an empirical coefficient derived by fit to species of several genera; and
c_{T2} is an empirical coefficient derived by fit to species of several genera.

$$C_f = \exp(\beta \cdot (T - T_S)) \tag{5}$$

where: C_f is the TERP or OVOC environmental correction factor;
β is an empirical coefficient derived by fit to species of several genera;
T is the leaf temperature (taken to be ambient temperature -- K); and
T_S is the normalizing temperature (303 K).

Table 1 presents the model parameters with their nominal value, range, and confidence interval [24, 27, 28, 29]. Although the model parameters are positively correlated, insufficient data were available to compute the correlation matrix. Therefore, in this analysis, the model parameters were treated independently for sampling.

TABLE 1. Biogenic Emissions Model Parameters And Uncertainty Values.

Parameter[1]	Nominal Value	Range	Confidence Interval
	Biogenic VOC Model Parameters		
α	0.0027	± 0.0015	1σ
C_{L1}	1.06	± 0.2	1σ
C_{T1}	95000	± 20000	1σ
C_{T2}	230000	± 150000	1σ
T_M	314	± 3	1σ
β	0.09	± 0.02	1σ
	Biogenic NO Model Parameters		
T_3	0.071	± 0.007	1σ
	Land Use Type: Grasslands		
T_2	8.8	± 4.4	1σ
T_1	0.66	± 0.33	1σ
	Land Use Type: Forests		
T_2	3.6	± 1.8	1σ
T_1	0.84	± 0.42	1σ
	Land Use Type: Wetlands		
T_2	4.4	± 2.2	1σ
T_1	0.92	± 0.46	1σ
	Land Use Type: Corn		
T_2	5.8	± 2.9	1σ
T_1	0.72	± 0.36	1σ
	Land Use Type: Cotton, Wheat, Soybeans		
T_2	2.9	± 1.45	1σ
T_1	1.03	± 0.515	1σ

[1]PAR (L), temperature (T), area of species coverage (A), and species/pollutant-specific emissions flux (E_S) are data inputs.

2.4 NITRIC OXIDE EMISSIONS ESTIMATES

BIOME is used to estimate NO emissions due to microbial activity in soils (e.g. microbial nitrification and denitrification processes [e.g. 46]). Microbial nitrification and denitrification processes occur in both agricultural soils and natural soils, but can be significantly enhanced in agricultural soils due to fertilization activity. In fact, studies have shown that NO emission fluxes from fertilized soils can approach those of urban areas [30, 31]. Biogenic NO emissions are estimated as:

$$E = E_S \cdot C_f \cdot A \tag{6}$$

where: E is the meteorological-specific NO emission rate ($\mu g \cdot hour^{-1}$);
E_S is the species-specific NO emission flux at 30°C ($\mu g \cdot ha^{-1} \cdot hour^{-1}$);
C_f is the hour-specific environmental correction factor;
A is area of species coverage (ha).

$$C_f = \exp(T_3 \cdot T_{soil}) \tag{7}$$

where: T_3 is an empirical parameter derived from emissions rate measurements ($°C^{-1}$);
T_{soil} is the soil temperature which is dependent on land use (°C).

$$T_{soil} = T_1 \cdot T + T_2 \tag{8}$$

where: T_1 is an empirical parameter derived from emissions rate measurements;
T is the ambient temperature (°C); and
T_2 is a land use-specific soil temperature adjustment factor (°C).

Table 1 presents the model parameters with their nominal values, ranges, and confidence intervals [24, 27, 28, 29].

2.5 MODEL INPUTS

The PAR (L) field was computed in two steps. First, the clear sky total radiation values were computed based on the work of Iqbal [32]. PAR, which falls in the visible spectrum of 400 nm to 700 nm, was taken as half of the total radiation. Second, the PAR field was scaled based on the spatially and temporally resolved cloud cover field. The development of the temperature (T) and cloud cover fields is discussed by Harley et al. [33].

The SoCAB species coverage (A) was derived from the Biogenic Emissions Landcover Database (BELD) and the LCC25 land coverage [34]. The BELD was created by combining the US Forest Service tree coverage (circa 1990), US Agricultural Census (1987), US urbanized areas (1990 Census Bureau), USGS AVHRR data (1990), and Environment Canada 1.1 kilometer land use data [35]. The BELD, which contains coverage estimates of 127 plant communities and species, was resolved to the county level in the United States and similar census tract levels in Canada. The LCC25 data set, which contains twenty-seven classes of species coverage for the United States, was used as a surrogate to allocate spatially data from the BELD.

Although their values involve some uncertainty, temperature, PAR, and species coverage were treated as certain values in this analysis. Research has indicated that newer PAR models are accurate to within ±9% to 11% [36]. If PAR varied by ±11%, the relative difference in the light correction factor (C_L) was computed to be less than 1%. If temperatures varied by ±10%, an 8% relative difference was computed for the temperature correction factor (C_T). Hence, the uncertainty in temperatures was a potential contributor to the overall uncertainty in the biogenic emissions estimates. However, as discussed later, the uncertainty in biogenic emissions factors overwhelmed the other sources of uncertainty; therefore, temperature and PAR were treated as certain data inputs in order to make the study more computationally tractable. Of note, as the uncertainty in biogenic emissions factors is reduced through additional measurements, the impact of uncertainty in the temperatures must be incorporated into the analysis. Future work will include the uncertainty in the temperatures.

The species coverage (A) presented an interesting problem in that no quantitative assessment of uncertainty in the data was found. Although species are correlated by location with other plant species, the correlation matrix was also unavailable. Further, the overall species coverage was spatially and temporally uncertain. As apparent from equations 1 and 6, large uncertainties in the species coverage result in large changes in the biogenic emissions estimates. Because valid species coverage uncertainty information was unavailable, the species coverage was treated as certain for the analysis. Additional research is required before uncertainties in the species coverage can be accurately incorporated into the analysis.

Table 2 lists the eighteen SoCAB plant species, as derived from the BELD, as well as the species-specific biogenic emission factors and uncertainty values. The nominal values for emissions factors are taken from Pierce and Geron [37], and the uncertainty values are taken from Guenther [27, 29] and Pierce [38]. As one can deduce upon review of Table 2, many plant species are grouped in the BELD. Benjamin et al. [39] identify at least 377 plant species in the SoCAB for which biogenic emissions factors can be assigned; however, it is not clear from Benjamin et al. [39] whether or not the 377 plant species have been inventoried (i.e. spatial extent and location). Table 2 and equations 1 and 6 show that the uncertainty in the biogenic emissions factors are the single, largest contributor to the overall uncertainty in biogenic emissions estimates. Future work will examine the extent to which the Benjamin et al. [39] data can be used in the uncertainty analysis.

2.6 ESTIMATING UNCERTAINTY IN BIOGENIC EMISSIONS ESTIMATES

Before the uncertainty analysis was conducted, the *Sampling Processor* (Figure 2) was added to BIOME. The *Sampling Processor* was built around the Iman and Shortencarier [40] program which can generate either Latin hypercube or random multivariate samples. Using the *Sampling Processor*, twenty-six samples were generated for each uncertain parameter -- seventeen model parameters (Table 1) and two data inputs (i.e. NO and VOC biogenic emissions factors -- Table 2). All samples were taken from a normal distribution. Twenty-six samples for each parameter were deemed sufficient based on the work Iman and Helton [41] which states that adequate distributions can be developed from a sample set, drawn by Latin hypercube sampling, which is comprised of 4/3 the number of parameters. Each combination of model parameters and model data were input to BIOME. The results of the multiple BIOME runs were compiled into cumulative distribution functions (CDF) from which uncertainty in the emissions estimates were calculated.

TABLE 2. SoCAB Species, Biogenic Emissions Factors, and Uncertainty Values.

Species	Area in Domain (ha)	Pollutant	Nominal Value (μg/ha/hr)	Range	Confidence Interval
		Rural Species and Plant Communities			
Barley	9,201	Isoprene	76000	± 50%	95%
		Monoterpenes	190000	± 50%	95%
		Other VOCs	114000	± 50%	95%
		Nitric Oxide	2567000	± 50%	95%
Cotton	8,049	Isoprene	76000	± 50%	95%
		Monoterpenes	190000	± 50%	95%
		Other VOCs	114000	± 50%	95%
		Nitric Oxide	2567000	± 50%	95%
Grass	1,245,386	Isoprene	562000	± 50%	95%
		Monoterpenes	1405000	± 50%	95%
		Other VOCs	843000	± 50%	95%
		Nitric Oxide	578000	± 50%	95%
Hay	129,311	Isoprene	378000	± 50%	95%
		Monoterpenes	945000	± 50%	95%
		Other VOCs	567000	± 50%	95%
		Nitric Oxide	128000	± 50%	95%
Miscellaneous Crops	516,514	Isoprene	76000	± 50%	95%
		Monoterpenes	190000	± 50%	95%
		Other VOCs	114000	± 50%	95%
		Nitric Oxide	128000	± 50%	95%
Scrub	1,263,005	Isoprene	378000	± 50%	95%
		Monoterpenes	945000	± 50%	95%
		Other VOCs	567000	± 50%	95%
		Nitric Oxide	578000	± 50%	95%
Scrub Woodland	322,406	Isoprene	27000000	± 50%	95%
		Monoterpenes	3490000	± 50%	95%
		Other VOCs	6510000	± 50%	95%
		Nitric Oxide	312000	± 50%	95%
Western Coniferous Forest	641,450	Isoprene	42700000	± 50%	95%
		Monoterpenes	11200000	± 50%	95%
		Other VOCs	13200000	± 50%	95%
		Nitric Oxide	45000	± 50%	95%
Woodland/ Cropland	31,239	Isoprene	25500000	± 50%	95%

Species	Area in Domain (ha)	Pollutant	Nominal Value (µg/ha/hr)	Range	Confidence Interval
		Monoterpenes	6630000	± 50%	95%
		Other VOCs	20530000	± 50%	95%
		Nitric Oxide	87000	± 50%	95%
Wetland Forest	97	Isoprene	38200000	± 50%	95%
		Monoterpenes	9230000	± 50%	95%
		Other VOCs	12320000	± 50%	95%
		Nitric Oxide	2000	± 50%	95%
Wheat	34,567	Isoprene	150000	± 50%	95%
		Monoterpenes	60000	± 50%	95%
		Other VOCs	90000	± 50%	95%
		Nitric Oxide	1925000	± 50%	95%
Western Woodlands	747,202	Isoprene	5250000	± 50%	95%
		Monoterpenes	2500000	± 50%	95%
		Other VOCs	3600000	± 50%	95%
		Nitric Oxide	45000	± 50%	95%
		Urban Species and Plant Communities			
BEIS2 Other Urban	1,038,008	Isoprene	112000	± 50%	95%
		Monoterpenes	281000	± 50%	95%
		Other VOCs	169000	± 50%	95%
		Nitric Oxide	116000	± 50%	95%
Scrub Woodland	36,544	Isoprene	27000000	± 50%	95%
		Monoterpenes	3490000	± 50%	95%
		Other VOCs	6510000	± 50%	95%
		Nitric Oxide	312000	± 50%	95%
Western Coniferous Forest	56,350	Isoprene	42700000	± 50%	95%
		Monoterpenes	11200000	± 50%	95%
		Other VOCs	13200000	± 50%	95%
		Nitric Oxide	45000	± 50%	95%
Woodland/ Cropland	961	Isoprene	25500000	± 50%	95%
		Monoterpenes	6630000	± 50%	95%
		Other VOCs	20530000	± 50%	95%
		Nitric Oxide	87000	± 50%	95%
Wetland Forest	4	Isoprene	38200000	± 50%	95%
		Monoterpenes	9230000	± 50%	95%
		Other VOCs	12320000	± 50%	95%

Species	Area in Domain (ha)	Pollutant	Nominal Value (μg/ha/hr)	Range	Confidence Interval
		Nitric Oxide	2000	± 50%	95%
Western Woodlands	64,768	Isoprene	5250000	± 50%	95%
		Monoterpenes	2500000	± 50%	95%
		Other VOCs	3600000	± 50%	95%
		Nitric Oxide	45000	± 50%	95%

2.7 AIR QUALITY MODELING

Air quality models, like the California/Carnegie Institute of Technology (CIT) air quality model, are used to determine the efficacy of emissions control strategies. CIT calculates the evolution of pollutants by accurately describing the physical and chemical processes that are responsible for the chemical transformation, transport, and fate of compounds in the atmosphere.

The CIT was used to quantify the response of the air quality field, particularly ozone and isoprene, to changes in the biogenic emissions fields, and to test how uncertainties in biogenic emissions affect the response of the ozone fields to emissions controls. Harley et al. [33] described the development of all the CIT inputs, except for the biogenic emissions estimates, for the 27-29 August 1987 episode. The current biogenic emissions estimates[1] were replaced with the biogenic emissions estimates that were developed using the more up to date methods previously described. A matrix of CIT control strategy assessment runs were made in which the anthropogenic emissions estimates were varied while the biogenic emissions estimates were held constant at the 2.5 percentile, 50 percentile, and 97.5 percentile levels as determined in the uncertainty analysis. The anthropogenic VOC emissions estimates were varied between 0% and 250% of the base case, and the anthropogenic NO_X emissions estimates were varied between 0% and 120% of the base case. The anthropogenic emissions estimate ranges were used to account for uncertainties that are believed to exist in the anthropogenic emissions estimates [43, 44, 45], as well as assess how emissions controls would impact air quality given the suspected uncertainty in the anthropogenic emissions estimates.

3. Results

The results of the biogenic uncertainty analysis are presented in Figure 3. The uncertainty ranges which are expressed as the 95% confidence interval for each pollutant are as follows: isoprene (Figure 3a), 0.03 to 3.7 times the expected value (1800 tons/day); nitric oxide (Figure 3b), 0.22 to 72 times the expected value (99 tons/day); other volatile organic compounds (Figure 3c), 0.54 to 1.7 times the expected

[1]Hereafter referred to as the California Air Resources Board (CARB) biogenic emissions estimates.

value (980 tons/day); and monoterpenes (Figure 3d), 0.54 to 1.8 times the expected value (700 tons/day).

Table 3 compares the CARB biogenic emissions estimates to biogenic emissions estimates developed in this study. The CARB biogenic emissions estimates include only isoprene and monoterpenes. This study suggests that the CARB monoterpene estimate may be significantly in error though the CARB isoprene emissions estimate is captured within the range of uncertainty that was computed in this study. This finding is consistent with the approach that was used to compute the CARB isoprene and monoterpene emissions estimates. That is, the CARB isoprene emissions estimates were based on limited measurements of ambient isoprene concentrations; however, little ambient monoterpene concentration measurements were available which makes such estimates highly uncertain.

Figures 4 through 7 show the spatial distributions of the daily biogenic emissions. Five important observations emerge upon review of Figures 4 through 7. First, the CARB isoprene and monoterpene emissions (Figures 4a and 7a respectively) do not cover the entire modeling domain. This is in part due to the fact that when the CARB biogenic inventory was constructed, the agencies were concerned primarily with the urban core emissions estimates.

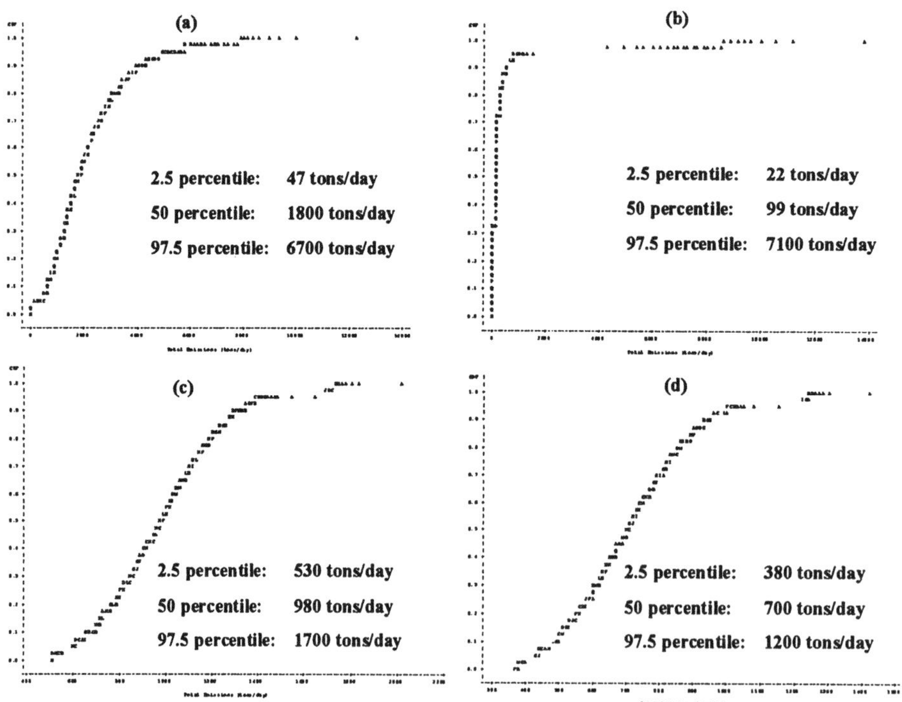

FIGURE 3. Cumulative Distribution of Biogenic Emission Estimates: (a) Isoprene, (b) Nitric Oxide, (c) Other Volatile Organic Compounds, (d) Monoterpenes

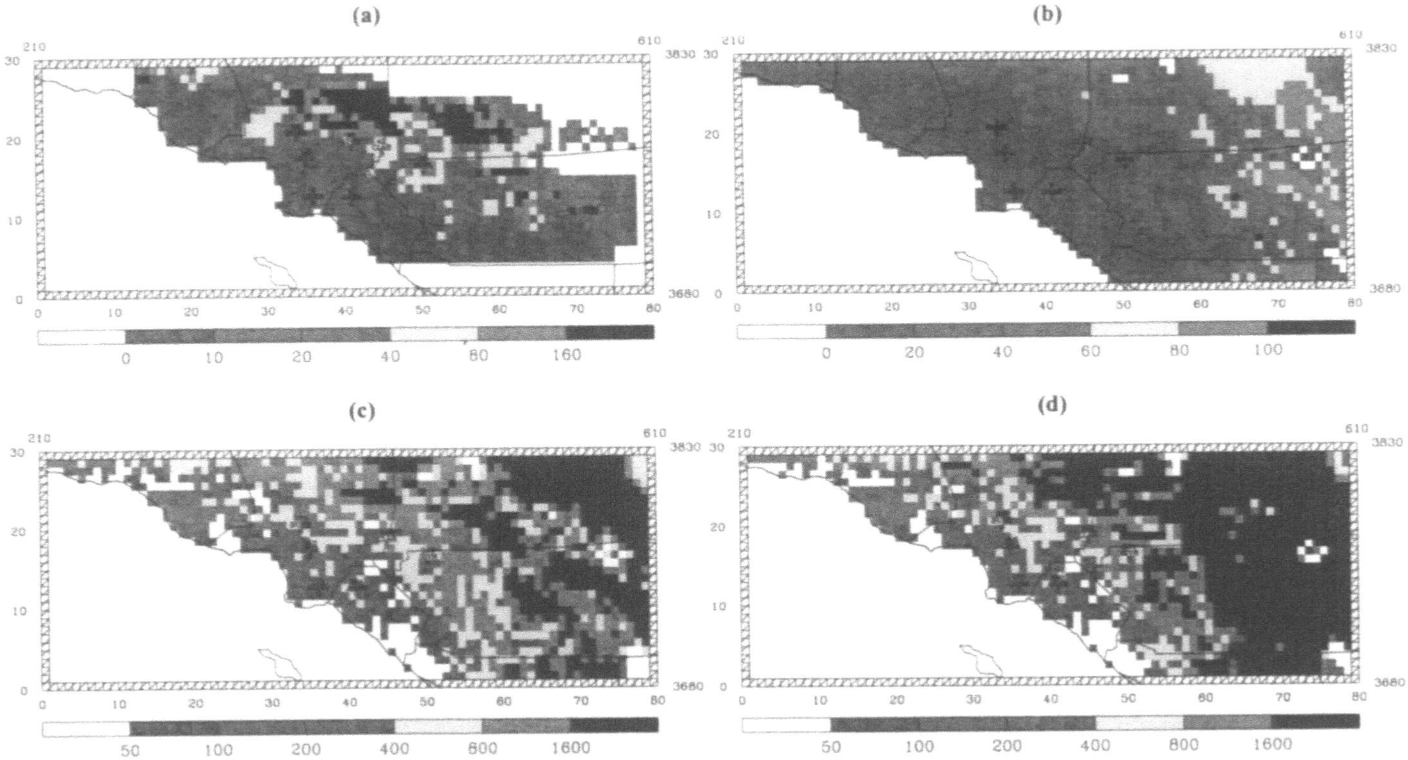

Figure 4. Spatial Distribution Of Isoprene Emissions (kg/day).

(a) CARB, (b) 2.5 Percentile, (c) 50 Percentile, (d) 97.5 Percentile

Selected observation stations are shown as a four character identifier (e.g. BURK). The value above the station is the predicted biogenic emission rate. The value below the station is the observed biogenic emission rate for which all values are zero since no biogenic emission rate measurements were made. The single, small cross-hair that is found in the eastern part of the domain identifies the maximum emission rate location.

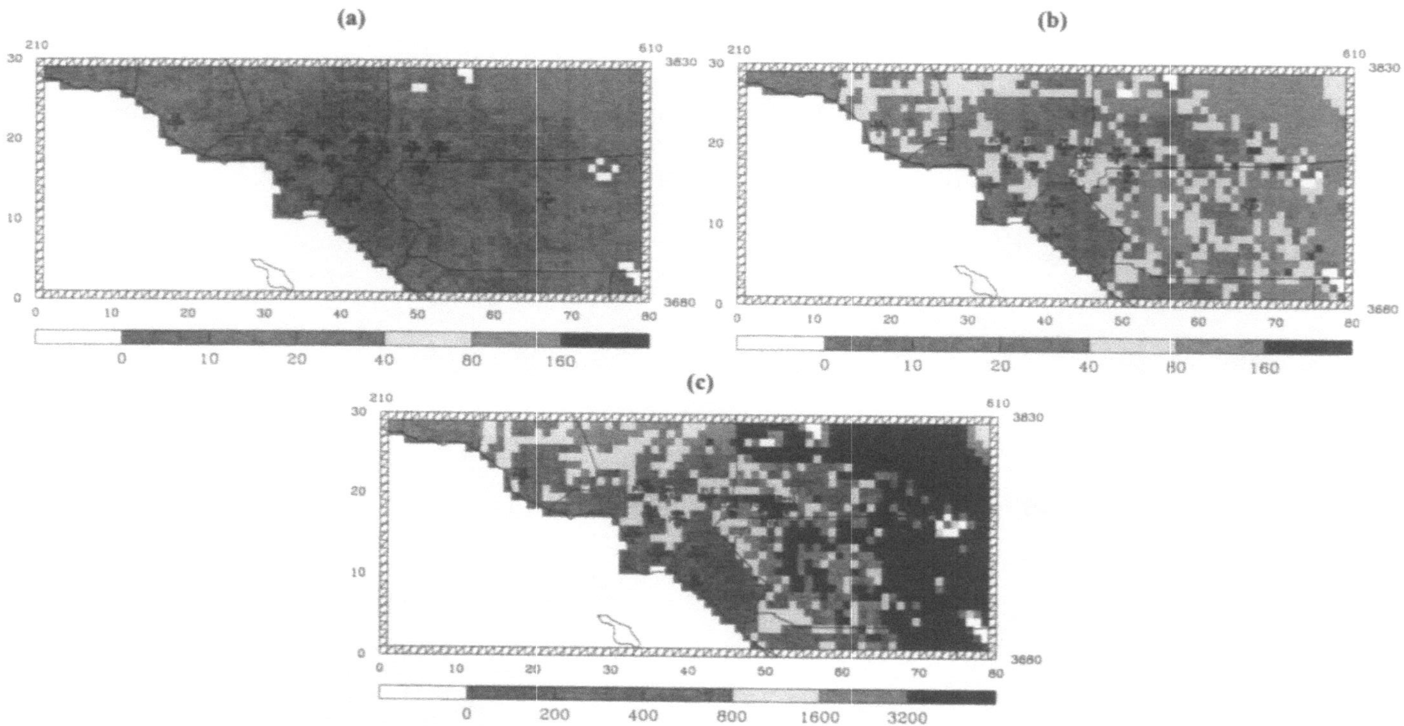

Figure 5. Spatial Distribution Of Nitric Oxide Emissions (kg/day).
(a) 2.5 Percentile, (b) 50 Percentile, (c) 97.5 Percentile

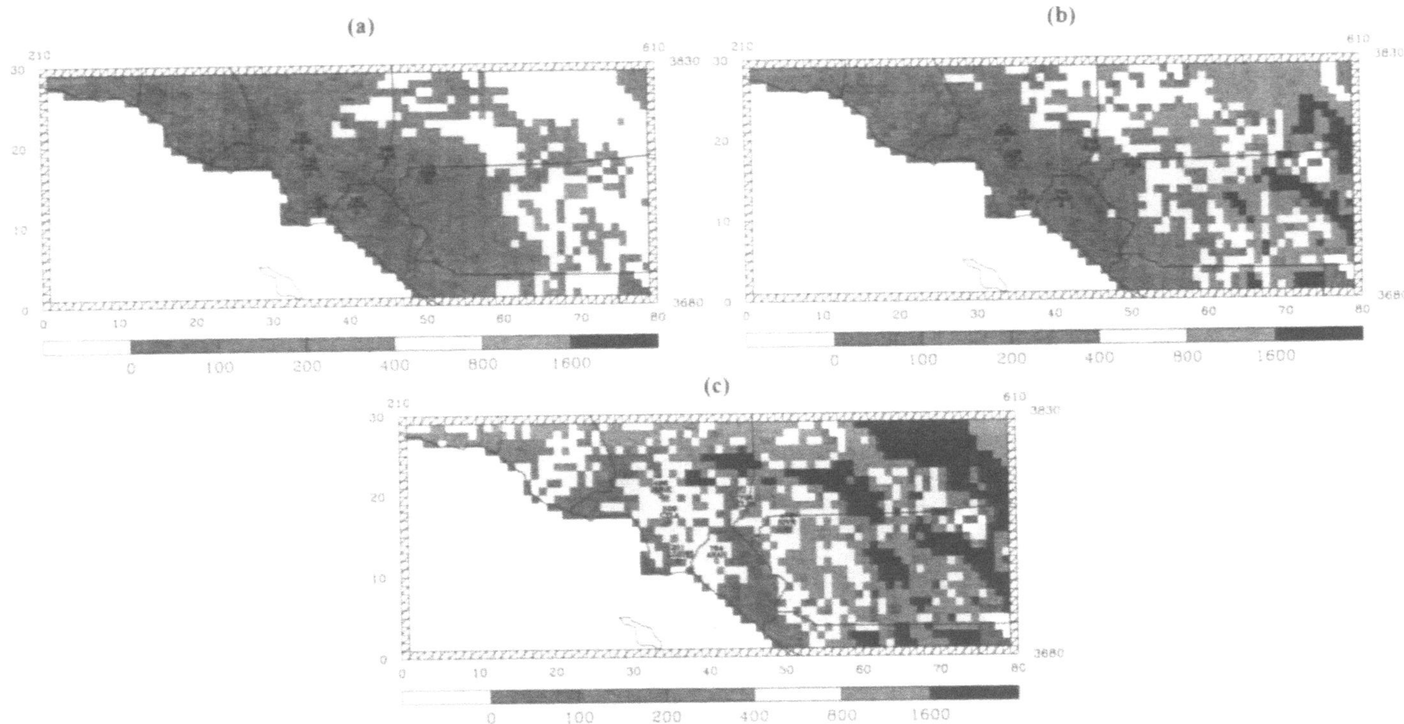

Figure 6. Spatial Distribution Of Other Volatile Organic Compound Emissions (kg/day).

(a) 2.5 Percentile, (b) 50 Percentile, (c) 97.5 Percentile

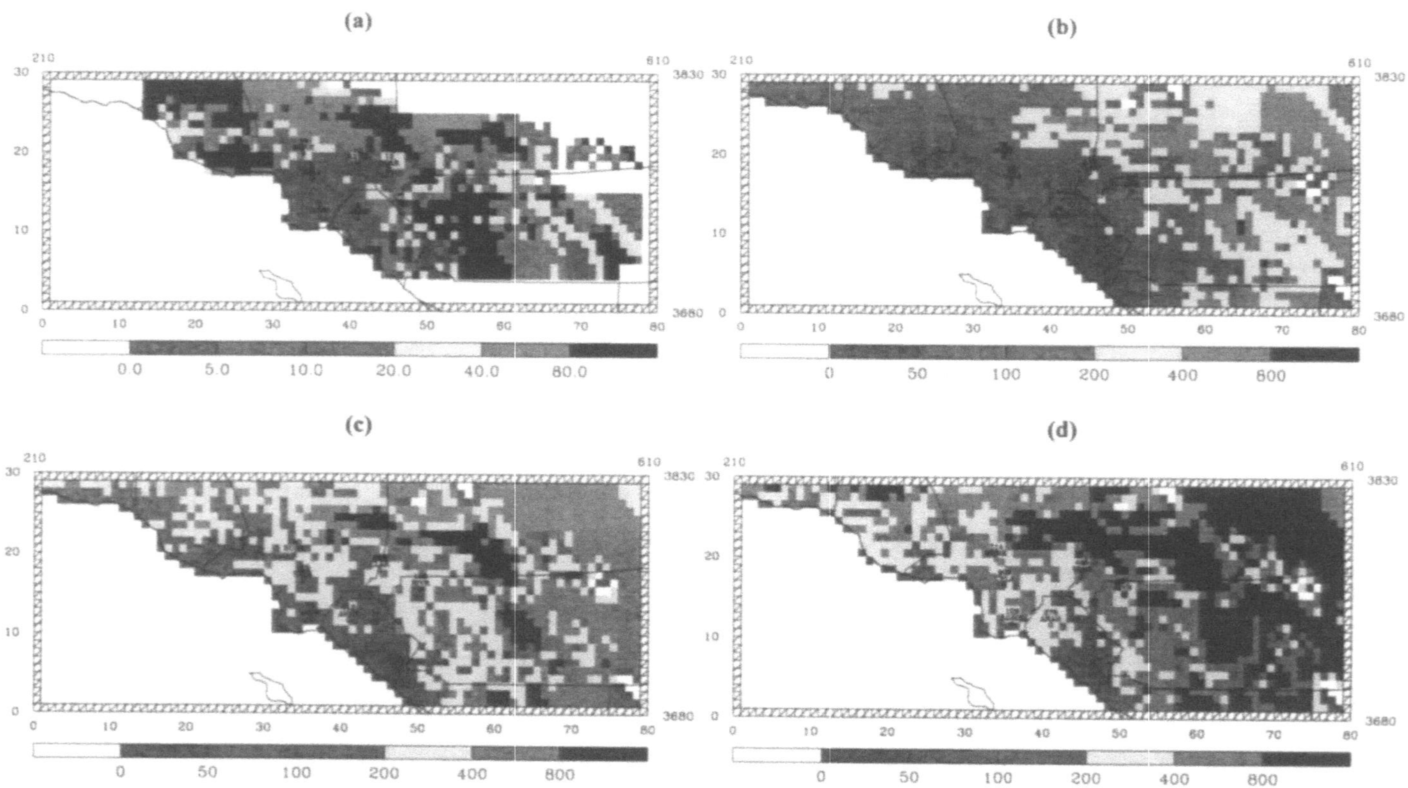

Figure 7. Spatial Distribution Of Monoterpene Emissions (kg/day).
(a) CARB, (b) 2.5 Percentile, (c) 50 Percentile, (d) 97.5 Percentile

TABLE 3. Comparison of Biogenic Emissions Estimates For August 29, 1987.

Pollutant (tons/day)	Biogenic Emissions Estimates			
	CARB	2.5 Percentile	50 Percentile	97.5 Percentile
Isoprene	60	47	1800	6700
Nitric Oxide	0	22	99	7100
Other VOCs	0	530	980	1700
Monoterpenes	67	380	700	1200

Second, no CARB biogenics exist for nitric oxide and other volatile organic compounds (Figures 5 and 6 respectively). This is because our realization of the magnitude and possible importance of biogenic nitric oxide has only recently become known. Further, the "other" volatile organic compound has been added to account for the oxygenated ten-carbon compounds which have different chemical reactivities than the monoterpenes. Third, the CARB isoprene emissions estimates may be in error by a factor of thirty, though as noted previously, the CARB isoprene emission estimate is captured within the range of uncertainty for isoprene. Note that ozone and isoprene observations do not support isoprene emissions above the median prediction as is discussed later. Fourth, the scale of the CARB monoterpene emissions is at least one order of magnitude less than the improved biogenic emissions estimates. This results from the use of increased biogenic emissions factors. Finally, the spatial structure appears to be consistent between the CARB and the improved biogenic emissions. This is encouraging since the CARB biogenics were developed based on entirely different data.

Figure 8 shows the spatial distribution of daily maxima ozone which results from CIT runs with the biogenic emissions. The results that are presented in Figure 8 reflect a factor of two increase in the anthropogenic VOC emissions to account for suspected bias in the anthropogenic VOC emissions estimates. This bias is primarily due to underestimated on-road mobile source emissions[43, 44, 45]. Review of Figure 8 reveals that as biogenic emissions estimates increase, the area of ozone exceedance also increases. In addition, the peak ozone value increases with increasing biogenic emissions estimates: CARB, 178 ppb; 2.5 percentile, 199 ppb; 50 percentile, 216 ppb; and 97.5 percentile, 432 ppb. Note that ozone observations in the far eastern part of the domain (e.g. Banning [BANN] and Palm Springs [PLSP] monitors -- Figure 8d) do not support biogenic emission levels much above the 50 percentile.

Table 4 presents the bias, gross normalized error, and normalized mean square error model performance statistics for ozone, isoprene, and nitrogen dioxide. Bias indicates the magnitude of the model's tendency to over- or under-predict the observed concentrations. Normalized gross error is a robust measure of overall model performance which provides a useful measure to compare model runs across alternative modeling scenarios. Normalized mean square error is also a robust measure of overall model performance where values less than 0.4 indicate good model performance. Review of Table 4 reveals that both the 2.5 percentile and 50 percentile biogenic emissions estimates result in about the same CIT model performance for ozone. Further, both the 2.5 percentile and 50 percentile biogenic emissions estimates result in

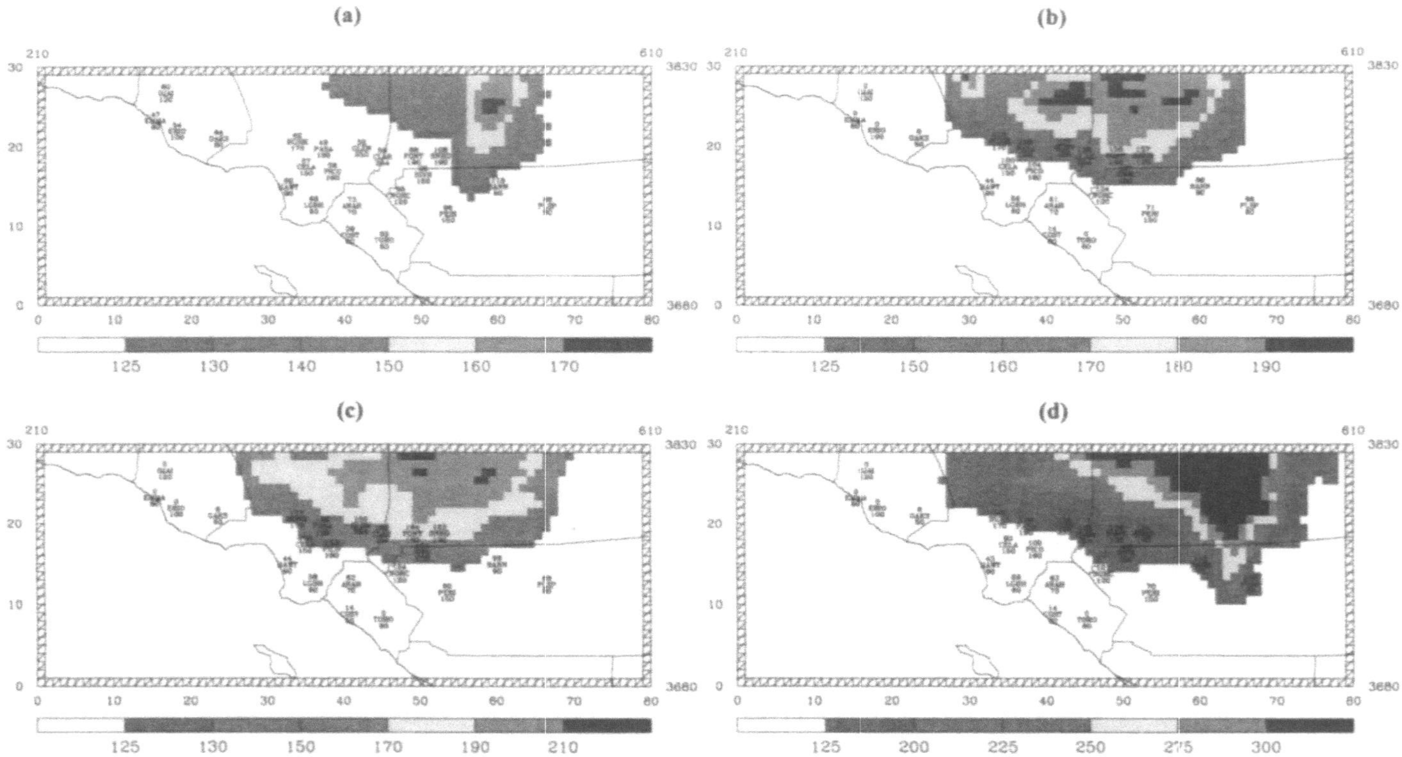

Figure 8. Spatial Distribution Of Daily Maxima Ozone (ppb).

(a) CARB, (b) 2.5 Percentile, (c) 50 Percentile, (d) 97.5 Percentile

(2x anthropogenic VOC emissions estimates)

Selected ozone monitors are identified with a four character abbreviation. The predicted daily maxima, four-cell weighted ozone average (ppb) is listed above the station, and the daily maximum (ppb) is listed below the station. The location of the predicted peak ozone is identified by a cross-hair. Stations with a predicted ozone value of zero are outside of the CIT computational domain.

better model performance for ozone than do the CARB biogenics. However, the median biogenic emissions estimates result in poor bias and gross error statistics for isoprene but a better normalized mean square error value where as the converse is true for the 2.5 percentile biogenic emissions estimates. These findings suggest that the isoprene emission estimate which will result in the best CIT model performance for both ozone and isoprene falls between the 2.5 percentile and 50 percentile values.

Finally, Figure 9 presents the ozone isopleths which result from the matrix of control strategy simulations. Note that the isopleths are based on the peak predicted ozone value.

Figure 9a shows the results of the matrix runs using the CARB biogenic emissions estimates. Figure 9b and Figure 9c result from the use of the 2.5 percentile and 50 percentile biogenic emissions estimates respectively. Results for the 97.5 percentile biogenics are not presented since all matrix runs resulted in peak predicted

TABLE 4. CIT Model Performance Statistics For August 29, 1987. (2x anthropogenic VOC emissions for all but CARB emissions).

Statistic	Biogenic Emissions Estimates			
	CARB	2.5 Percentile	50 Percentile	97.5 Percentile
		Ozone		
Bias (%)	-32	-25	-21	-35
Gross Normalized Error (%)	44	42	42	56
Normalized Mean Square Error	0.7	0.3	0.3	0.6
		Isoprene		
Bias (%)	-41	-62	730	810
Gross Normalized Error (%)	60	78	740	820
Normalized Mean Square Error	14	36	4.5	4.5
		Nitrogen Dioxide		
Bias (%)	-4	-32	-30	41
Gross Normalized Error (%)	48	46	46	87
Normalized Mean Square Error	0.3	0.4	0.4	0.4

ozone values above 300 ppb. Comparison of Figure 9a to Figure 9b shows that although the isoprene emissions are lower at the 2.5 percentile level than the CARB estimates, sufficient monoterpenes and other VOCs are present to alter the effectiveness of VOC emissions controls. That is, under the CARB biogenics scenario, it is possible to attain the ozone NAAQS with VOC-only emissions controls (Figure 9a), but it is unlikely that VOC-only emissions controls will result in attainment under the 2.5 percentile biogenics scenario (Figure 9b). The effectiveness of VOC emissions controls further decreases as biogenic emissions are increased (Figure 9c).

4. Discussion

Biogenic emissions estimates are highly uncertain due primarily to the lack of more complete information concerning biogenic emissions factors. However, as shown in Table 4, model predictions which use the biogenic emissions estimates greater than the 50 percentile are not supported by the isoprene and ozone observations. This suggests that the uncertainties are overestimated at the high-end of the biogenic emissions estimate distributions in this domain. Although the CARB isoprene emissions estimates are greater than the 2.5 percentile biogenic emissions estimates, the 2.5 percentile biogenic emissions estimates result in an area of predicted ozone exceedance that is greater than the area of predicted ozone exceedance with the CARB biogenic emissions estimates (Figure 8). This is due to both the greater monoterpene emissions estimates

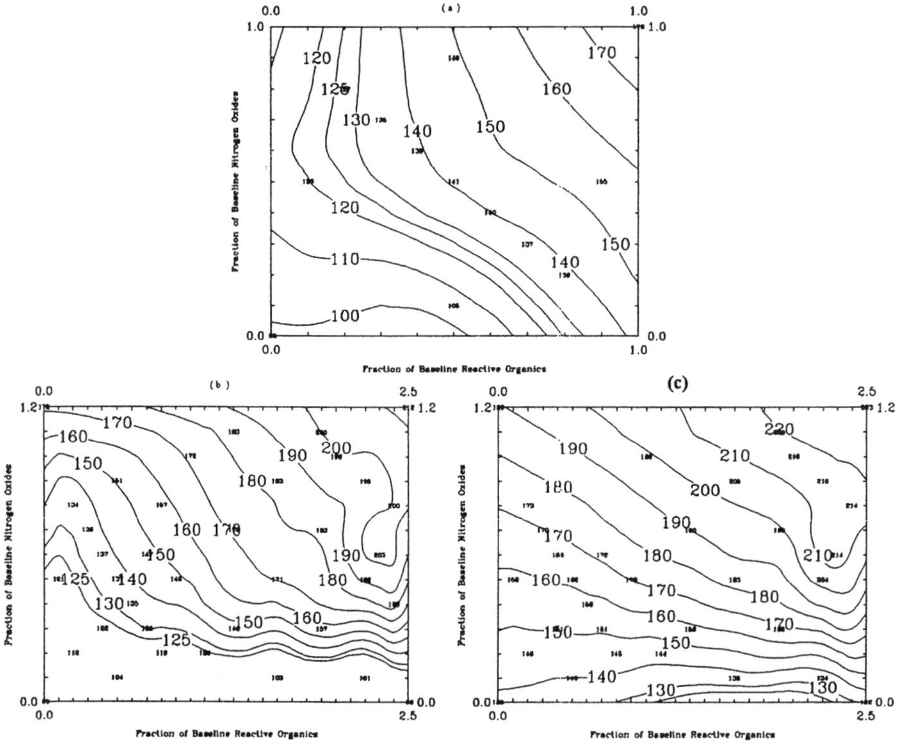

Figure 9. Ozone Isopleths.

(a) CARB, (b) 2.5 Percentile, (c) 50 Percentile

at the 2.5 percentile level, and the fact that the other VOCs and nitric oxide emissions are estimated by the biogenic emissions model but are not in the CARB biogenic emissions estimates. The results of this study suggest that biogenic emissions do impact emissions control strategy selection in the SoCAB. If biogenics are actually emitted at the 2.5 percentile level, the emissions controls which are suggested in the 1994 AQMP (77% NO_X and 69% VOC controls) are likely to be sufficient (Figure 9b) for the SoCAB to attain the current ozone. However, if biogenic emissions are actually higher than the 2.5 percentile levels, as suggested by the study results, the 1994 AQMP emissions controls are likely insufficient to meet the current ozone NAAQS. Based on the findings of this study, more stringent NO_X controls are probably necessary to meet the current ozone NAAQS. This contradicts recent findings by the SCAQMD which state that stringent VOC emissions controls will help the SoCAB to attain the ozone NAAQS. Further, if the SCAQMD continues to pursue this emissions control strategy, scarce resources are likely to be depleted with little return in improved air quality.

5. Conclusion

Improved biogenic emissions estimates for the 27-29 August 1987 SoCAB ozone episode were prepared using state-of-the-science biogenic emissions modeling techniques. The uncertainties in the biogenic emissions estimates were found to be very large; however, the biogenic emissions estimates that were greater than the 50 percentile biogenic emissions estimates were not supported by the ozone and isoprene observations. The improved biogenic emissions estimates had a significant impact on the air quality modeling results. The predicted peak ozone and the predicted ozone exceedance area increased over the CARB emissions estimates. The impact that uncertain biogenic emissions estimates have on the choice of emissions control programs in the SoCAB was also significant. Biogenic emissions were likely emitted between the 2.5 percentile and 50 percentile levels in which case the emissions controls that were mandated in the 1994 AQMP will probably be insufficient to meet the one hour ozone NAAQS. Furthermore, NO_X emissions controls were predicted to be more effective in meeting the one hour ozone NAAQS. Finally, it was shown that the isoprene predictions from the lower end of the uncertainty range agreed well with isoprene observations. This suggests that the current biogenic emissions model can be used successfully in air quality modeling studies though further testing and evaluation in other domains is warranted.

5.1 FUTURE WORK

The biogenic emissions factors are based almost exclusively on rural biomass which is typically greater than urban biomass. Therefore, improved urban biomass values will likely result in lower urban biogenic emissions estimates. As previously noted, the biogenic VOC model parameters are positively correlated; however, the model parameters were treated as uncorrelated in this study. Addition of the correlation matrix will likely increase the mean and increase the range of uncertainty. Finally, a cost analysis should be conducted to determine what the fiscal impact will be of choosing an improper control strategy.

6. Acknowledgements

This work was partially supported through a National Science Foundation Fellowship and the Georgia Institute of Technology President's Fellowship. The author also acknowledges the support of Dr. Alec Guenther of the Atmospheric Chemistry Division, National Center for Atmospheric Research (NCAR), Mr. Thomas E. Pierce of the Atmospheric Characterization and Modeling Division of NOAA (currently on assignment to the Office of Research and Development, U.S. EPA), and Dr. Eric Williams of the Aeronomy Laboratory, NCAR.

7. References

1. NRC, 1991. Rethinking the Ozone Problem in Urban and Regional Air Pollution. National Research Council, National Academy Press, Washington, D.C.
2. Seinfeld, J., 1986. Atmospheric Chemistry and Physics of Air Pollution. Wiley Interscience, John Wiley & Sons, New York, NY.
3. Finlayson-Pitts, B. J. and J. N. Pitts, 1986. Atmospheric Chemistry. Wiley Interscience, John Wiley & Sons, New York, NY.
4. Mlot, C, 1995. "A Clearer View of Why Plants Make Haze." *Science*, Volume 268, May
5. Pierce, T. E., B. K. Lamb, and A. R. Van Meter, 1990. "Development of a Biogenic Emissions Inventory System for Regional Scale Air Pollution Models." Paper 90-94.3. In the Proceedings of the 83rd Annual Meeting of the Air and Waste Management Association, 24-29 June, Pittsburgh, PA.
6. Lamb, B. K., D. Gay, H. Westberg, and T. E. Pierce, 1993. "A Biogenic Hydrocarbon Emissions Inventory for the U.S.A. Using a Simple Forest Canopy Model." *Atmospheric Environment*, 27A: 1673-1690.
7. Pier, P. A., 1995. "Isoprene Emission Rates from Northern red Oak Using a Whole-Tree Chamber." *Atmospheric Environment*, 29 (12): 1347-1353.
8. Winer, A. M., J. Arey, S. M. Aschmann, R. Atkinson, W. D. Long, C. L. Morrison, and D. M. Olszyk, 1992. "Emission Rates of Organics from Vegetation in California's Central Valley." *Atmospheric Environment*, 26A: 2647-2659.
9. Guenther, A., R. Monson, and R. Fall, 1991. "Isoprene and Monoterpene Emission Rate Variability: Observations with Eucalyptus and Emission Rate Algorithm Development." *Journal of Geophysical Research*, 96:10799-10808.
10. Tingey, D. T., M. Manning, L. C. Grothaus, and W. F. Burns, 1980. "Influence of Light and Temperature on Monoterpene Emission Rates from Slash Pine." *Physiological Plantarum*, 65: 797-801.
11. Tingey, D. T., M. Manning, L. C. Grothaus, and W. F. Burns, 1979. "Influence of Light and Temperature on Isoprene Emissions from Live Oak." *Physiological Plantarum*, 47: 112-118.
12. Zimmerman, P. R., 1979. "Tampa Bay Area Photochemical Oxidant Study. Determination of Emissions Rates of Hydrocarbons from Indigenous Species of Vegetation in the Tampa/ St. Petersburg, Florida Area." EPA 904/9-77-028, United States Environmental Protection Agency, Atlanta, GA.

13. Winer, A. M., L. Chinkin, J. Arey, R. Atkinson, J. Adams, J. Karlik, 1995. "Critical Evaluation of a Biogenic Emission System for Photochemical Grid Modeling in California." Final Report to the California Air Resources Board, Contract Number 93-725, Sacramento, CA.
14. Wilkinson, J. G., T. E. Pierce, R. A. Emigh, 1995. "An Intercomparison of Biogenic Emissions Estimates from BEIS2 and BIOME: Reconciling the Differences." For presentation at the 1995 AWMA Emissions Inventory Conference, RTP, NC.
15. CFR, 1995. *Code of Federal Regulations* (40 CFR Part 91), Office of the Federal Register, National Archives and Records Administration, U.S. Government Printing Office, Washington, D.C.
16. SCAQMD, 1996. Draft 1997 Draft Air Quality Management Plan. http://www.aqmd.gov/aqmp/97draft/execs.html
17. SCAQMD, 1994a. Air Quality Trends, 1976-1993. Appendix II-B, September 1994 Air Quality Management Plan, South Coast Air Quality Management District, Diamond Bar, CA.
18. SCAQMD, 1994b. Summary of Air Quality, 1990-1993. Appendix II-A, April 1994 Air Quality Management Plan, South Coast Air Quality Management District, Diamond Bar, CA.
19. SCAQMD, 1994c. Ozone Attainment Demonstration for the South Coast Air Basin. Appendix V-A, September 1994 Air Quality Management Plan, South Coast Air Quality Management District, Diamond Bar, CA.
20. Chameides, W., R. Lindsay, J. Richardson, and C. Kiang, 1988. "The role of biogenic hydrocarbons in urban photochemical smog: Atlanta as a case study," *Science*, 241, 1473-1475.
21. Tingey, D., 1981. "The effect of environmental factors on the emission of biogenic hydrocarbons from live oak and slash pine." Atmospheric Biogenic Hydrocarbons. edited by J. Bufalini and R. Arnts, pages 53-72, Butterworth, Stoneham, MA.
22. Guenther, A., P. R. Zimmerman, P. C. Harley, R. K. Monson, R. Fall, 1993. "Isoprene and Monoterpene Emission Rate Variability: Model Evaluations and Sensitivity Analysis." *Journal of Geophysical Research*, Volume 98, Number D7:12609-12617.
23. 23. Geron, C. D., A. B. Guenther, T. E. Pierce, 1994. "An improved model for estimating emissions of volatile organic compounds from forests in the eastern United States." *Journal of Geophysical Research*, Volume 99, Number D6:12773-12791.
24. Williams, E. J., A. Guenther, and F. C. Fehsenfeld, 1992. "An Inventory of Nitric Oxide Emissions from Soils in the United States." *Journal of Geophysical Research*, Volume 97, Number D7:7511-7519.
25. Wilkinson, J. G., C. F. Loomis, R. A. Emigh, D. E. McNally, T. W. Tesche, 1994. "Technical Formulation Document: SARMAP/LMOS Emissions Modeling System (EMS-95)." Final report prepared for the Lake Michigan Air Directors Consortium (Des Plaines, IL) and the Valley Air Pollution Study Agency (Sacramento, CA).
26. Fall, R. And R. Monson, 1992. "Isoprene emission rate and intercellular isoprene concentration as influenced by stomatal distribution and conductance." *Plant Physiology*, 100, pages 987-992.

27. Guenther, A., 1997. Personal Communication. January 6.
28. Williams, E. J., 1997. Personal Communication. January 7.
29. Guenther, A., 1995. Personal Communication. December 14.
30. Shepherd, M. F., S. Barzetti, and D. R. Hastie, 1991. "The production of atmospheric NO_X and N_2O from fertilized agricultural soil." *Atmospheric Environment*, 25A:1961-1969.
31. Williams, E. J., D. D. Parrish, M. P. Buhr, and F. C. Fehsenfeld, 1988. "Measurement of Soil NO_X Emissions in central Pennsylvania." *Journal of Geophysical Research*, Volume 93:9539-9546.
32. Iqbal, M., 1983. An Introduction To Solar Radiation. Academic Press, New York, 202-210.
33. Harley, R. A., A. G. Russell, G. J. McRae, L. A. McNair, D. A. Winner, M. T. Odman, D. Dabdub, G. R. Cass, and J. H. Seinfeld, 1992. "Continued Development of a Photochemical Model and Application to the Southern California Air Quality Study (SCAQS) Intensive Monitoring Periods: Phase I." Final Report to the Coordinating Research Council Project: SCAQS - 8.
34. Kinnee, E., C. D. Geron, T. Pierce, and T. Birth, 1995. "Creation of a land use inventory for estimating biogenic ozone precursor emissions in the United States." Submitted for publication in *Ecological Applications*.
35. Pierce, T. E., 1996a. "Documentation for /beis2/SOS/." ftp://monsoon.rtpnc.epa.gov/pub/beis2/SOS/AAREADME.
36. NREL, 1994. "Estimating the Uncertainty of Solar Radiation Data." National Renewable Energy Laboratory, http://solstice.crest.org/renewables/solrad/method-3.html.
37. Pierce, T. E. and C. D. Geron, 1996. The personal computer version of the Biogenic Emissions Inventory System (PCBEIS2.2). AAREADME, anonymous ftp at monsoon.rtpnc.epa.gov located in /pub/beis2/pcbeis22.
38. Pierce, T. E., 1996b. Personal Communication. November 21.
39. Benjamin, M. T., M. Sudol, L. Bloch, and A. M. Winer, 1996. "Low-Emitting Urban Forests: A Taxonomic Methodology For Assigning Isoprene And Monoterpene Emission Rates." *Atmospheric Environment*, Volume 30, Number 9:1437-1452.
40. Iman, R. L. and M. J. Shortencarier, 1984. "A FORTRAN 77 Program And User's Guide For The Generation f Latin Hypercube And Random Samples For Use With Computer Models." Prepared for the Division of Risk Analysis and Operations, Office of Nuclear Regulatory Research, U.S. Nuclear Regulatory Commission, Washington, D.C. NUREG/CR-3624, SAND83-2365, Sandia National Laboratories, Albuqerque, NM.
41. Iman, R. L. and J. C. Helton, 1985. "A Comparison of Uncertainty and Sensitivity Analysis Techniques for Computer Models." Prepared for the Division of Risk Analysis and Operations, Office of Nuclear Regulatory Research, U.S. Nuclear Regulatory Commission, Washington, D.C. NUREG/CR-3904, SAND84-1461, Sandia National Laboratories, Albuquerque, NM.
42. Winer, A. M., D. R. Fitz, and P. R. Miller, 1983. "Investigation of the Role of Natural Hydrocarbons in Photochemical Smog formation in California." Report to the California Air Resources Board, Contract Number A0-056-32, Statewide Air

Pollution Research Center, University of California at Riverside.

43. Lurmann, F. W., H. H. Main, 1992. "Analysis Of The Ambient VOC Data Collected In The Southern California Air Quality Study." Final Report to the California Air Resources Board, Contract Number A832-130.
44. Pierson, W. R., A. W. Gertler, and R. L. Bradow, 1990. "Comparison of the SCAQS Tunnel Study with Other On-Road Vehicle Emissions Data." *Journal of the Air and Waste Management Association*, Volume 40:1495-1504.
45. Ingalls, M. N., L. R. Smith, and R. E. Kirksey, 1989. "Measurement of On-Road Vehicle Emission in the California South Coast Air Basin - Volume I: Regulated Emissions." Report to the Coordinating Research Council, Atlanta, GA.
46. Aneja, V. P., 1994. "Workshop of the Intercomparison of Methodologies for Soil NO_X Emissions: Summary of Discussion and Recommendations." *Jounal of the Air and Waste Management Association*, 44: 977-982.

AIR POLLUTION SPACE-TIME ANALYSIS

F. FERREIRA, J. SEIXAS, C. NUNES, and J.P. SILVA
Faculdade de Ciências e Tecnologia
Universidade Nova de Lisboa
Quinta da Torre, 2825 Monte de Caparica, PORTUGAL

Abstract

The environmental decision-making process is preceded by data analysis, which is conditioned by a specific temporal and spatial scale. This paper aims to present a methodology that integrates the time-space framework of air quality data to infer the temporal pattern and spatial variability that could be interpreted for environmental decision purposes. Variograms that accommodate time and space lags were used for the analysis and proved to be effective. Temporal and spatial trends were found for data collected on an hourly and daily basis and its environmental meaning is discussed. Visualization of spatial patches of air pollution in Lisbon during a working day is performed through the use of an image processing technique, named morphing.

1. Introduction

Environmental management must include three sequential tasks: data capture, data analysis and decision making. Since any environmental issue is dynamic by nature and manifests itself on a certain spatial scale, the process of environmental data capture and analysis has to accommodate its twofold nature, the spatial and the temporal dimension.

Ideally, environmental monitoring should be proceeded on a continuous-space and continuous-time frame, considering the appropriate spatial and temporal scale for the phenomena under study. Point source observations provide only small, sometimes inadequate clues to a much larger problem. Appropriate exploratory data analysis must answer questions about patches, defined as some spatial extension of significant data, besides point data, as well as questions regarding meaningful periods of time.

The goal of this paper is to present a methodology to assess simultaneously the spatio-temporal patterns of air quality at the city of Lisbon, in order to assist the environmental management decision-making.

I. Linkov and R. Wilson (eds.), Air Pollution in the Ural Mountains, 341–350.

2. Environmental Data Analysis

Quite different approaches are being used to deal with the spatial nature of the data, either from a geostatistical, or from a modeling perspective, or simply from spatial measures, either global or local. The underlying rational in each of these methods is that the computed parameters express a kind of ìnatural characteristicî of a spatially contiguous set of pixels [1]. Although the individual point values may vary, the pattern can be distinctive.

Exploratory spatial data analysis (ESDA) include traditional techniques for visualizing spatial data such as the display of box plots by distance bands or selected regions, or maps of locations of data statistical properties, such as post plots of the quartiles of data values. More sophisticated visual tools include, among others, the spatial lag pies, the spatial lag scatterplots, and pocket plots, being these tools inappropriate for large data sets and for directional analysis. Recent spatial indicators accommodate these features, being some examples, the Gi statistics, the local indicators of spatial association, and the box plot of variogram clouds suggested by Getis and Ord [2], and Anselin [3,4].

Spatial patterns can be described quantitatively in terms of the semivariance function, which is based on the idea that statistical variation of data is a function of distance. Although the variogram modeling has been used as a basis for spatial interpolation, it can also be adopted as an exploratory analysis tool, being its parameters reliable for the interpretation of environmental issues. For a depth revision of these techniques, see Isaacs and Srivastava [5], Cressie [6].

The issue of spatial data analysis has been considered from a static perspective, as illustrated from the above revision. The inclusion of time in spatial data analysis has been proposed rather on a statistical modeling approach, than as an exploratory tool. In fact there is an ever-increasing number of papers about space-time models and analysis, being the generic problem of space-time geostatistical data related with estimation. In these models, there is usually an assumption that the large-scale spatial and temporal components do not interact or that the small-scale components are homogeneous, which allows more precise estimates of model parameters.

The consideration of time within the spatial data analysis framework has been proposed by several authors, either within the variogram modeling [7,8], or within the spatial adaptive filtering based on a moving-window scheme [9]. Considering the data time range, one can state that the probability of the data to be more alike at successive time slices is higher than for distance-lag time slices. The thesis for this paper is that animated visualization techniques provides the basis for integrating time in spatial data analysis in a well-understandable basis.

3. A New Methodology for Space-time Air Quality Data Analysis

The sampling scheme of air quality data usually includes a set of monitoring stations at specific points, measuring pollutant concentration on a quasi-continuous fashion. Therefore, air quality monitoring is extremely exhaustive in the time domain but sparse in the space one. Some experiments on air quality analysis have been unsuccessful if only the spatial dimension of the data is considered. The lack of data spatial structure can be due to the very dynamic behavior of the air quality variables at each monitoring station, which is influenced by environmental factors such us wind patterns, building architecture, and air pollutant sources characteristics. However, for each station one can expect a temporal pattern, and the assessment of the similarities and/or differences among them can highlight some smoothing of the spatial dimension of the data.

Let us denote by Z(x,t) the air quality concentration at a given time t at the monitoring station located at x, usually defined by a two-coordinate pair on a plane reference system. Adopting the rational beyond spatial analysis, one can state that observations close in time tend to be more similar than those further apart, at least when considered the quasi-continuous sampling time domain. Thus, for a specific location (x_0) in the space domain, the semivariance of z (x_0,t) can be computed, considering pairs of data at successive increasing temporal lags N(d), and using the following expression:

$$\gamma(d)=\frac{1}{2N(d)}\sum_{t=1}^{N(d)}\left[z(x,t)-z(x,t+d)\right]^2 \tag{1}$$

If there is a semivariance function, the variogram parameters can reveal some meaningful temporal patterns of air quality, mainly the time range. This analysis has to be conducted for all the spatial locations available for the study. The temporal range (d_t) concluded from the above model, and for the condition of its similarity for all the spatial locations, can now be incorporated for the spatial variability assessment, by the expression:

$$\gamma(h)=\frac{1}{2d_t N(h)}\sum_{x=1}^{N(h)}\sum_{t=1}^{d_t}\left[z(x,t)-z(x+h,t)\right]^2 \tag{2}$$

Using that approach, time stationarity is assured for the spatial analysis of air quality data which gives a major probability of success on the use of the variogram for the spatial analysis of air quality data. The integration of the time domain in the computation of the traditional semivariance has been suggested by some authors [7,8].

4. Application

The proposed methodology was applied for the city of Lisbon, and for the NO_2 pollutant, which derives mainly from the traffic activity. Figure 1 shows the spatial distribution of the air quality monitoring stations referenced on a 10 m panchromatic SPOT image.

FIGURE 1. Air quality monitoring stations and main traffic paths
Referenced on the SPOT image

Two different NO_2 data sets were analyzed, namely the hourly average from April 18, 1995, to assess the air quality behavior during a working day, and the daily averages from January to October, 1995, to identify weekly patterns of air pollution. The results will be presented in the next two sections.

4.1. HOURLY-ANALYSIS

Urban air pollution is a serious environmental management problem. Due to the wide range of air pollutants concentrations for the different monitoring stations, the semivariance values are very different, which make difficult to find the existence of similar patterns among those stations. A better comparison is obtained if the semivariance values are divided by the variance of the data, as shown in Figure 2(a), which presents the temporal variogram for the NO_2 hourly data. This experimental temporal variogram is used only for time patterns interpretation, and not for modeling purposes. A time-range of six hours can be identified. This trend is similar for all the monitoring stations as a consequence of the rush hour period during a working day.

The time range found in the experimental temporal variogram was used to compute the spatial variogram. According to expression (2), the experimental semivariance values were calculated considering the square differences at each lag distance station pair, summed for six consecutive hours. The choice of the data series of six hours was random, and the same time range was considered for all monitoring stations, due to the similar temporal pattern.

The experimental variogram presents a spatial range around 3.5 Km, where data are spatially correlated following a spherical model, presented in Figure 2(b).

$$\gamma(h) = 4500 + 4241.2\left(1.5\frac{h}{4000} - 0.5\frac{h^3}{4000^3}\right), h < 4000 \tag{3}$$

(a) (b)

FIGURE 2: NO_2 hourly data from Lisbon: (a) temporal, and (b) spatial variograms.

This might be related with the similarities between traffic patterns among the different air quality monitoring stations in Lisbon during the day. For kriging interpolation the experimental variogram was fitted to the spherical model, according to expression (3).

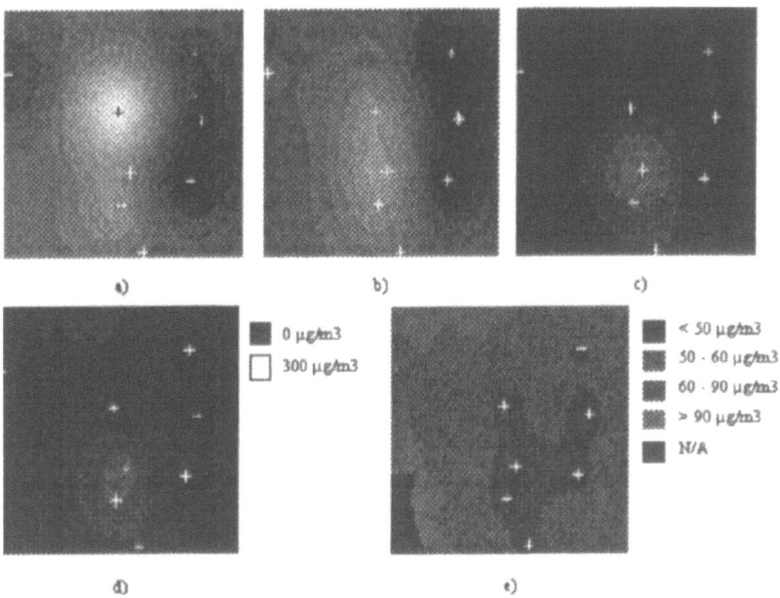

FIGURE 3: Map of the kriged NO2 values (a-d) and correspondent standard deviations(e).

The kriged NO_2 concentrations as well as its standard deviation were estimated for a 10 km grid. The spatial resolution was set at 50m in both the northings and the eastings from the sampling grid.

Figure 3 displays examples of air quality isosurfaces for the city of Lisbon, referring to the morning rush hour (7:00 (a) and 8:00am(b)), to lunch-time (2:00pm, (c)), and to steady-traffic conditions (9:00pm, (d)). It is quite visible the movement of higher concentrations towards South along the day, which is related with higher traffic from Lisbon entrances to downtown. The presence of more extreme air pollution patches during the morning rush hour when compared with the evening, is due to the wind speed at 18 April, which was about 8 km/h in the morning and about 27 km/h during the afternoon and evening, providing better dispersion conditions. Figure 3 (e) shows the kriged standard deviations for NO_2.

4.2. DAILY-ANALYSIS

The temporal pattern of the daily NO_2 data was assessed using the daily averages between January 1, 1995 and October 31, 1995. Figure 4 (a) shows the experimental temporal variogram for five monitoring stations. As before, the semivariance values were divided by the data variance, in order to get better comparisons. A very similar trend among those stations was found, and two plateaus (2 and 7 days) can be identified, probably corresponding to two different temporal patterns. In fact, considering the daily air pollution behavior at the city, the time range of 2 days should be associated with weekends, while the time range of 7 days should be associated with the whole week. Therefore, two spatial semivariance functions were computed in order to identify the corresponding experimental variograms. As for the previous case of the hourly data, a random sample of two consecutive (weekends) and five consecutive (working-day) daily values, were taken into account. Figure 4 (b) displays the spatial variogram for the weekends.

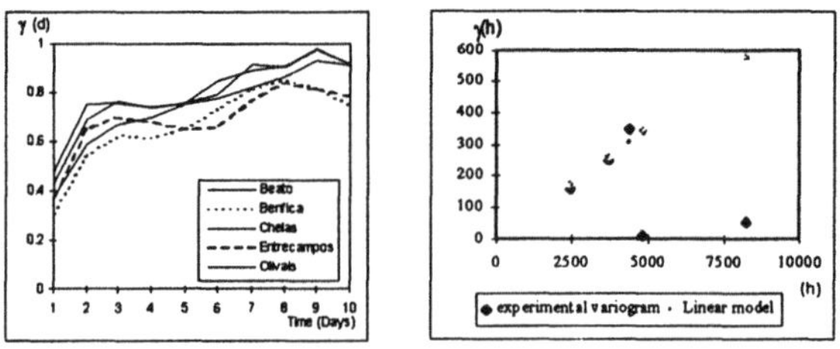

Figure 4: NO_2 daily data of Lisbon, (a) temporal variogram and (b) spatial variogram for the weekends.

For both cases, the experimental variograms show the first three points aligned until a spatial distance of about 4.5 km. In terms of environmental science, this means that the dispersion factors influence similarly the NO_2 daily values, either weekend days, or working days, resulting in a similar spatial variability structure. The two extremely low values presented at the weekend variogram, is associated with the monitoring stations located at the suburb areas, which at the weekend days, have similar air pollutant levels (weekend familiar activities) but are far way from each other. It is the case of Benfica and Entre Campos (h=4800 m, g(h)=11.5) and Benfica e Olivais (h=7800 m, g(h)=25.6).

Linear models were fitted to the experimental variograms, namely for the weekends ($\gamma(h) = 0.075 * h, h < 4000$) and working-week ($\gamma(h) = 0.051 * h + 90, h < 4500$) patterns. The kriged NO_2 concentrations as well as its standard deviations were estimated for a 10 km grid. As for the hourly case, the spatial resolution was set at 50m in both the northings and the eastings from the sampling grid.

4.3. VISUALIZATION OF AIR POLLUTION DATA

Temporal-spatial analysis could be enriched by means of visualization. The integration of time in scientific visualization schemes is always performed by an animated sequence of frames. If the data from each time period is correlated, the resulting animation show smooth evolution. Unusual changes are then jarring and then readily apparent [10]. Therefore, and within the context of this work, once environmental data has been stated on a spatial framework, the temporal tracking of significant patches can be assessed using animation techniques.

Since, one can assume that observations close together in time are more alike that those further apart, a more realistic temporal tracking of spatial data can be performed by using a morphing algorithm [11]. Morphing algorithms rely on mesh points to define the correspondence points between two images. Usually, the mesh point selection is a time-consuming but conceptually easy task. When applied to air quality isocontours, the mesh point selection appears more difficult, since one has to recall for exogenous variables such as wind direction, to assist on the correspondence points between the two images. The morphing process then warps and fades from one image to the other, using the mesh of correspondence points to define the warping function. With well-placed mesh points, this process generates a natural sequence of intermediate images and ìnaturalî transitions according to the criterion of human perception can be produced. Morphing methods provide a better sense of time continuity because of the smoothing behavior obtained between two consecutive time frames, and because it can generate any number of frames, which increases the time resolution.

For clearness of the visualization of NO_2 pollution patches, a threshold value was calculated, considering the whole time period. Then, a reclassification of the kriged images were performed in order to get the regions above that threshold, which represent polluted areas, with interest for air quality analysis. The visualization of the polluted patches over Lisbon was performed for a working day, based on a subset of the hourly kriged maps.

The interpretation of the Lisbon-air-quality movies referring to a working day reveals that the paths of the NO_2 patches are roughly associated with the traffic entries and downtown traffic jam. The presence of more extreme air pollution patches during the morning rush hour when compared with the evening, is due to higher the wind speed at 18 April throughout the afternoon.

In terms of implementation issues, image analysis was performed using the IDRISI 4.1 software, which is a Geographic Information System software. Variogram analysis as well as kriging interpolation was run on Visual Basic, within the Microsoft Excel® software. The morphing sequences were produced using Elastic Reality™ 1.0.

5. Conclusions

A methodology to identify and integrate the temporal trends of air pollution data, in the assessment of its spatial variability patterns, based on variograms, was presented. Hourly NO_2 data leads to a clear temporal trend relating with the urban traffic behavior during a working-day. Similar trends were found for the seven monitoring stations, which reveals common temporal traffic patterns to the several stations in the city. Temporal patterns similarity was also identified for the NO_2 daily data among the monitoring stations. However, within this time frame, two variogram structures were found, referring to the weekends and to the working-week. Integrating the significant temporal trends over space, temporal-spatial variograms show that in some situations, such as the one presented, a linear trend can be found. However, it is also demonstrated that each time and spatial scale selected may lead to different interpretations of the data, and consequently to different air quality related decisions.

6. Ackowledgments

We would like to thank the Comiss„o de Gest‚o do Ar de Lisboa who provide us the air pollution data. We also thank CNIG for the SPOT image of Lisbon.

7. References

1. DeJong, S. M., and Burrough, P. (1995). A Fractal approach to the classification of mediterranean vegetation types in remotely sensed images, *Photogrametric Engineering & Remote Sensing*, **61**, **8**, pp. 1041-1053.
2. Getis, A. and J. K. Ord, (1992) The Analysis of Spatial Association by Use of Distance Statistics, *GeographycalAnalysis*, **24**, 3, pp. 189-206.
3. Anselin, L. (1993) Exploratory Spatial Data Analysis and Geographic Information Systems, *DOSES Workshop on SpatialAnalysis and GIS*, Lisboa.
4. Anselin, L. (1994) Local indicators of spatial association-LISA, *Research Paper 9331*, Regional Research Institute, West Virginia University.

5. Isaacs, E. and Srivastava., R. (1989) *An Introduction to Applied Geostatistics,.* Oxford University Press, New York.
6. Cressie, N., (1991) *Statistics for Spatial Data*, John Wiley & Sons, Inc., New York.
7. Soares, A., T·vora, J., Pinheiro, L., and Almeida, J. (1993) Predicting Probability Maps of Air Pollution Concentration: A Case Study on Barreiro/Seixal Industrial Area, *Geostatistics Troia¥92*, Ed. Amilcar Soares, Klywer Academic Publishers, Holland, pp.625-636.
8. Miller, E. (1996), Towards a Four-Dimensional GIS: Four-Dimensional Interpolation Utilizing Kriging, GIS Research UK, University of Canterbury.
9. Seixas, J. (1996) Exploratory spatial-time data analysis: the case of desertification, accepted for the 2nd ESF-NSF GISDATA Summer Institute, Berlin, 24 July-1 August.
10. Becker, R., A., Eick, S. G. , Wilks, A. R. (1995) Visualizing Network Data, *IEEE Transactions on Visualization and Computer Graphics* **1**, 1, pp. 16-28.
11. Wolberg, G.(1988) *Digital Image Warping,* IEEE Computer Society Press Monograph.

AUTOMATED INFORMATION SYSTEM FOR DECISION MAKING SUPPORT DESIGNED TO REDUCE ENVIRONMENTAL CONSEQUENCES OF INDUSTRIAL ACCIDENTS

S. V. PAVLOV, S. V. PLEKHANOV
The Institute of the Problems of Applied Ecology and Natural and Mineral Resources Utilization 12/1, 8 Marta street, Ufa, 450005, RUSSIA

Failures in industrial operations or in transportation have a considerable impact on the environment. Some examples include failure of oil pipelines and explosions in the industrial installations. Such failures are especially frequent in industrial zones saturated with dangerous objects. The damage caused depends on the speed of the administrative decision making and liquidation of the emergency. To reduce time, automated emergency systems for decision-making support are required.

A decision-making support system for the operative-dispatching service is being developed in the Republic of Bashkortostan, Russia. It will be used by the Ministry of Emergency Situations and Environmental Safety that is a coordinating organization for emergency liquidation. The purpose of this information system is to reduce the time between an emergency occurrence and the beginning of restoration, the acceleration of restoration work, and, as the result, the reduction of environmental consequences. This time reduction is achieved by automation of the following stages of the decision-making process:

- automated data registration is provided at the stage of receiving the incoming information;
- data visualization and automated search is provided at the stage of information analysis;
- automated message dispatch via communication networks is provided at the stage of decision realization.

The structure of the information system is shown on the Figure. The information system consists of four main subsystems:

- incoming information registration;
- information search;
- emergency services database management;
- hazardous industrial objects database management.

I. Linkov and R. Wilson (eds.), Air Pollution in the Ural Mountains, 351–354.

The registration subsystem provides the current data on emergencies and registration of incoming and outgoing messages. The place, date, time, type and a brief description of the emergency; the time of receiving the information and the source of the information are stored into the database. If additional messages received, they are recorded into the database and linked to the main message. The subsystem can generate generalized reports about the situation and detailed reports about a specific emergency. Old information can be transferred from the main database to the archive or deleted.

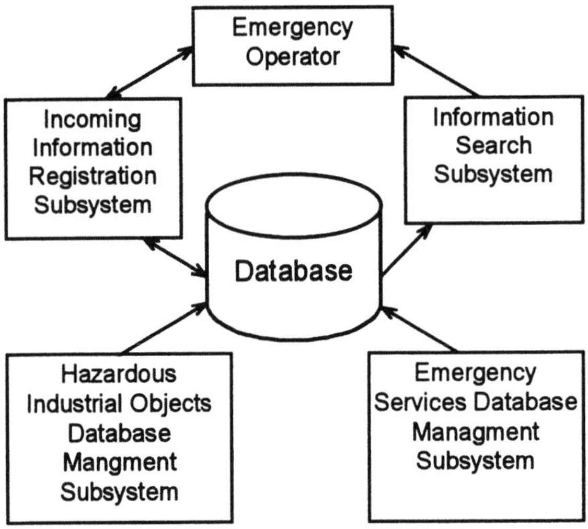

FIGURE 1. The Structure Of The Information System

The information search subsystem provides an ability to search information on emergency services and hazardous industrial objects. Hazardous objects can be searched by their location and by their environmental impact. This combined search permits a quick identification of the emergency locations by the analysis of the incoming flow of spatial information about accidents.

The emergency services can be searched by their location, available equipment and facilities, and also by their speed of response and their range. This search allows to make a reasonable and quick team selection for restoration work according the accident type and place. For the chosen team or organization, a list of telephones or other accessible communication channels can be found in the database.

The database management subsystem provides an ability to store and update the information about emergency services and hazardous industrial objects. Location of the enterprises and services, their personnel and telephone numbers are stored in the database. The speed of response and range, list of the available equipment and facilities are specified for each emergency service. For each hazardous industrial object the types of possible environmental impact are specified.

An addition of a Geographical Information System module is planned for linking the data to an electronic map of the republic. That will enable the analysis of the geographical location of objects.

This information system allows to improve the quality of the decisions related to the emergency liquidation by:

- automated registration and processing of the incoming and outgoing information; facilitatation of the data visualisation; analyses of the restoration work and preparation of reports
- increasing the speed and reliability of the identification of emergency objects and the forecasting of scales and the consequences of a failure by automated search of the information on probable emergency sources;
- quick selection of the nearest facility with the required specialized equipment for restoration work by an automated search of the information on emergency services.

The information system is installed in the Ministry of Emergency Situations and Environmental Safety. Currently it includes information on more than a hundred emergency services and hazardous objects. A simplified version of the emergency services subsystem are realized as an independent software system "Emergency Services Register" for use by local administration and government bodies.

HAZARD RANKING FOR INDUSTRIAL AREAS

V.E. GVOZDEV
Ufa State Aviation Technical University,
450000, K.Marks str., 12, Ufa, RUSSIA

A.M. SHAGIAKHMETOV
Institute of Problem of Apply Ecology and Natural Resources Use,
450000, 8-March str., 10/1, Ufa, RUSSIA

The Republic of Bashkortostan has developed extensive oil and chemical processing industry. Many of the ecologically dangerous enterprises are located in cities (Ufa, Sterlitamak and Salavat) and therefore pose a risk to the population. Prevention of emergencies has become extremely important due to the increased number and scale of accidental situations in the RB. The Ministry of Ecology and Emergencies responsible for public safety and risk management requested that all industrial cities be divided in zones ranked by their relative hazards.

This paper maps the chemical risk zones for the city of Ufa, the capital of the Republic of Bashkortostan with population of more than 1,000,000. The city map is presented on Figure 1. The highest population density is in the central and SouthWest areas. The main hazardous objects are listed in Table 1.

TABLE 1. Major hazardous enterprises in Ufa.

#	Name	Chemical	Industry
1	"Himprom"	chlorine	Chemical
2	"NUNPZ"	nitrogen	Oil-processing
3	"UNPZ"	nitrogen	Oil-processing
4	"Ufaneftehim"	chlorine	Oil-processing
5	"Ufaorgsintez"	nitrogen	Chemical
6	City sewerage system	chlorine	Sewerage
7	"UMPO"	nitrogen	Machinery engineering
8	"Bashkirriba"	nitrogen	Food storage
9	"UMKK"	nitrogen	Food processing and storage
10	"Ufamolzavod"	nitrogen	Food processing and storage
11	"Bashhladokombinat"	nitrogen	Food storage
12	"Severnii vodozabor"	chlorine	Water purification
13	"Gormolkombinat"	nitrogen	Food storage
14	"Ugnii vodozabor"	chlorine	Water purification
15	"Severno-Kovshovii vodozabor"	chlorine	Water purification

I. Linkov and R. Wilson (eds.), Air Pollution in the Ural Mountains, 355–356.

For each source of hazard, the area at risk was calculated using predictive models for accidental resuspension of toxic pollutants. This methodology was developed by the Russian Scientific-Research Institute of Civil Protection and Emergencies and is based on Gaussian dispersion modeling. The methodology is suitable for modeling spread of gaseous, vaporous or aerosol pollutants. The input data for calculating areas at risk are location of source, type of pollutants stored at facility, its amount, landscape variables and meteorological conditions (such as direction and speed of wind, temperature of air and so on).

The Geographical Information System ArcView^{R} 2.0 (GIS) was used to overlay the individual hazards and to map the combined risk zones (Figure 1). Eight hazard categories were distinguished. The most dangerous zones are influenced by seven and eight different hazard sources (black areas on the map). Unfortunately, these zones are located in the northern region of the city (center of the map) and have the highest population density.

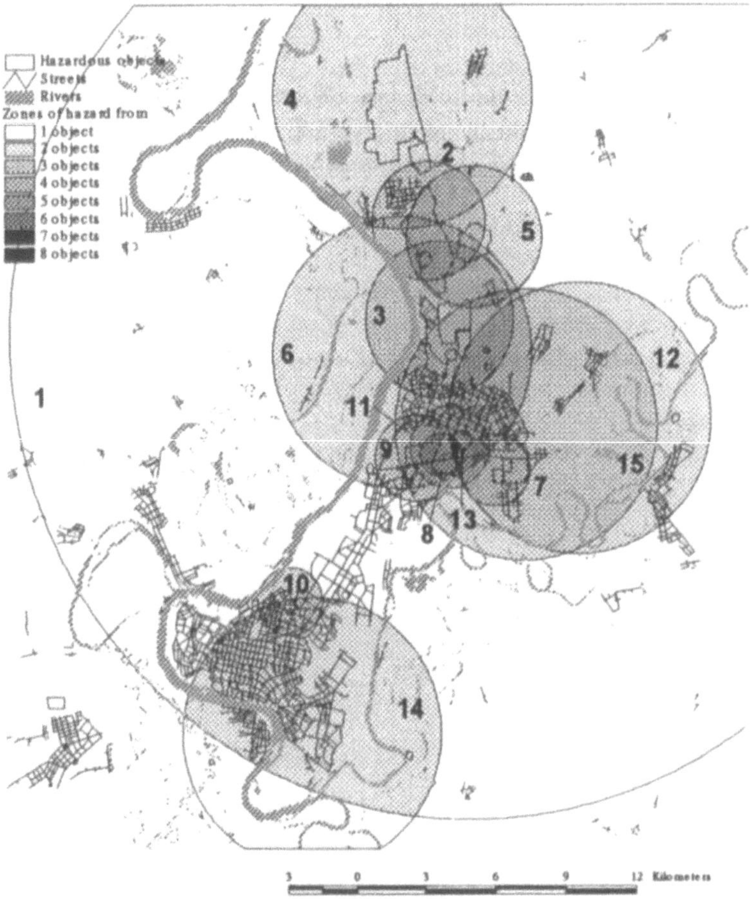

FIGURE 1. Ufa: City map and areas at risk from the industrial enetrprises listed in Table 1.

ASSESSMENT OF SOURCES OF HEAVY METALS AND SO_X EMISSIONS IN THE SOUTHERN URAL MOUNTAINS AND RECOMMENDATIONS FOR EMISSIONS ABATEMENT

Report of the Working Group on Policy Issues[1]

M.J. ELLENBECKER
University of Massachusetts Lowell, Department of Work Environment, One University Ave, Lowell, MA, USA

B. MOREL
Carnegie Mellon University, Department of Engineering and Public Policy, Pittsburgh, PA 15213. USA

I.P. MILBORROW
School of Social Sciences, Univeristy of Bath, Claverton Down, Bath, BA2 7AY, UK

1. Statement of Purpose

To identify key environmental policy issues in the Ural mountains, particularly concerned with air pollution. Discussions between Western and Russian experts aim to formulate a strategy to address the serious environmental issues of the Region.

2. Summary of Discussions

The Western experts asked the Russian representatives to identify the most pressing environmental problems in the Ural region, especially those associate with the Magnitogorsk Iron and Steel Works (MMK) operations. Initially, the discussions were somewhat stinted. However, progress did take place and the following issues were discussed.

a) the levels of heavy metals in Magnitogorsk and the Southern Ural region and their effects on the various receptors.
b) the pollution monitoring that exists in Magnitogorsk and the Southern Ural region. The pollutants for which data are available are ammonia, dust, SO_2, H_2S and SO_x.

[1] Members of Working Group:: M.J. Ellenbecker (Co-Chairman), B. Morel (Co-Chairman), I.P. Milborrow (Rapporteur), T Belotserkovets, A Djabasov, S Pantus, S Plekhanov, G Gostiaeva, N Andrenova, A Revunova

I. Linkov and R. Wilson (eds.), Air Pollution in the Ural Mountains, 357–359.

c) the levels of pollution 'norms' that exist for the Ural region. These norms are set by Federal bureaus and firms incur fines if these are exceeded. The absolute levels of these norms were not established.
d) the most efficient control options to abate pollution at the Magnitogorsk plant.
e) the availability of funds to facilitate the upgrading of production technologies in order to abate pollution from steel making operations.
f) the possibility of applying some form of cost-benefit analysis to the operations of the Magnitogorsk plant.
g) the Western experts raised concern about possible bureaucracy that could hinder access to the relevant data for assessment. The local Russian representatives reassured the group on this issue and noted that preliminary co-operation with local authorities had already taken place.
h) the Western experts stressed the need to formulate a structured approach to our work. Clarity and vision are essential in order to make applications to foundations or NGOs for funding to carry out further research, training etc. It would be valuable to show the Russian representatives how to write and submit a proposal. The Western experts stressed that NATO was keen to assist practically in the Ural region and that further initiatives were possible.

Discussions were held with Dr Kostiukov from MMK and he prepared a list of objectives and submitted it to the group. The issue here was pollution control technologies. Three proposals were outlined:

- A project to install clean-up technologies for the control of inorganic emissions (e.g. iron oxides) while pouring pig iron into the converter. The estimated cost of the project was US$2 million.
- The reconstruction of valve boxes in dust collectors and new seals for coke oven doors on four coke ovens. The estimated cost of the project was US$2 million. This project would be more important from a health perspective than the converter project since the emissions are much more hazardous (e.g. benzene, phenols, ammonia, aromatic hydrocarbons).
- To obtain equipment for the chemico-analytical control of impurities arising in the production of tin plate. A chrome-film analyzer and surface oil detector were mentioned as possible solutions. No cost estimates were given.

The Western experts believe that this list should be regarded as a preliminary and tentative. There are undoubtedly other environmental problems at MMK that need to be considered. Further liaison with MMK will take place in order to obtain a greater understanding of the process technologies and the associated environmental burdens.

3. Work Plan

The following work plan was devised and suggested as a framework to create a strategy document.

A. Define the problems to be considered

B. Assess the availability of data (environmental and economic)
C. Prioritize and formulate key indicators (5) that characterize the problems
D. Define 5 corresponding goals or endpoints
E. Consider and propose various strategy options
F. Prepare proposals to achieve the stated goals

4. Toward the Preparation of a Research Plan

A number of members of the group decided to commence some work on a project to reduce emissions of heavy metals and SO_x emissions. As a first step it was agreed that a detailed inventory of emitting sources and a characterization of the corresponding technologies in the Magnitogorsk and Southern Urals should be undertaken. This would enable us to assess the degree of the environmental problem and would fulfil tasks (A) and (B) described in Section 3 above. On the basis of this, a decision could be taken on which plants (sources) could merit further investigation for a detailed case study. The members of the group who agreed to do the work were M Ellenbecker, B Morel, I Milborrow and T Belotserkovets. The tasks agreed for each member are detailed below.

Russian Experts (led by T Belotserkovets, Institute of Terrestrial Ecology, Chelyabinsk)

First, prepare a comprehensive list of sources of heavy metal and SO_x emissions. This would identify the location of polluting industries and the level of emissions (kg/year) in the Magnitogorsk and Southern Ural Region. An inventory of monitoring stations would also be prepared. Second, the technical characteristics of the major polluting sources would be reported (e.g. what technologies are used in different processes and the age of the equipment). The nature of emissions would be documented (e.g. particle size, gas concentrations, temperature, volume flow of exhaust gas). Economic characteristics of sources would be considered. Data would be needed on employee levels and wage costs, production levels and costs, firm profitability etc.

Western Experts

Preliminary investigations will be undertaken to assess the feasibility of undertaking the above outlined work. This will include networking in the US and Europe and identifying other members of the proposed research team. The group suggests that contact with Dr Linkov should be maintained with a view to future liaisons with Russian experts.

5. Conclusions

On the basis of the above, a decision will be taken as to whether to proceed with further work.

Part 4

Air Pollution around the World: Problems and Solutions

THE CONTAMINATION AND RECOVERY OF NATURAL ECOSYSTEMS BY SMELTING AND MINING ACTIVITIES AT SUDBURY, ONTARIO

T.C. HUTCHINSON and D. GUNDERMAN (SARGENT)
Environmental and Resource Studies Program
Trent University, Peterborough, Ontario, CANADA

Abstract

The mining and smelting of nickel, copper, cobalt, iron, and other precious metals at Sudbury, Ontario resulted over many years, in the largest point source of sulphur dioxide in the world. The smelters have run since the early years of the 20th century and continue to be major producers of nickel and copper. Emissions as high as 3.2 million tons of SO_2 per year were occurring by the late 1960's. The area around the smelters is one of thin, podsolic soils originally covered with forest. A very large number of lakes also occur in the area amongst the bedrock of the Pre-Cambrian Shield.

The damage to the forests, the soil microbial flora and fauna, and to the soils was massive, affecting an inner zone of several thousand square kilometres. The major phytotoxic factor was sulphur dioxide and sulphuric acid aerosol, but large, toxic accumulations of heavy metals occurred in the soils and the leachates acidified and added metals to the aquatic systems. Many lakes became devoid of fish and of zooplankton. Acid and metal-tolerant varieties of algae, of soil fungi, of grasses and trees were evolved.

Since 1972, a 381 m. high Superstack has been in operation. At the same time, a small but very polluting smelter at Coniston was closed. Regional air quality has greatly improved. The paper focuses on the extent and pattern of response of the terrestrial and aquatic systems to the aerial contamination, and then on the substantial recovery in ecosystem function and biodiversity since clean up began.

1. The Development of Mining and Smelting at Sudbury and the Associated Regional Pollution

Sudbury is presently a city of about 85,000 people located on the acidic Pre-Cambrian Shield about 400 km north of Toronto. It is famous as the centre of one of the largest mining and smelting complexes in the world. It has also been infamous as an area of major air pollution and resultant destruction of the forests and contamination of the water bodies. The nickel, copper and iron ore bodies which now comprise the Sudbury Basin were discovered in the 1880s. The Basin itself is a geological anomaly stretching about 160 km along its southwest/northeast axis and about 70 km along its northwest/southeast axis. Around the rim

I. Linkov and R. Wilson (eds.), Air Pollution in the Ural Mountains, 363–373.

of the Basin, there are raised hills while the centre of the Basin is lower giving an overall ellipsoidal saucer shape. It is believed to have been caused by meteorite impact of such force that the earth's crust was broken and magmas were released to the surface. The cooling and crystallization of these magmas resulted in the very rich nickel, copper, iron ore bodies which have been mined intensively. Following the initial discoveries, a large number of mines were located around the perimeter of the Basin and a small roast yard and smelter was established at Copper Cliff as early as 1888, [14]. At first, the major interest was in the copper ores, while the nickel with which it was associated and which was in greater quantity, was initially considered a nuisance from which the copper needed to be separated. By 1890 however, new discoveries of the value of nickel in hardening steels resulted in a major focus on the nickel also.

In the early days of Sudbury mining, much of the ore was taken by train and ship to Swansea in south Wales and somewhat later, to the United States for smelting. From 1903 approximately, major smelters were developed at Copper Cliff and in 1913, a further smelter was developed at Coniston. A third smelter was opened at Falconbridge in 1928.Along with the early smelters, open roast yards were used for the initial reduction of the sulphur content of the ore before it went for smelting. The ore contained up to 24% sulphur in the form of pyrite with much higher percentages of sulphur occurring in another iron sulphide form, pyrrhotite. At the roast yards, raw ore was burned after an initial ignition with logs cut from trees in the area. See Symington and Hutchinson in this volume. [13]

The roast yards drove off sulphur dioxide in large quantities and during their life, from 1888 to 1929 , were estimated to have discharged 10 million tonnes of S02 at ground level into the Sudbury environment, [11] The ore, after burning for several months, was then taken to the smelters for further processing using air and oxygen as a gas. The sulphur-dioxide produced from the smelting was discharged through smoke stacks. Over time, taller and taller smoke stacks were built, from 30 metres in height to 120 to a final Super Stack of 381 metres at the Copper Cliff smelter, which went online in 1972. These smelters and roast yards also allowed aerial discharge of dust and particulates, containing metals, especially Ni, Cu, Co, and Fe as well as metalloids such as As and Se. The sulphur dioxide, in particular, was phytotoxic and killed the vegetation over a large area. It also severely acidified the surface soils and increased the solubility of toxic elements in the soil and mobilized potentially toxic elements such as Al, [8]. There are literally thousands of lakes within 100 km of Sudbury and many of these also became acidified with loss of their fish populations, changes in water chemistry and effects carried on down through the food chain including the elimination of molluscs, zooplankton, and even of insect larvae. Numerous complaints from the farmers and from local people, drew attention to potential human health effects and effects on their living environment.

The smelters became infamous for the magnitude of their sulphur fumigations. These by the 1960s were adversely affecting an area of approximately 5,300 sq. km., within which at least once a year, a fumigation occurred which was damaging to all vegetation. Towards the centre of the area, such damaging fumigations occurred many times per year. Sulphur emissions increased through 1920s to the 1970, which reached a peak production of approximately 3.2 million tonnes per year in 1968-70. By this time, the initially forested landscape had been devastated over a very large area with the trees killed and with particularly denuded landscape occurring around each of the three smelters. The three smelters were

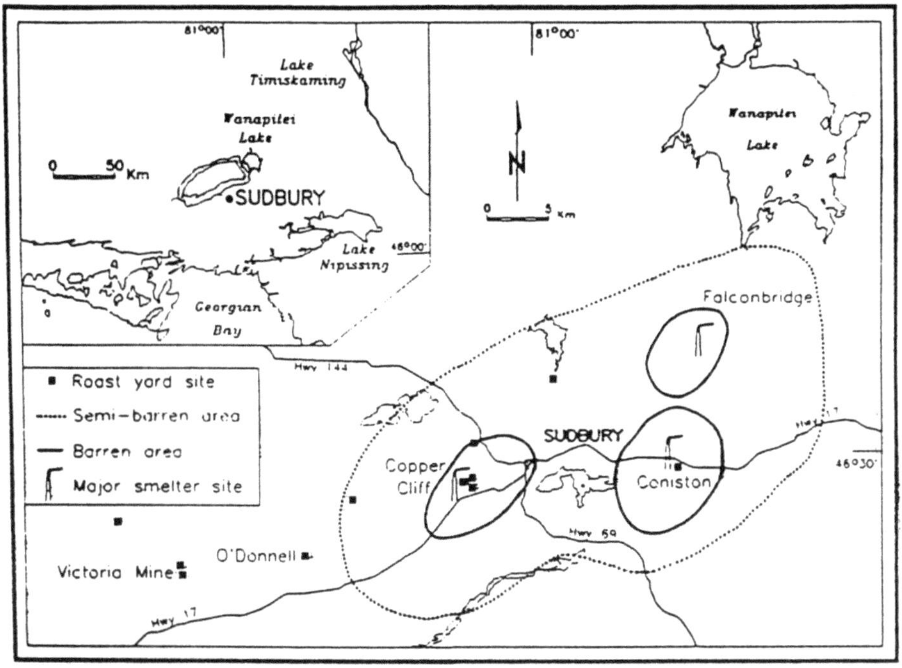

FIGURE 1. The location of the Sudbury Basin and the major sites of roasting and smelting activity and Struik's (1973) "zones of site and vegetation stability" based on air photographs.

located five km west of Sudbury (Copper Cliff), 12 km northeast of Sudbury (Falconbridge), and 8 km east of Sudbury (Coniston) Fig. 1. The town of Sudbury, which was the headquarters for all of this mining activity and for trade and commerce, thus lay within the triangle of the three smelter locations and was subjected to some degree of air pollution on an almost daily basis. The other small towns in the vicinity were also regularly fumigated especially when static weather occurred and during temperature inversions. In 1972, the new Super Stack opened at Copper Cliff and the Coniston smelter was closed. For the first time in many years, the air quality began to improve. Also, while the mining and smelting are still the major sources of employment in the Sudbury region, there has been a gradual but extensive diversification of industry and of commerce. This has coincided with a substantial reduction in the number of workers employed by the two major smelting companies ie. International Nickel Company and the Falconbridge Nickel Company but without massive unemployment and economic disruption of the city and its people.

The town of Sudbury has transformed itself from a place often depicted as one of the pollution capitals of the world to an area in which much re-greening has already occurred and in which natural and assisted/managed recover and rehabilitation has taken place. This latter has been a community effort which has involved thousands of school children as well an unemployed persons, the mining companies, local and regional governments, and job creation schemes from the provincial and federal governments. Sudbury, over the past 25 years, has been transformed along with its surroundings. [6] The mining and smelting towns of the Urals, it can be hoped and expected, will be able to make the same transformations in the near future.

The Sudbury smelters, like the steel mills at Magnitgorsk, became major producers of metals on a world scale. By the 1960s, 60% of the world's nickel outside of the Soviet Union was produced at Sudbury. It was also a major producer of copper, iron, cobalt, gold, silver, platinum, thallium, osmium, selenium, arsenic and sulphur. From an environmental perspective however, it was the sulphur dioxide emissions which became the dominant factor. 2.6 million tonnes of SO_2 were produced in 1960 and this increased to 3.2 million tonnes by 1970. This represented 10% of all sources of sulphur dioxide in North America including coal burning, and 25 % of all smelter sources in North America. Most of this was being discharged by 1972 through one Superstack. The quantities of SO_2 being generated were far in excess of the company's ability to produce economic byproducts. Large quantities of sulphuric acid were produced for use in the chemical and plastics industry and in production of superphosphate fertilizers, up to 2500 tonnes per month.

2. Effects on Terrestrial Vegetation

The original vegetation of the Sudbury region was that of a mixed hardwood, conifer forest. The dominant trees were white pine (Pinus strobus), red maple (Acer rubrum), red oak (Quercus rubra), with a ground cover of shrubs such as Kalmia polifolia, and several species of blueberry eg. Vaccinium angustifolium and V. myrtilloides. The sulphur dioxide first killed all the coniferous species, so that by 1970 white pines had been eliminated within a radius of 15 km around each of the three smelters. As fumigations continued, the death of trees spread from the hilltops down into the valleys and simultaneously into the shrub layers of ground flora. Mosses and especially lichens were particularly susceptible, [4,1]. This death of the forest vegetation allowed major soil erosion to take place. These soils were shallow to start with, perhaps only 15 cm in depth, but over time almost all of this soil was lost from the slopes, being washed into the lakes and streams and also into low lying marshes and wetlands. The particulate accumulation of heavy metals must have begun in the surface soils around the roast yards and the smelters from the early days. Surface soil accumulations in excess of 3,000 ppm of both copper and nickel occurred. In addition, accumulations of several hundred ppm of cobalt also occurred. The highest concentrations were in the surface soils, decreasing with depth in the soil profile, [8]. The concentrations also decreased in a linear manner from the point sources. We calculated from measurements of deposition rates at different distances made in the period 1975 to 1978 that more than 98% of the total sulphur emissions were transported greater than 100 km from the smelters and therefore, contributed substantially to regional air pollution [5]. Indeed, under certain meteorological conditions the plume from the Sudbury stack has been detected across the Atlantic, a distance in excess of 1,000 km and sensitive sulphur dioxide instruments in Greenland indicated elevated levels from which back trajectories suggested Sudbury as a source, 3,200 km away. While most of the sulphur dioxide was carried long distances from Sudbury, nevertheless a substantial amount was converted to sulphate and deposited more locally and dry deposition of sulphur dioxide also occurred. This deposition was substantially increased during rain events and when the air was cleansed by snow falls. Such rain and snowfall was very acid, with the pH of rain falling through the smoke plume being 2.8. [9]

Since smelting began, more than 100 million tonnes of sulphur dioxide have been discharged into the environment. In addition, tens of thousands of tonnes of copper, iron, cobalt and nickel have also been discharged [11]. In the period 1960-72, 40,000 tonnes per year of particulate was emitted. From 1973- 81, 15,000 tonnes per year of particulate were emitted. The projection for 1995 was that particulate emissions would be down to 1,500 tonnes. Fifty to sixty percent of the particulate fell within 50-60 km of the smelters [5]. The composition of the particulate material has changed over time as the quality and chemistry of the ore being used has altered. In the early years, nickel predominated over copper but for the period 1973-81, the emissions were as follows: 1800 tonnes/year Fe, 700 tonnes/year Cu, 500 tonnes/year Ni, 200 tonnes/year Pb, 100 tonnes/year As. The total suspended particulate levels for nickel were 50-80 μg/m2 [11].

By 1972, the central industrial area around the smelters was described as an industrial barren. It was almost entirely devoid of vegetation except that for <u>Vaccinium</u> on sandy area and, in wetter areas, the cattail (<u>Typha angustifolium</u>) and the grass (<u>Phragmites communis</u>) still survived. A moss (<u>Pohlia nutans</u>) was found to be remarkably tolerant of extremes both of acidity and of heavy metals. It occurred in the forests many kilometres from the smelters as a minor constituent of the forest floor but was found in the most contaminated Sudbury conditions, an opportunity to occupy large areas and presumably benefiting from the almost complete lack of competition from other plant species, combined with its remarkable pre-adaptation to the adverse soil conditions.

3. Aquatic Impacts

The substantial erosion and runoff, especially of metal contaminated silt and clay, caused increased rates of sediment deposition and accumulation of severely metal contaminated sediment on the bottom of the lakes and streams. In two lakes near the Coniston smelter (Alice and Baby Lake), nickel and copper levels exceeded 3,000 ppm in the upper sediments and in Baby Lake, nickel concentrations in sediments at 5 cm depth were 8,500 ppm, copper levels peaked at 6,000 ppm in these same sediments at a shallower depth i.e. more recent deposition [15]. In these same two lakes, in the period 1968-72, all zooplankton and phytoplankton had been eliminated. Some life still existed in the sediments themselves, notably the presence of red chironomids [10]. In Baby Lake, the pH of the water column had been reduced from 7.0 to about 4.0. A few unicellular algae cells still existed at the surface of the sediment and these were, inevitably, acid tolerant and heavy metal tolerant, [12]. To detect these algal cells, it was necessary to filter 75 litres of water from the lakes. Fish populations had been eliminated at an early stage. Dixit et al. [3] described the changes in the diatom assemblages over time in the Sudbury lakes and from these assemblages they inferred the pH of the lake water at the time the diatoms were deposited into the sediments. In one example , that of Clearwater Lake, the pre-industrial inferred pH for the period 1780 to 1880 was around about 6.0. It is then inferred to have fallen over the next 100 years to a pH of 4.7 and has since recovered to 5.5, Dixit et al. [3].

4. Recovery of the Terrestrial and Aquatic Ecosystems since 1972

Following the building of the superstack with its effective dispersion of SO_2 out of the Sudbury area, together with the closure of the Coniston smelter and the reduction overall of SO_2 emissions, the chemistry of the lakes began to change. Fig. 2 shows this change (recovery) for sulphate, pH, nickel and copper contents for Baby Lake, situated very close to the Coniston smelter. pH recovered from 4.0 to 7.0 within 15 years. Sulphate levels were the key factor in this, decreasing from 80 mg.l to 20 by 1987. Copper and nickel concentrations, which had been severely toxic to phytoplankton, dropped to non-lethal levels (Fig. 2).

Following along with these changes in water chemistry has been a re-invasion or re-establishment of biota and a food chain. By 1987, 26 algal species were found in Baby Lake and 41 in Alice Lake next to it. In the 1960s, only five algal species were found and these were highly metal tolerant and present in very small quantities [7]. Zooplankton also began to re-appear, including rotifers. By 1987, 13 zooplanktonic species were back in Baby Lake and 15 in Alice Lake. Several of these are well known to be acid tolerant species. In more recent times, from 1990 to the present, some fish species have re-invaded, including sticklebacks and sunfish. Bioassays for the lake water toxicity, using Chlorella vulgaris as a test organism, have shown that the water in Baby Lake is now a much more supportive medium for algal growth then it was in 1968, [7].

In the terrestrial barren areas around Coniston, major changes in soil chemistry have also occurred. Soil pits were taken in 1972 and then again in 1992 from immediately adjacent to them. Soils were collected at sites of increasing distance out from Coniston to 52 km. Samples were also collected down the pit from surface soils of 0-1 cm, through 1-5 cm to 5-10 cm depth. When these soils were chemically analysed using exactly the same methods as used in 1972, striking changes were found in pH, Ni, Cu, Co and in Al. Basically, the degree of acidity and metal toxicity had greatly reduced in the 20-year interval, due to a major reduction in SO_2 and in acidic inputs. Much of the copper and nickel and H ions have been leached through the soils and have flowed into the rivers and drainage areas. Such changes in the soils are illustrated in Figs. 3 and 4. Data for H+ ion concentrations and for nickel only are shown but are typical results. While major reductions in soil levels of nickel and H ion concentrations are found, the reduction is also seen throughout the 10 m soil profile, but with greatest effects in the surface layer. The sites out to 19 km show chemical recovery while the most distant site at 50 km shows a small increase in H+ ion (as well as in Al). This distant effect is believed due to the long distance transport of particulates from the new Superstack, which are beginning to acidify soils at great distances from the source, while the inner zone recovers.

In the Coniston Valley, acid and metal tolerant grasses began to colonise very rapidly after the smelter closed [2]. Since then, on the rocks and on the bare soils, some lichens have re-colonized. Tree seedlings are growing in amongst these grasses at sites from 3-15 km from the closed smelter. Many trees which survived the worst pollution periods are beginning to re-grow, while others have remained in a curious "bonsai" state. The whole area is re-greening in a quite spectacular manner. New communities of higher plants are developing in the inner areas.

In summary, it can be said that despite the Sudbury area being one of the most polluted landscapes in North America as recently as 25 years ago, the reduction in sulphur and

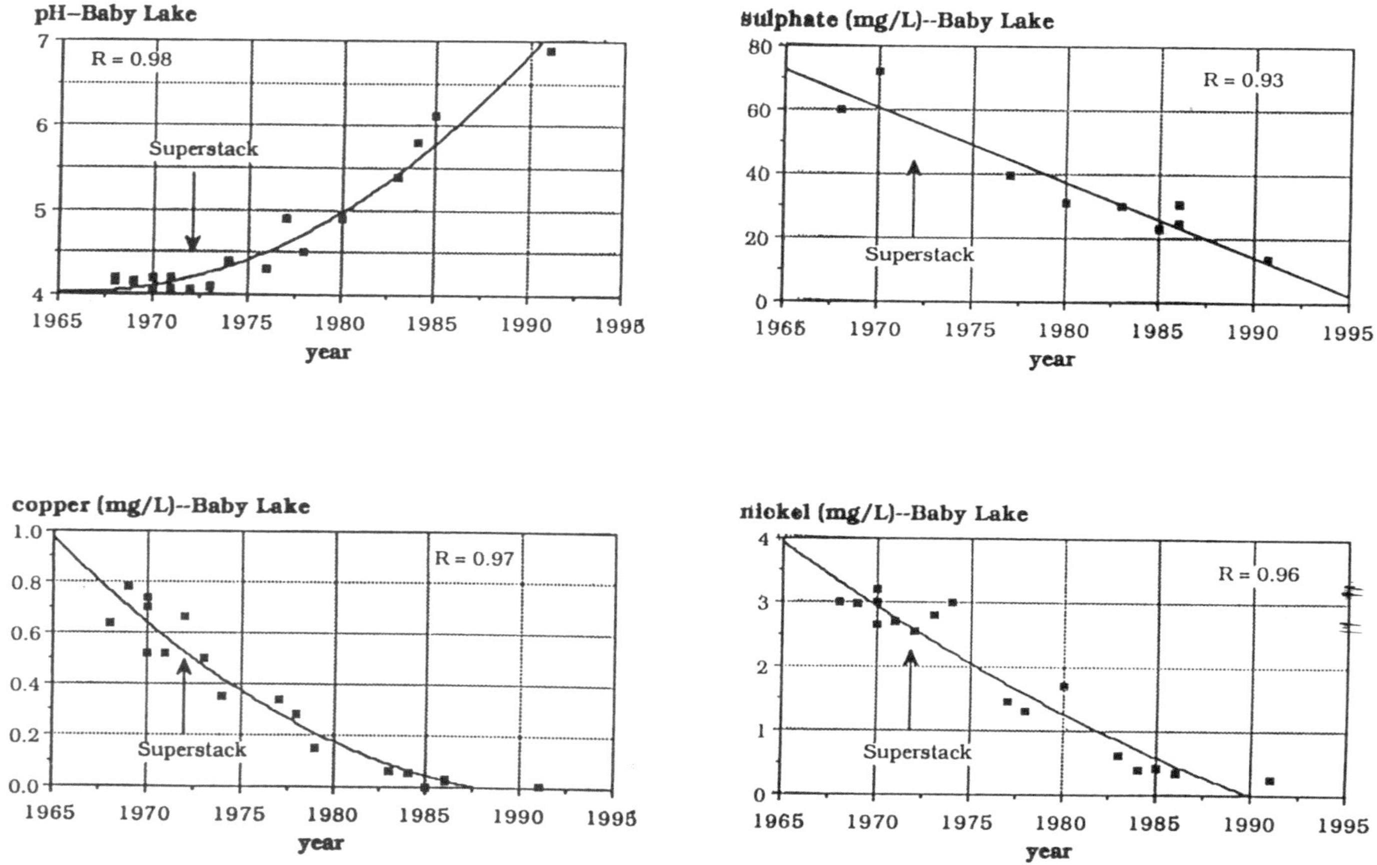

FIGURE 2. Water chemistry of Baby Lake in period 1968-1992

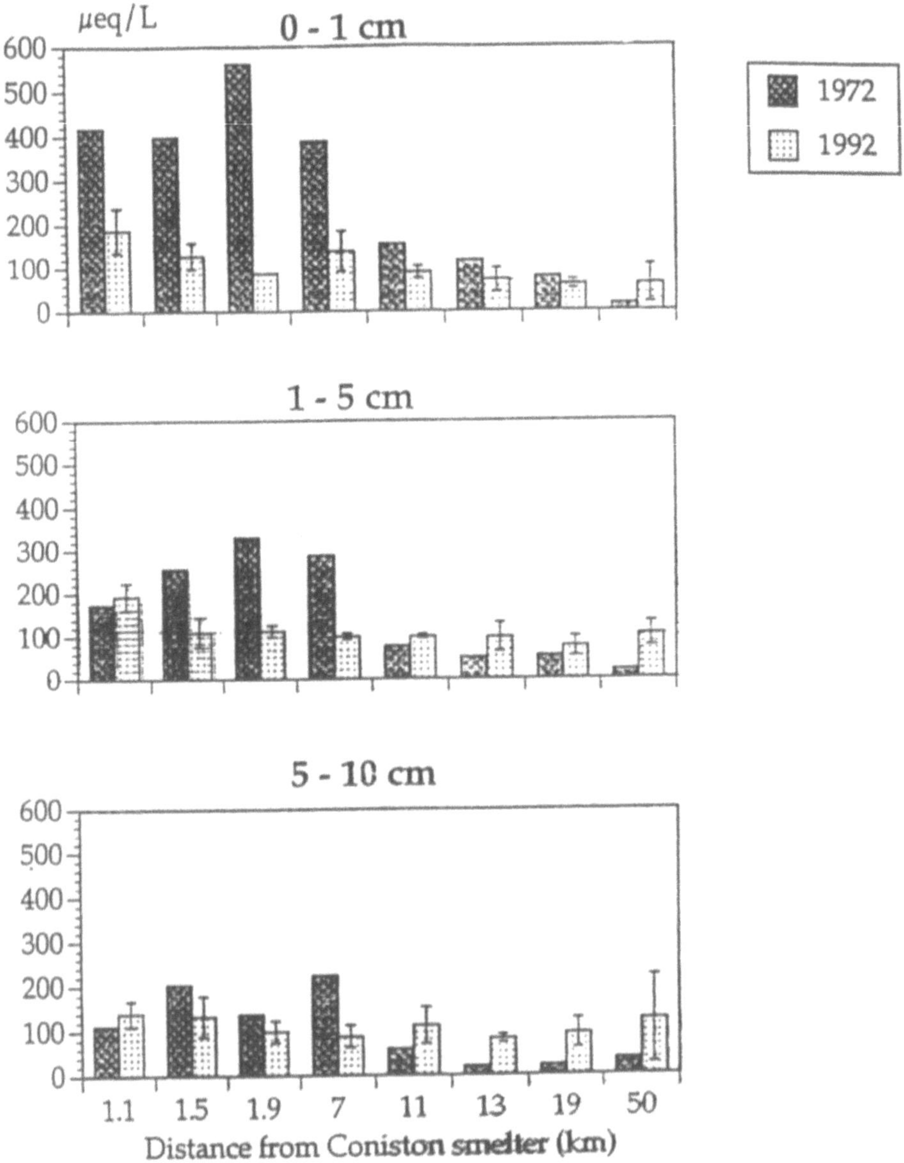

FIGURE 3. Changes in the concentration of hydrogen ions found in the 0-1, 1-5, and 5-10 cm soil depths compared to the same sites analyzed in 1972. 1992 data is represented by means $\pm$ standard deviations (n=3) and 1972 data is represented using means only (n=3).

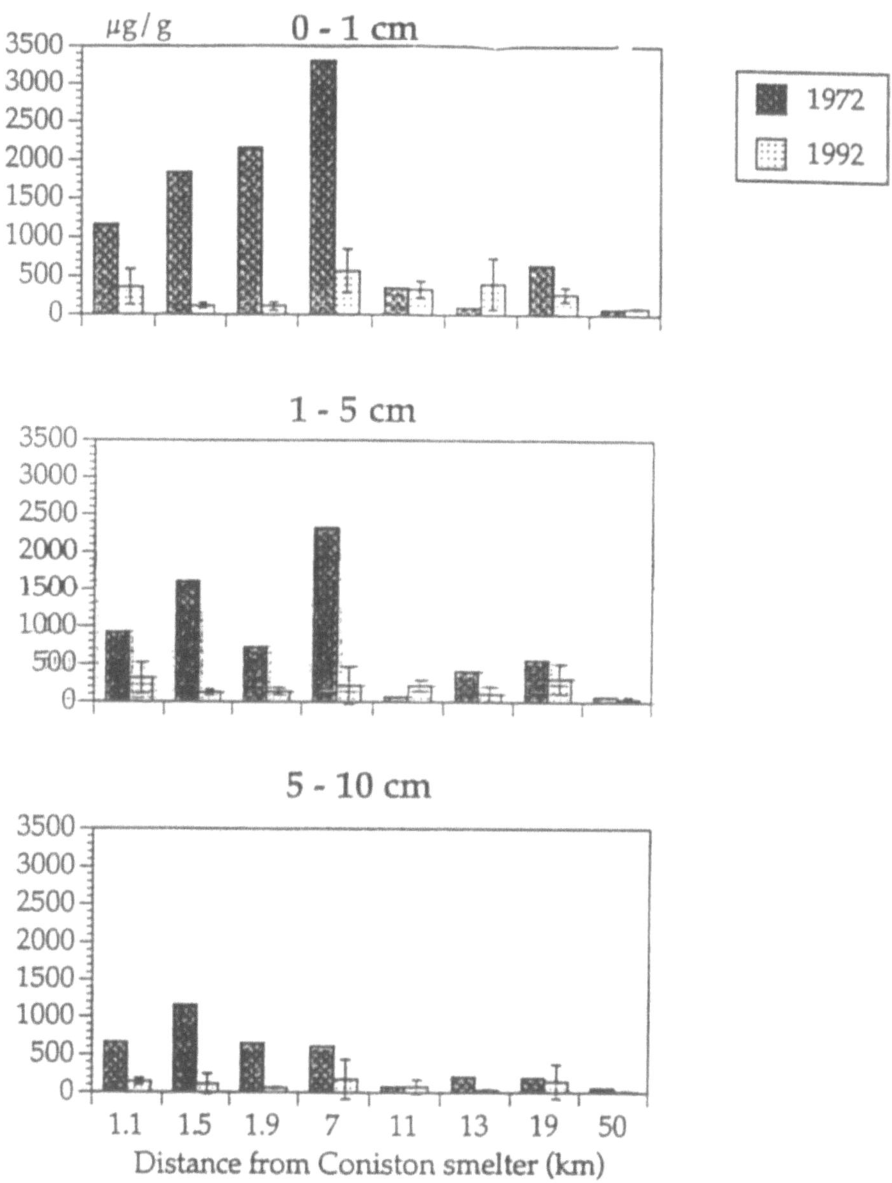

FIGURE 4. Changes in the total concentration of nickel found in the 0-1, 1-5, and 5-10 cm soil depths compared to the same sites analyzed in 1972. 1992 data is represented by means ± standard deviations (n=3) and 1972 data is represented using means only (n=3).

particulate emissions have allowed a major improvement to take place. Chemical changes have driven this recovery which is apparent in the re-establishment of phytoplanktons, zooplankton and fish, while on land the forests are beginning to re-invade and be re-established. Acid and metal tolerant species and races have been an important intermediary in this process. The human environment is now much improved with a reduction in health risks to both children and adults. Deliberate management interventions along roadways have also improved the aesthetics of the area. Sudbury was one of the cities to be given a UN re-greening award at the 1992 Rio Conference on the Environment. The ability of its ecosystems to recover from massive pollution insult lends hope that those in the polluted cities of the Urals will be able to do the same.

5. References

1. Beckett, P. J. (1995). Lichens, sensitive indicators of improving air quality. In Restoration and Recovery of an Industrial Region. Ed. J. M. Gunn. pp 81-92. Pub. Springer-Verlag.
2. Cox, R.M. and T. C. Hutchinson. (1979). Metal co-tolerances in the grass Deschampsia cespitosa. Nature, London 279:231-233.
3. Dixit, S. S., A. S. Dixit, J. P. Smol and W. Keller (1995). Reading the records stored in the lake sediments. In Restoration and Recovery of an Industrial Region. Ed. J. M. Gunn. pp 33-46. Pub. Springer-Verlag.
4. Freedman, B. and T. C. Hutchinson. (1980). Long term effects of smelter pollution at Sudbury, Ontario on surrounding forest communities. Can. J. Bot. 58:2123-2140.
5. Freedman, W., and T. C. Hutchinson. (1980a). Pollutant inputs from the atmosphere, and accumulations in soils and vegetation near a nickel-copper smelter at Sudbury, Ontario. Can. J. Bot. 58:108-132.
6. Gunn, J. M. (1995). Progress in restoring the smelter-damaged landscape near Sudbury, Canada. In Restoration and Recovery of an Industrial Region. Pub. Springer-Verlag.pp 358.
7. Havas, M., D. G. Woodfine, P. Lutz, K. Yung, H. MacIsaac and T. C. Hutchinson. 1995. Biological recovery of two previously acidified, metal-contaminated lakes near Sudbury, Ontario, Canada. Water, Air and Soil Pollution 85:791-796, 1995.
8. Hutchinson, T. C. and L. M. Whitby (1976). Heavy metal pollution in the Sudbury mining and smelting region of Canada. I. Soil and vegetation contamination by nickel, copper and other metals. Environ. Conservation. 1:123-132.
9. Hutchinson, T. C., and L. M. Whitby. (1977). The effects of acid rain-fall and heavy metal particulates on a boreal forest ecosystem near smelting region of Canada. Water, Air and Soil Pollution, 7:421-38.
10. Hutchinson, T. C. and M. Havas (1986). Recovery of previously acidified lakes near Coniston, Canada following reductions of atmospheric sulphur and metal emissions. Water, Air and Soil Pollution. 28: 319-333.
11. Potvin, R. R. and J. J. Negusanti (1995). Declining industrial emissions, improving air quality, and reduced damage to vegetation. In Restoration and Recovery of an Industrial Region. Ed. J. M. Gunn. pp 51-66. Pub. Springer-Verlag.

12. Stokes, P.M., T. C. Hutchinson and K. Krauter. (1973). Heavy metal tolerance in algae isolated from contaminated lakes near Sudbury, Ontario. Can. J. Bot. 51:2155-2168.
13. Symington, M. S. and T. C. Hutchinson (1998) Retrospective analysis of the environment and community associated with the O'Donnell Roast Bed, Sudbury, Ontario, Canada. In Air Pollution in the Urals. Ed R. Wilson and I. Linkov. Proc. NATO Advanced Research Institute, held May 1997 at Magnetigorsk, Russia.
14. Winterhalder, K. (1995). Early history of human activities in the Sudbury area and ecological damage to the landscape. In Restoration and Recovery of an Industrial Region. Ed. J. M. Gunn. Pub. Springer-Verlag.
15. Woodfine, D. G. (1998). Doctoral thesis. Chemical and biotic recovery of metal contaminated and acidified lakes in the Sudbury region. Canada. Trent University, Peterborough, Ontario, Canada. Watershed Ecosystem Graduate Program.

RETROSPECTIVE ANALYSIS OF THE ENVIRONMENT AND COMMUNITY ASSOCIATED WITH THE O'DONNELL ROAST BED, SUDBURY, ONTARIO, CANADA

M.S. SYMINGTON
Canadian Heritage and Development Studies Graduate Program
Trent University, Peterborough, Ontario, CANADA

T.C. HUTCHINSON
Environmental and Resource Studies Program
Trent University, Peterborough, Ontario, CANADA

Abstract

The world's largest open ore roasting bed operated at O'Donnell, west of Sudbury, Ontario, Canada, from 1916 to 1929. It processed the sulphur-rich, copper-nickel ores of the Canadian Copper Company prior to smelting. By the time it was closed, the ground level fumigations of sulphur dioxide had devastated vegetation for miles around. Roast yard workers were located at the townsite of O'Donnell which was located a mile downwind. Interviews with former O'Donnell residents have been conducted in an attempt to document the social impacts the roast yard had upon this community. Sixty-eight years later, the roast bed itself is still almost devoid of vegetation. However, re-colonization and recovery elsewhere has allowed a forest, albeit unusual in species composition, to grow almost to the edge of the roast bed. The primary colonizers are metal-tolerant grasses, sedges and mosses. The surface soils of the recovered forest are still strongly acidic as is the roast bed itself (pH 3.9-4.2). Concentrations of nickel and copper in the roast bed still exceed several thousand μg/g, while several hundred μg/g Ni and Cu occur in surface forest soils to 300m from the roast bed edge. The social and environmental impacts of the land use changes of the O'Donnell roast bed area from the early 1900's are described.

1. History of Roasting

Roasting was a primary smelting activity in which the mined nickel-copper sulphide ore was roasted to reduce its' sulphur content. Wood was piled to form a thick base upon which green ore was placed. After lighting, the wood burnt out after approximately 60 hours leaving the ore in vigorous combustion, continuing to burn for

I. Linkov and R. Wilson (eds.), Air Pollution in the Ural Mountains, 375–381.

up to 7 months [5]. The O'Donnell Roast Bed was 2,300 m long with approximately 250,000 tons of ore being roasted at one time, generating intense ground level fumigations of sulphur dioxide and sulphuric acid aerosol, as well as particulate matter.

The first roast beds were located in the village of Copper Cliff. The fumes were intense and created great discomfort for its residents. The new location of O'Donnell (20km outside of the town) was chosen in an attempt to minimize complaints from farmers who were demanding compensation for their crops being damaged by the roasting process.

To provide housing for the workers employed to run the roast bed, the village of O'Donnell was built 1.6 km downwind. Since the roast bed burned year round, there was little relief from noxious fumes for the residents of the village of O'Donnell. In 1916, at the establishment of O'Donnell there were as many as 30 families and many more single men working at this roasting yard. The town consisted of many family homes, a one room school house, a butcher's shop, a general store and post office, more than 4 boarding houses which housed the single men of O'Donnell. In 1918, O'Donnell was mechanized and as a result, 10 times the amount of ore was roasted at any one time. This mechanization had a dramatic impact on both the size of workforce in that it was reduced from >200 to approximately 60, as well as increasing the environmental impact on the area surrounding the roast bed.

Interviews with former O'Donnell residents, all aged between 80-95 years of age in 1996, revealed the following health and environmental conditions:

- -residents experienced nose bleeds on a daily basis
- -fumes were so severe that men, on their way to work in the morning constantly had to lay on the ground to breathe fresh air, in order to be able to continue to work.
- -metal chains, which were set up around the roast bed so workers could find their way along, had to be frequently replaced due to the corrosive action of the acidic sulphur fumes coming off the roast bed.
- -there was not a blade of grass within miles of the roast bed
- -the school was closed 50% of the time because the sulphur fumes were so dense that the students were unable to see the black board.
- -the sulphur fumes killed germs and as a result the residents viewed themselves as the healthiest people in the area. However, many workers quit after a few days because of the conditions.

Despite the extreme conditions, all of the former residents interviewed have fond memories of their time at O'Donnell.

2. Environmental Recovery

Since the roast bed's closure in 1929, there has been remarkable recovery of the surrounding ecosystem. The forest has recovered naturally to the edges of the historical site of the roast bed. Understanding these recovery mechanisms can provide invaluable information towards rehabilitating similarly stressed environments. Soil chemistry of the area surrounding the roast bed shows elevated levels of heavy metals 68 years after the environmental stress was alleviated (Table 1).

TABLE 1. Surface soil chemistry and pH. Concentrated nitric acid digest, analysed with ICP-MS. n=5. (Modified from Hutchinson and Symington, [4]).

Site	pH	Ni (μg/g)	Cu (μg/g)	Co (μg/g)	Ca (μg/g)	Cd (μg/g)	Mg (μg/g)
Roast Bed (vegetated)	4.0	2069	1728	58	5	0.31	8
Roast Bed (non-vegetated)	3.8	6672	3648	167	4	0.41	4
Roast Bed (D. Cespitosa)	4.4	8885	10051	299	21	1.06	8
10 metres from Roast Bed	3.6	217	308	14	6	0.2	7
30 metres from Roast Bed	4.1	355	499	12	5	0.21	3
50 metres from Roast Bed	4.4	115	149	5	4	0.13	1
100 metres from Roast Bed	4.3	352	388	11	3	0.31	1
300 metres from Roast Bed	3.8	283	338	8	3	0.67	1

Surface soil chemistry shows a decrease in metal contamination as distance increases from the roast bed. On the roast bed itself concentrations exceed several thousand ppm for Ni and Cu, with Co levels also being well above background, and within 10 metres drop to a few hundred for both Ni and Cu, and to below 15 for Co, with a pH of 3.8-4.4. Three sites were chosen on the roast bed itself to explore the differences (if any) between the vegetated and non-vegetated sites. Most (99%) of the roast bed is devoid of vegetation. The levels of Mg at the two vegetated sites on the roast bed were two times the concentration found on the non-vegetated site. The *D. cespitosa* site has both elevated Cd and Ca as well as very high nickel and copper levels.

Soil profiles in 5 cm depth increments show the mobility of heavy metals to a depth of 35 cm. In general, metals were more tightly bound to surface soils, presumably a function of the greater amount of organic material at the surface (Figure 1). The profiles for nickel and copper show the surface retention of these two metals, while the profiles for cobalt show a greater movement of Co down through the profile. This behaviour is particularly evident at a distance of 300 m from the roast bed.

Background levels of copper and nickel for this area and soil type are 20-40 μg/g Ni and 30-50 μg/g Cu as cited by Bowen [1]. As is clearly shown, the metals in the soils surrounding the roast bed are still elevated well above background. These same surface soils were then assessed as to their potential toxicity to plant growth using a bioassay technique in which seeds of radish variety Cherry Belle were germinated and grown for 4 weeks on soils collected to a depth of 10 cm.

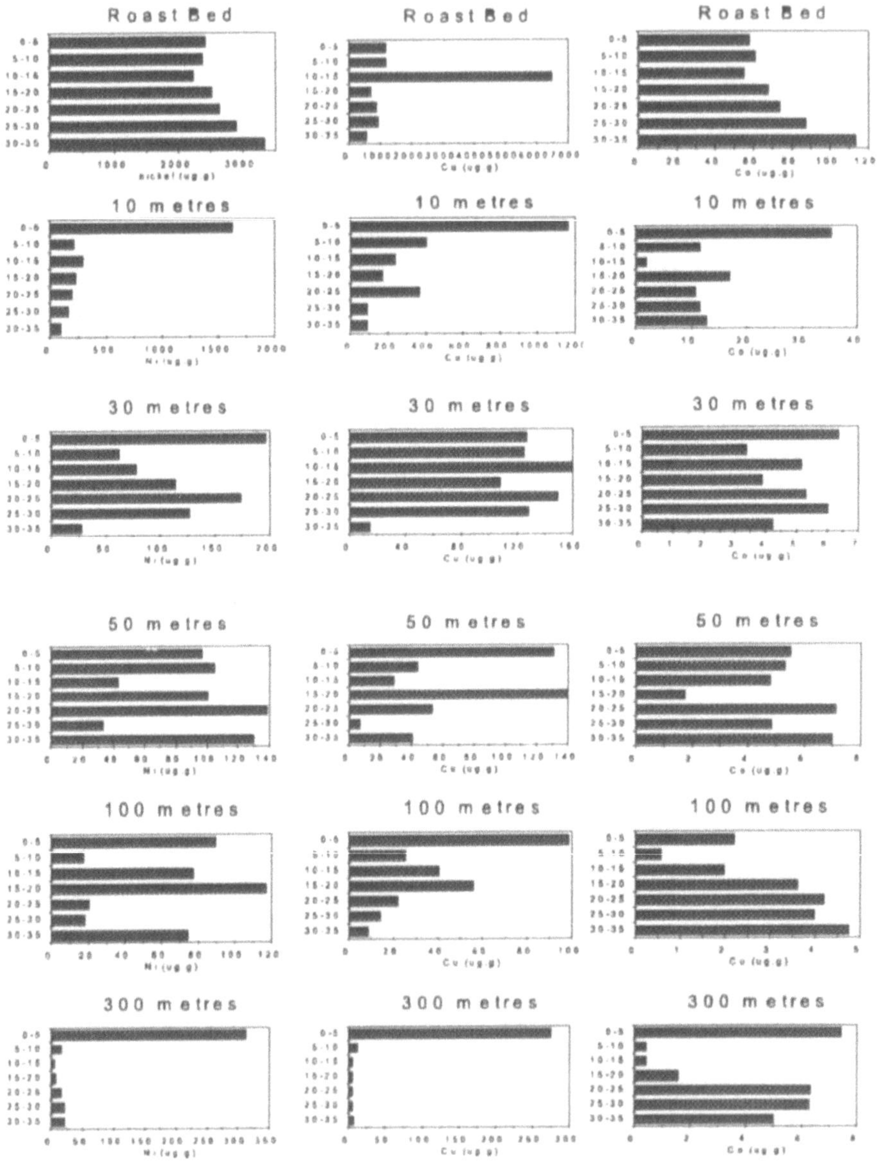

FIGURE 1. Soil profiles for Ni, Cu, and Co (μg/g) from sites at set distances from the Roast Bed. Samples analysed by ICP-MS (n=1).

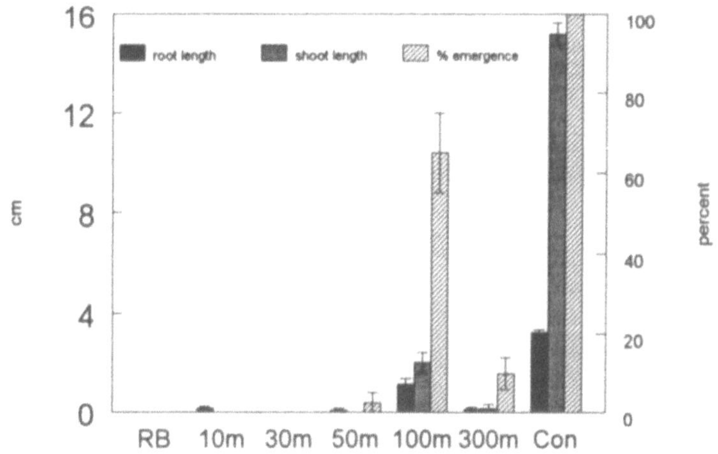

FIGURE 2. Bioassay of the present day roast bed soils using radish variety Cherry Belle.

The seedlings were grown in plastic pots under greenhouse conditions. It is apparent (Figure 2), that despite the soils surrounding the roast bed being now naturally vegetated, the soils are still highly toxic to a moderately sensitive species such as radish *Raphanus sativus.*

There was little to no growth for radish in soils out to 300m away from the Roast Bed. There is significant emergence at 100m, which is the wettest site and perhaps has had more leaching than the other sites. The control (CON) on the far right gives an indication of the lack of growth at the other sites.

Despite it having been shown that the soils surrounding the roast bed are still toxic, there are tolerant plant species that are able to colonize and thrive in these harsh conditions of high levels of heavy metals and low pH (Table 2).

A marked transition in composition of the vegetation occurs between sites within 50 m of the roast bed and that at greater distances. In spite of these harsh conditions (i.e. low pH and toxic metal concentrations), tolerant species such as the grass *D. cespitosa* (see [2], and [3]) the moss *Pohlia nutans* and the sedge *Carex* spp. are able to become established on the actual roast bed and may then play a critical role in this ecosystem.

These species may provide the initial organic material needed for less tolerant species to colonize. In addition these tolerant species could have an impact on soil chemistry (Table 1). There is a species transition that occurs at approximately 50m

TABLE 2. O'Donnell roast bed percent cover of ground vegetation - 1994. Based on 10 one metre square quadrates placed randomly at each site. Modified from Hutchinson and Symington [4].

Species	Distance from Roast Bed (m)									
	0	0	10	30	50	80	100	150	200	300
Pohlia nutans	15	1	52	93	26	-	-	-	-	-
Agrostis scabra	9	-	2	3	-	-	-	-	-	-
Carex spp	3	-	2	1	-	-	-	-	-	-
Deschampsia cespitasa	-	9	-	-	-	-	-	-	-	-
Betula papyrifera	-	-	7	1	5	2	1	1	-	1
Kalmia polifolia	-	-	1	1	43	84	67	3	4	33
Polytrichum communis	-	-	-	1	41	22	1	-	1	-
Deschampsia flexuosa	-	-	-	-	-	1	1	1	-	12
Ledum groenlandicum	-	-	-	-	-	-	-	62	12	-
Fragaria virginiana	-	-	-	-	-	-	-	2	14	-
# lichen species	1	-	1	1	1	1	5	2	3	7
# vascular species	2	2	7	6	5	14	18	29	30	25
% Bare ground	65	90	-	3	-	-	-	-	-	-

from the more tolerant colonizers to the less tolerant shrubs and grasses as can be seen in Table 2. The number of lichen species increases from 1 on the roast bed to 7 at a distance of 300m from the roasting site. The percentage of bare ground drops essentially to zero as soon as you move away from the roast bed itself and the number of vascular species peaks 200 m from the roast bed.

In conclusion, the area surrounding the roast bed still has phytotoxic conditions due to a strong acidity combined with excessive concentrations of copper and nickel. Despite this, a forest now grows on these soils, developed over a 60 year recovery period. This certainly includes a re-developed soil microbial flora and fauna. On the roast bed itself, plants have re-established at metal concentration in excess of 8000 μg/g Ni and 10 000 μg/g Cu and a pH of 4.4. Understanding the mechanisms these recovery processes are important in order to accelerate restoration efforts of similarly stressed ecosystems. The roast bed provides a unique outdoor laboratory to investigate the processes of natural recovery of a severely contaminated ecosystem.

3. Acknowledgements

Funding for this work was provided by Natural Sciences and Engineering Research of Canada (NSERC). International Nickel Company for access to the site. Thanks to Eric P.S. Sager for invaluable help in the lab and field.

4. References

1. Bowen, H.J.M. (1966) *Trace elements in Biochemistry*. Academic Press, London and New York. 241 pp.
2. Cox, R.M. and T.C. Hutchinson (1980) Multiple-metal tolerances in the grass *Deschampsia cespitosa* from the Sudbury smelting area. *New Phytol.* 84: 631-647.
3. Cox, R.M. and T.C. Hutchinson (1981) Multiple and co-tolerance to metals in the grass *Deschampsia cespitosa*: adaptation, preadaptation and 'cost'. *Journal of Plant Nutrition.* 3 (1-4), 731-741.
4. Hutchinson, T. C. and M.S. Symington (1997). Persistence of metal stress in a forested ecosystem near Sudbury, 66 years after closure of the O'Donnell roast bed, in the *Journal of Geochemical Exploration*, Mining and Metals in the Environment eds. R.J. Allan and W. Salomons. 58: 323-330.
5. Royal Ontario Nickel Commission (1917). *Report of the Royal Ontario Nickel Commission with Appendix.* No. 2960. Printed and published by A.T. Wilgress, Printer to the King's Most Excellent Majesty.

HEAVY METAL ATMOSPHERIC DEPOSITION AROUND AN IRON SMELTER COMPLEX STUDIED BY THE MOSS BIOMONITORING TECHNIQUE

M.V. FRONTASYEVA
Frank Laboratory of Neutron Physics, Joint Institute for Nuclear Research
141980 Dubna, Moscow Region, RUSSIA

E. STEINNES
Department of Chemistry, Norwegian University of Science and Technology
N-7034, Trondheim, NORWAY

Abstract

The moss *Hylocomium splendens* was used as a biomonitor to study changes in the atmospheric deposition of metals around an iron smelter complex in Northern Norway between 1989 and 1993. Samples representing the last three years' growth of moss, collected on the ground at 25 sites located 1-25 km from the source area, were analyzed by instrumental neutron activation analysis using epithermal neutrons. A total of 38 elements were determined, including most of the heavy metals known to be released into the air from this kind of industry. A re-orientation of the industrial production occurred around 1990; the most important developments being the closing of an iron smelter and the starting-up of a new ferrochrome smelter. Accordingly, the Cr deposition in the vicinity of the smelter increased by about a factor of 100, while the Fe deposition decreased by almost a factor of 10. Other elements showing a decrease over this period were Mo, Sr, Se, Zr, Re, Hf, Ta, Th, and U, while there was an appreciable increase for Ni, W, and Au. Most other elements, including V, Mn, Co, Cu, Zn, As and Sb, had similar levels as before. This technique should be well suited to study the metal deposition around Magnitogorsk and other industrial sources in the Ural region.

1. Introduction

Mosses can be favorably used as biomonitors of the atmospheric deposition of trace elements because they lack a root system and therefore depend on supply of chemical

I. Linkov and R. Wilson (eds.), Air Pollution in the Ural Mountains, 383–389.

substances derived from above. Analysis of naturally growing mosses to monitor heavy metal deposition is a well-established technique [1-2] and has been shown to be useful in studies on a local [3-5] as well on a large [6-8] scale. Studies of a limited number of elements using atomic absorption spectrometry have been reported most frequently, but also multi-element techniques, such as instrumental neutron activation analysis (INAA) [8-11] and inductively coupled plasma mass spectrometry (ICP-MS) [12] have been employed. INAA using epithermal neutrons was shown to be feasible for the determination of more then 40 elements [11], including some elements not previously reported in moss studies.

Among the local pollution problems studied using moss analysis is metal deposition in the vicinity of iron and steel manufacturing industries [3,13]. In 1989, the moss technique was used to investigate the deposition of selected heavy metals near an iron smelter complex at Mo, a small industrial town in northern Norway [14]. For certain elements, very high deposition rates were observed. In 1993 after a re-orientation of the industrial production, the most important parts of which was to close a large iron smelter and replacing it with the production of ferro-alloys, the study was repeated. As the application of INAA in moss monitoring studies related to iron industries had not been previously reported, it was decided to submit the moss samples collected from the two different years to INAA in the IBR-2 reactor at Dubna, characterized by a very high ratio of epithermal to thermal neutrons, in order to study simultaneously how the deposition patterns had developed for a large number of elements.

2. Experimental

Samples of the feather moss *Hylocomium splendens*, which grows on organic topsoils in forests and other natural environments, were collected at 25 sites situated 1-25 km from the smelter complex. The network of sampling sites is shown in Fig.1. The sampling was carried out according to a procedure described in detail elsewhere [7]. For each site 5-10 subsamples were taken within a 50×50 m area and these were combined in the field. The unwashed samples were air-dried at 30° C and extraneous plant material was removed. The three youngest fully developed segments of each *Hylocomium* plant were taken for analysis. No further homogenization of the samples was performed [10]. Disposable polyethylene gloves were used during all handling of samples.

The procedure employed for activation, radioactivity measurement and data processing are described elsewhere [15].

3. Results and Discussion

Mean values and ranges for 38 elements in moss samples collected at the same 25 sites in 1989 and 1993, respectively, are given in Table 1.

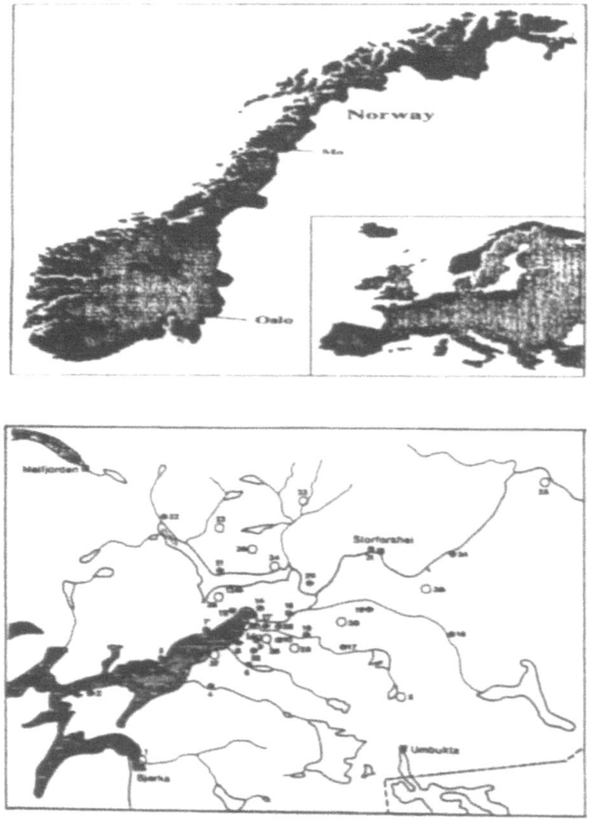

Fig. 1 Location of study area and individual sampling points (originally 39 points were sampled as shown, but only the 25 marked with filled circles were included in this work).

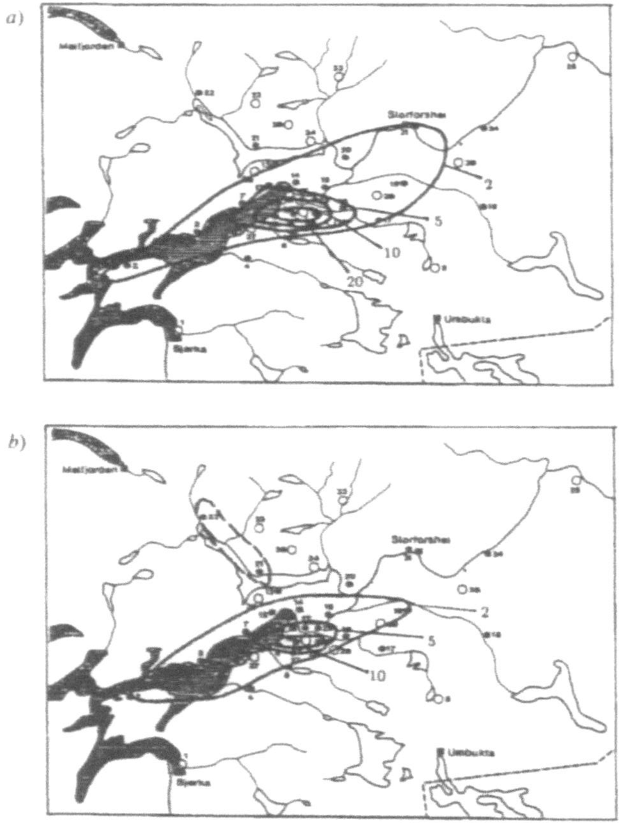

Fig. 2 Isopleths showing deposition pattern of V around the Mo iron smelter complex relative to background level, constructed for the (*a*) 1989 and (*b*) 1993 sampling data. See text for details.

TABLE 1. Summary of results obtained for 38 elements in 25 moss samples from the vicinity of an iron smelter complex.

	1993		1989	
Element	Mean (ppm)	Range (ppm)	Mean (ppm)	Range (ppm)
Na	310	127-732	294	93-635
Mg	2015	781-6520	1861	556-4230
Al	1186	293-4460	1244	243-3100
Cl	201	88-375	294	50-1110
K	3470	1640-4770	3845	1930-7160
Ca	1903	995-4830	2871	1450-6740
Sc	0.20	0.03-0.70	0.41	0.06-1.41
V	4.10	0.79-19.0	5.72	1.05-31.0
Cr	390	4.3-3900	11.7	0.5-50
Mn	325	61-917	384	89-1460
Fe	2440	300-8700	12 280	700-72 100
Co	0.68	0.07-3.0	0.61	0.06-2.2
Ni	2.84	< 0.5-15.6	1.69	< 0.5-6.96
Cu	8.23	3.5-20.3	12.3	2.9-29.0
Zn	88	16-542	99	31-397
As	0.37	0.06-1.56	0.62	0.06-2.20
Se	0.31	0.16-0.75	0.47	0.21-1.17
Br	6.36	3.5-11.4	6.94	3.6-12.2
Rb	12.5	3.3-26.4	17.2	6.7-46.2
Sr	8.9	< 4-52	19.9	< 4-50
Zr	3.0	< 2-24	11.0	< 2-46
Mo	0.16	< 0.08-0.76	0.87	< 0.08-3.85
Ag	0.057	<0.03-0.18	0.059	< 0.03-0.16
Sb	0.166	< 0.03-0.69	0.250	< 0.05-0.76
I	1.82	< 1.0-3.3	2.26	<1.0-4.3
Cs	0.28	0.05-0.89	0.37	< 0.05-1.03
Ba	26.7	15-57	33.1	12-83
La	0.45	< 0.10-1.38	0.69	< 0.10-2.87
Sm	0.139	0.04-0.45	0.33	0.05-1.34
Eu	0.016	< 0.01-0.13	0.031	< 0.01-0.12
Tb	0.020	< 0.005-0.096	0.019	< 0.005-0.067
Yb	0.032	< 0.010-0.120	0.069	< 0.010-0.230
Hf	0.097	< 0.04-0.36	0.179	< 0.04-0.71
Ta	0.011	< 0.003-0.038	0.043	< 0.003-0.180
W	1.68	< 0.6-9.4	1.71	< 0.6-6.4
Au	0.0042	0.0006-0.037	0.0002	< 0.0001-0.010
Th	0.114	0.02-0.47	0.267	0.04-1.10
U	0.035	0.03-0.22	0.143	<0.03-0.51

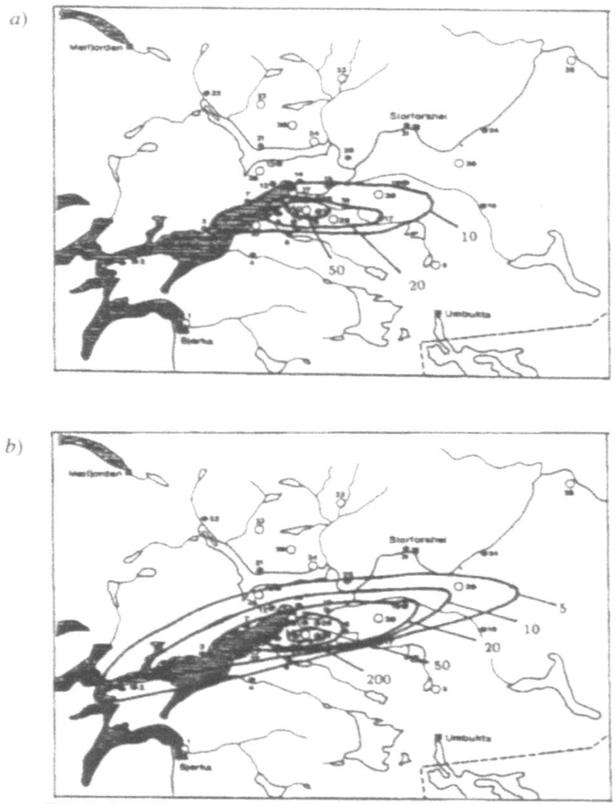

Fig. 3 Isopleths showing deposition pattern of Cr around the Mo iron smelter complex relative to background level, constructed for the (*a*) 1989 and (*b*) 1993 sampling data.

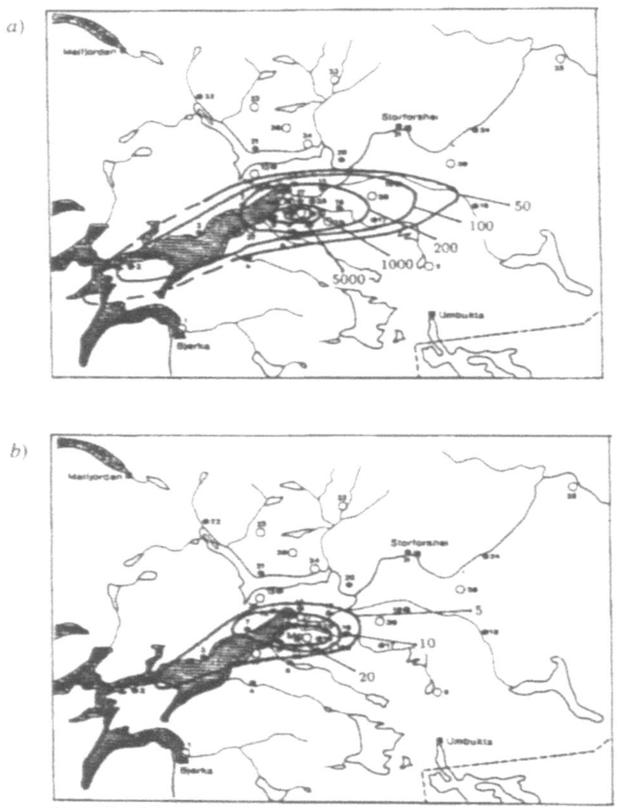

Fig. 4 Isopleths showing deposition pattern of Fe around the Mo iron smelter complex relative to background level, constructed for the (*a*) 1989 and (*b*) 1993 sampling data.

The spatial distribution of multi-site pollution data is most often expressed in the form of isopleths based on observed single values. In the present work an alternative approach is used. Background levels of the individual elements in *Hylocomium splendens* in the actual region of Norway are established on the basis of data from nationwide deposition surveys in 1985 [10] and 1990 [16]. The observed values from the present work are divided by the corresponding background values and the resulting ratios, expressing the enrichment factors over the background levels, are used as a basis for isopleth construction instead of the primary concentration values. Examples of such isopleths are shown for 3 elements that are strongly enriched in the area around the smelter complex in Fig. 2 (V), Fig. 3 (Cr) and Fig. 4 (Fe).

Chromium shows the most dramatic difference between the samples collected in 1989 and the samples collected in 1993, with an increased deposition by a factor of 100, which is clearly associated with the new ferrochrome plant started after 1989. Also for Ni, W and Au (not shown) there has been a fairly clear increase in the moss metal-levels in the vicinity of the industrial complex. The deposition of Fe, however, has decreased substantially after the closing of the iron smelter. A similar decrease is observed for Mo, and to a lesser extent for Sr and some of the refractory lithophilic elements (Sc, Zr, Re, Hf, Ta, Th, U). Most of these elements are enriched in moss samples near the industry, which indicates that there must be a mineral dust problem, possibly associated with the handling of iron ore, and perhaps also disposal of slag from the smelting processes.

For V the changes in industrial activity have not had any strong influence on the deposition levels. A further discussion of the temporal and spatial pattern of each element and their relations to details in the local industrial development is beyond the scope of this paper.

In conclusion, ENAA appears to be a very useful technique in studying air pollution problems associated with iron and steel industries, employing moss as a biomonitor. The most significant metal pollutants associated with such industries can thus be determined. This technique should thus be well suited to study the metal deposition around Magnitogorsk and other industrial centers in the Ural region.

4. References

1. Rühling, A. and Tyler, G. (1973) Heavy metal deposition in Scandinavia. *Water, Air, Soil Pollut.*, 2, 445-455.
2. Rasmussen,L. and Johnsen, I. (1976) Uptake of minerals, particularly metals, by epiphytic *Hypnum cupressiforme*. *Oikos,* 27, 483-487.
3. Ellison, G., Newham, J., Pimchin, M.J. and Tompson, J. (1975) Heavy metal content of moss in the region of Consett (North East England). *Environ. Pollut.*, 11, 167-174.
4. Andersen, A., Hovmand, M.F. and Johnsen, I. (1978) Atmospheric heavy metal deposition in the Copenhagen area. *Environ. Pollut.*, 17, 133-151.

5. Folkeson, L. (1981) Heavy-metal accumulation in the moss *Pleurozium schreberi* in the surroundings of two peat-fired power plants in Finland. *Ann. Bot. Fenn.*, 18, 245-253.
6. Gydesen, H., Pilegaard, K., Rasmussen, L. and Rühling, A. (1983) *Moss Analyses used as a Means of surveying the Atmospheric Heavy-metal deposition in Sweden, Denmark and Greenland in i980.* Bulletin SNV PM 1670, Swedish Board of Enviromental Protection, Stockholm, 44 pp.
7. Rühling, A., Rasmussen, L., Pilegaard, K., Mäkinen, A. and Steinnes, E. (1987) *Survey of Atmospheric Heavy Metal Deposition in Nordic Countries in 1985.* Report NORD 21, Nordic Council of Ministers, Copenhagen, 44 pp.
8. Steinnes, E., Rambæk, J.P. and Hanssen, J.E. (1992) Large scale multi-element survey of atmospheric deposition using naturally growing moss as biomonitor. *Chemosphere*, 25, 735-752.
9. Steinnes, E. (1980) Atmospheric deposition of heavy metals in Norway studied by analysis of moss samples using neutron activation analysis and atomic absorption spectrometry. *J. Radioanal. Chem.*, 58, 387-391.
10. Steinnes, E., Hanssen, J.E., Rambæk, J.P. and Vogt, N.B. (1994) Atmospheric deposition of trace elements in Norway: Temporal and spatial trends studied by moss analysis. *Water, Air, Soil Pollut.*, 74, 121-140.
11. Frontasyeva, M.V., Nazarov, V.M. and Steinnes, E. (1994) Moss as monitor of heavy metal deposition: Comparison of different multi-element analytical techniques. *J. Radioanal. Nucl. Chem.*, 181, 363-371.
12. Steinnes, E., Johansen, O., Røyset, O. and Ødegård, M. (1993) Comparison of different multielement techniques for analysis of mosses used as biomonitors. *Environ. Monit. Assess.*, 25, 87-97.
13. Pilegaard, K. (1979) Heavy metals in bulk precipitation and transplanted *Hypogymnia physodes* and *Dicranoweisia cirrata* in the vicinity of a Danish steelworks. *Water, Air, Soil Pollut.*, 11, 77-91.
14. Steinnes, E., Andersson, E.M. and Sjøbakk, T.E. (1992) *Environmental Monitoring in Rana. Heavy Metal Contamination in the Terrestrial Environment.* Report 495/92, Norwegian State Pollution Control Authority, Oslo, 18 pp.
15. Frontasyeva, M.V. and Steinnes, E. (1995) Epithermal neutron activation analysis of mosses used to monitor heavy metal deposition around an iron smelter complex. *Analyst,* 120, 1437-1440.
16. Berg, T., Røyset, O., Steinnes, E. and Vadset, M. (1995) Atmospheric trace element deposition: Principal component analysis of ICP-MS data from moss samples. *Environ. Pollut.*, 88, 67-77.

EXPOSURE TO LEAD AND ITS EFFECT ON CHILDRENS' HEALTH IN BELOVO (METALLURGICAL PLANT)

B.A. REVICH
Centre of Demography and Human Ecology, Institute of Forecasting of Russian Academy of Sciences, 32 Krasikova ul., Moscow, 117418, RUSSIA.

A.A. BYKOV and S. LYAPUNOV
Plekhanov Economic Academy, Stremiannyi per. 28, Moscow, 113054, RUSSIA.

I.SERYOGINA
Geological Institute, Russian Academy of Sciences, RUSSIA

A.PRIKHOZHAN
Psychological Institute, Russian Academy of Education, RUSSIA

M. SOBOLEV
Medical Academy of Saint-Petersburg, RUSSIA

1. Introduction

The purpose of this study is to assess the effects of environmental lead contamination on children's health.

In the course of this work the following problems were tackled:

1. assessment of environmental lead contamination in the vicinity of Belovo Metallurgical Plant (Kemerovskaya Oblast, Eastern Siberia)
2. assessment of lead concentration in children's blood, hair and teeth;
3. estimation of children's mental status;
4. assessment of different routes of the child exposure to lead;
5. assessment of effectiveness of different measures aimed at reducing lead contamination and children's exposure to lead.

The Belovo Metallurgical Plant was built in 1930. It is situated on the northeastern outskirts of the town. The principal raw material used here is zinc sulfide concentrate with a content of zinc sulfide up to 70% and that of lead sulfide - up to 1.3%. The plant produces raw zinc, zinc powder, white zinc paint and sulfuric acid. Since the time of its commissioning, the Plant atmospheric emissions have been

I. Linkov and R. Wilson (eds.), Air Pollution in the Ural Mountains, 391–404.

measured in hundreds of tons of lead as well as zinc, cadmium, sulfur dioxide, nitric oxides, hydrogen sulfide, soot and other pollutants. However, data on the amount of *Pb* emissions are available only from 1983.

Since 1972 there has been a gradual decrease in the amount of atmospheric emissions from the Plant due to the introduction of new technological processes and the commissioning of local dust precipitators.

Recognizing a high level of environmental contamination with lead and other heavy metals the Plant management systematically moved the inhabitants outside the protection zone which over 1 km extends according to the current legislation. All in all, 360 households had been moved, but till today nearly 3 thousand people still reside inside the zone, not to mention 2 nursing schools, a secondary school and the Radio Engineering Plant. There are a lot of private houses with plots of land where people grow potatoes, vegetables and fruits.

2. Sampling Methodology

Soil. Soil samples were taken by the 'envelope' method. 110 samples of authentic soil and 10 ones of the averaged processed soils from kitchen gardens, including those of the crops, were selected.

Drinking water. 7 samples of drinking water have been taken.

Air. A 7-day 24-hour air sampling was performed at three stationary sites at a distance of 0.5, 1.5 and 3.5 km from the Plant. Air samples were drawn through the aerosol filters.

Plant foods (vegetables). Vegetables (potatoes, beet, carrot, and tomatoes) were sampled with soil samples over the examined plot. 30 samples of vegetables were selected.

Indoor dust deposits and painted surfaces. Indoor dust deposits were collected by dislodging or brushing them away into bags from the places of their concentration. Paint sampling was carried out by means of removal of thin pieces of enamels and pigments from the painted surfaces. 23 samples of indoor dust and 8 of paint samples have been selected.

Blood. Venous blood was drawn by disposable syringes into special containers filled with dry heparin manufactured by *"Becton Dickinson Vacutainer Systems."* Blood storage and transportation was done inside refrigerators. 91 blood samples have been selected.

Hair. Hair samples weighing up to 0.5 g were cut from the occipital part of the scalp. 67 hair samples have been selected.

Teeth. Samples of teeth were collected by a dentist at the children's polyclinic. 15 samples of teeth have been selected.

3. Analytical methods.

For the determination of lead in the samples of soil, paint, and ambient air collected on aerosol filters X-ray fluorescence (XRF) analysis was used. The analysis was made by means of an X-ray fluorescence analyzer ARF-6 (Russia) with the wave length dispersion (XRF-WD). The determined lead level was 0.1 - 3000 ppm.

For soils the detection limit for lead equals 1 ppm. The achieved detection limit for lead in an aerosol filter makes up 0.1 μg with an exposure time of 400 sec. Detection limit for lead in paint equals 0.01 mg/cm^2 with an exposure time of 400 sec.

For lead determination in the samples of drinking water, dust and human tissues (blood, hair, teeth) and also in plant samples method of flame atomic absorption spectrophotometry (FAAS) was used.

Samples of drinking water and dust were analyzed by means of a direct introduction of the prepared specimen into the acetylene-air flame. Carl Zeiss spectrophotometer model AAS-3 (Germany) was used.

Samples of human tissues (blood, hair, teeth) and plant were analyzed by FAAS using microcolumns with diethylenetriamintetraacetate (DETATA-sorbent) for on-line flow sorption preconcentration of lead (FI-FAAS). "Quant-AFA" spectrophotometer (Russia) combined with PC-controlled flow-injection system (AKVITA, Russia) was used.

All samples for analysis were digested by mixture of chloric and nitric acids before the lead measurements. The detection limit for lead by the method of direct atomic absorption analysis equals 0.5 μg/mL. Detection limit for lead in the method of FAAS with on-line flow sorption preconcentration was 0.005 μg/ml.

The accuracy of the analytical methods that were used was determined by the analysis of international standard reference materials. Table 1 lists the results of lead determination in some standard samples.

4. Children examination methods.

To determine the major factors that contribute to the accumulation of lead in children's blood and hair, the families of the examined children were interviewed. The questionnaire included 22 questions. After the preliminary analysis of the questionnaire data the following indicators were used for the subsequent statistical analysis: school location; child's age; place of residence; period of stay at a given address; date of building the house; availability of a carpet or wall - to - wall carpeting in a child's room; proximity of the house to a highway: up to 50 m, between 50 and

TABLE 1. Lead concentration in standard samples determined by *XRF (WD)* and FI-FAAS methods

Sampling method	Soil, mg/kg XRF				Coal ash, mg/kg XRF	
Standard	GBW07402	GBW07406	GBW07306	GBW07309	ZUK-1	SRM-1633b
Manufacturer	NRCCRM				Russia	NIST
Certificate	20	314	23	285	16	
Autopsy data	21	310	25	270	16	71
Sample method	**Blood, μg/dL FI-FAAS**					
Standard	194	492	594	1494		
Manufacturer	**CDC-BLLRS**					
Certificate	0.4	14.9	7.1	4.5		
Autopsy data	<0.5	14+1.8	9.2+1.5	4.2+1.0		

200 m and over 200 m; feeling of proximity to a highway; availability of a garden or a kitchen - garden situated within town limits or out of town; share of food products grown in one's garden or kitchen - garden in the family annual ration; occupational exposure of parents to lead; mother's education; smoking of parents and the number of smoked cigarettes.

For the assessment of children's mental status, two groups of indices were used - those reflecting the condition of higher mental functions and those testifying to the level of manifested anxiety. The assessment of higher mental functions is based on the Bruininks - Oseretsky test [1] and consists of 70 tasks, thus making it possible to detect abnormalities of these functions which define neurological dynamics, higher regulation and certain cortical function disturbances. The testing results were evaluated against the following score scale: 1 point - no disturbances or mild disturbances in performing a test; 2 points - medium disturbances; 3 points - serious disturbances manifested by the structural changes in the observed function; and 4 points - inability to perform a test when a child does not understand the task and requires neurological examination and follow-up.

The assessment of anxiety was made with the use of children's version of the Manifest Anxiety Scale (CMAS) developed by a group of US psychologists (A. Castaneda, B.R. McCandless, D.S. Palermo, 1956) on the basis of the Manifest Anxiety Scale (MAS) for adults (J. Taylor, 1951, 1953). The above test was adapted for use in Russia by A. M. Prikhozhan, Doctor of Psychological Sciences. In connection with this test, anxiety means a generalized chronic emotional experience under mental and somatic stress, and this sort of experience has nothing to do with any kind of critical situation. The questionnaire comprises 53 questions: 42 of them are put to clarify various indices of anxiety and the rest 11 ones are aimed at finding out an inclination of a respondent to have biased results under the influence of a social desirability factor, i. e. striving to give a 'correct' or a 'good' answer. The questionnaire was filled in by the children of Belovo attending the schools that are located in districts with various lead levels of environmental pollution. In the most contaminated district, 117 children of Group 1, including 73 girls and 44 boys, and 17 children of Group 2, including 11 girls and 6 boys, were interviewed respectively.

The obtained results were presented in the form of points and as a normalized scale index from 1 to 10 with an average value of 5.5 and a standard deviation of 2.25.

Statistical analysis. Multiple regression analysis was made according to the *Systat - 4* program.

5. Research Findings

5.1 AIR

Late in the 1980s an extremely high *Pb* concentrations reached 6 $\mu g/m,^3$ 20 times higher than the maximum allowable concentration (MAC) in the ambient air in Belovo. Owing to measures taken to reduce the amount of emissions and as a result of a sudden curtailment of the production, in 1994 - 1995 an atmospheric concentration of lead at a distance of 1 km from the Plant decreased down to 0.66 - 2.34 $\mu g/m^3$ which is still higher than the MAC.

In May 1996 when the Plant was shutdown, lead concentration in the air was controlled also. Average daily concentration in the air was much lower than in previous years and varied between 0.001 - 0.006 $\mu g/m^3$ at the posts outside the Plant (1.5 - 3.5 km). The maximum atmospheric lead concentrations (0.12 - 0.24 $\mu g/m^3$) were determined at a distance of 0.5 km from the Plant. Most probably, it is connected with deflation of its dumping sites and surface contamination of its territory in a hot, dry and windy weather. A model calculation of the air lead concentration at a distance of 1 km from the operating Plant demonstrated that with existing level of technology and cleaning of its emissions, it can reach 0.8 $\mu g/m^3$, i. e. almost 3 - times higher than MAC.

5.2 SOIL

The danger for children's health of lead contaminated soil may be considered both from the point of view of its harmful effect on young children due to polluted playgrounds and territories of their residence, and possible lead ingestion with agricultural crops. Lead concentration in the urban soil and kitchen - gardens is found in the range of 30 - 3000 mg/kg and 36 - 130 mg/kg respectively. The areas 15 km away from the town were considered to be a local background. In these locations the soil lead concentrations make up, on the average 30 mg/kg, i. e. 4.3 times lower that a standard maximum level of 130 mg/kg for this type of soil (neutral loams).

The pattern of lead contamination over an urban territory was found to be quite variable the measurement showed in Figure 1:

1. In 8% of the territory lead *concentrations* exceed 1000 mg/kg. These were areas around the industrial zone. There are no residential districts here.
2. In 10% of the territory the soil lead concentrations were observed in the range of 400 - 1000 mg/kg. Over this area which is actually the protection zone there are also some houses.
3. In 21% of the territory the soil lead concentrations were observed in the range of 130 - 400 mg/kg, i. e. exceeding the allowable level of soil contamination. This area has uneven boundaries, occupying the central part of the newly - built town and some districts of the old part of town with household plots of land. Two secondary schools included into the study are also situated over this territory.
4. In 61% of the territory the soil lead concentrations were observed in the range of 30 - 130 mg/kg. These are mainly south - western and western outskirts of the town, including newly - constructed residential micro - districts where one of the examined secondary schools is also located.

5.3 INDOOR DUST.

Lead concentration in the dust (23 samples) collected inside the school premises (350 - 990 mg/kg) were 3 - 5 times higher than in the residential houses (61 - 180 mg/kg). Higher dust lead concentrations are typically observed on the upper floors of the houses. But the highest lead concentration in indoor dust was found in a school situated at 3.5 km distance from the Plant. Most probably it is related to the mechanism of lead transport with fine aerosols less than 2.5 μm.

5.4 PAINTED SURFACES

The concentration of lead was determined in the paint used for painting the walls, floors and ceilings of the schools (8 samples of paint). The highest lead concentration of 4600 mg/kg was determined in red flooring paint. Recalculation of lead concentrations per a unit of surface lead loading area showed up 0.1 - 0.5 mg/cm^2 which is lower than the US standard (1mg/cm^2).

5.5 VEGETABLES

The lead concentration was determined in 30 samples of vegetables (potatoes, carrot, beet, tomatoes, cabbage) grown in a kitchen - garden at a distance of 0.5 - 1.0 km from the Plant territory where the soil contained 96 - 130 mg/kg of lead and at a distance of 3.0 - 3.5 km from it the concentration of lead was 36 -71 mg/kg. The concentration of lead in raw root - crops (potatoes, carrot, beet) and raw vegetables were 0.01 - 0.19 mg/kg and 0.03 - 0.09 mg/kg respectively, with MAC = 0.5 mg/kg. A maximum lead concentration was found in carrot samples (0.11 - 0.19 mg/kg). But the decrease of lead concentration in vegetables depending on the proximity of a kitchen - garden to the Plant could not be established. A relatively low accumulation of lead is associated with a barrier effect of the plant root and the fixation of its mobility complexes by

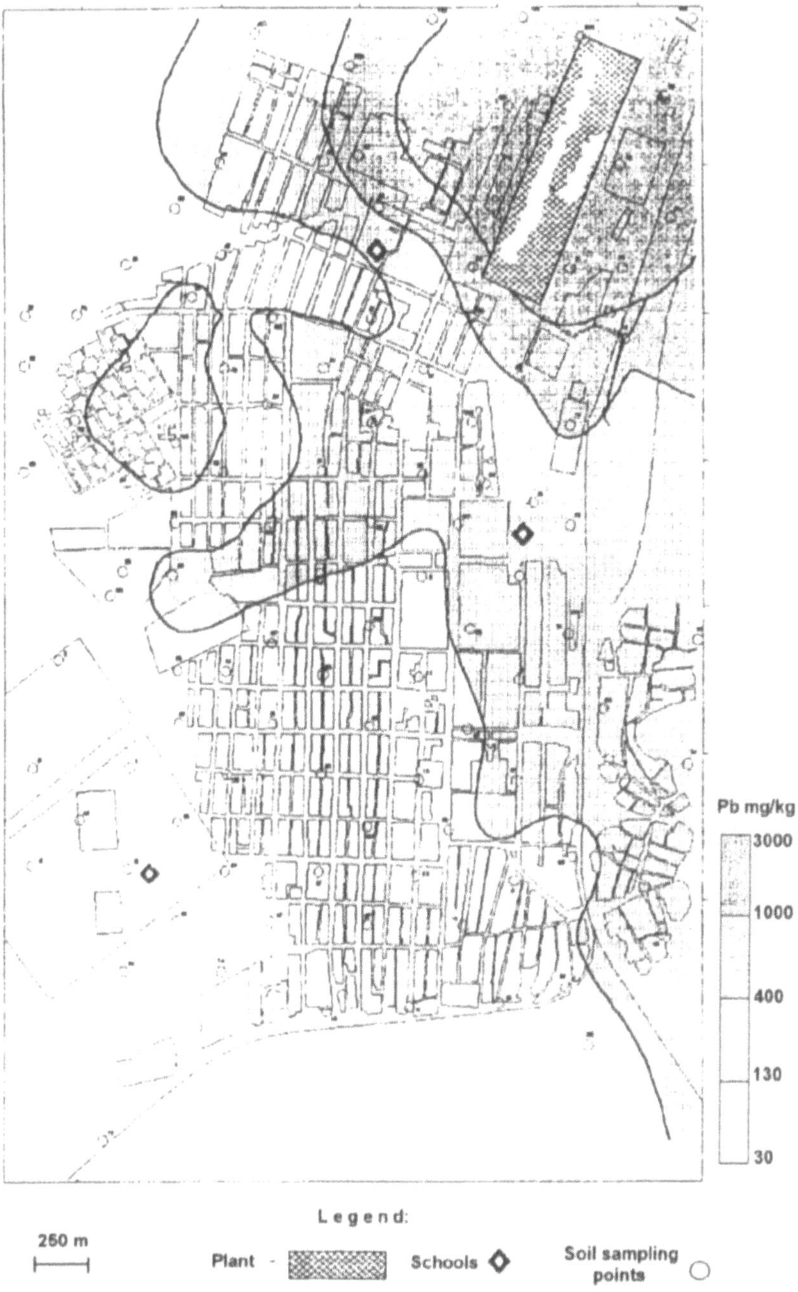

FIGURE 1. Soil lead concentration in Belovo

humus. Some time ago the same situation in Belovo was described by the experts of the Novosibirsk Institute of Soil Science and Agricultural Chemistry of the Siberian Branch of the Russian Academy of Sciences [4]. In Belovo daily ingestion of lead with vegetables by a child makes up 1.5 - 3.2 μg, thus not exceeding a daily adolescent normal rate of 4 -5 μg.

5.6 DRINKING WATER

A concentration of lead in the samples of drinking water withdrawn from the main municipal watersheds and water supply network in different parts of town (7 samples) does not exceed 1 μg /L, with MAC = 30 μg/L.

5.7 BLOOD

Blood lead level in children higher than 15 μg/dL is a sign of biochemical changes in a child's body (Inorganic Lead, 1995). The US Centers for Disease Control (CDC have developed a more detailed danger scale of lead accumulation in children's blood. The level below 9 μg/dL is considered as normal; between 10 and 19 μg/dL - as slightly increased when there can be behavioral and educational problems in children; and between 20 and 44 μg/dL - as increased when treatment is indicated (Lead [2]).

Initial studies related to lead in human tissue of the Belovo population were implemented in 1989 [5]. Blood lead levels in the neonates and women in labor made up 25.3 ± 2.8 μg/dL and 23.4 ± 3.5 μg/dL respectively which are much higher than the permissible ones. A relationship was also observed between the concentration of lead in mother's and infant's blood and in mother's blood and her breast milk. During the given examination blood lead levels were determined in 91 children, aged 7 - 8, living in different districts - near the Plant, downtown and in the newly - built districts at 3.5 - 4.0 km distance from the Plant. Blood lead levels in children varied between 0.5 and 39 μg/dL with an average concentration of 9.9 μg/dL (Table 2) which means that in 42% of the examined children these levels exceeded the allowable limit of 10 $\mu g/m^3$ and in 4% of them they were above 20 μg/dL.

TABLE 2. Blood and hair lead levels in children of Belovo, 1996

Statistical parameter	Pb in blood, μg/dL	Pb in hair, ppm
Number of children	91	66
Arithmetical mean	9.9	4.5
Average error	0.5	0.6
Standard deviation	5.2	4.9
Geometric mean	8.5	
Geometric mean error	3.3	

Since in recent years there has been a decrease in the amount of lead emissions of the Plant, consequently blood lead levels in children are also decreased compared to previous years.

5.8. HAIR

An average lead concentration in hair of 67 children made up 4.5 ± 0.6 μg/g. Statistical differences in blood and hair lead levels in children living in different districts of Belovo were not found. This could probably be explained by population mobility within a small town.

No significant correlation between blood and hair lead levels has been detected. To find out any relationship between blood and hair lead levels and a number of features were included in the questionnaire used. Multiple regression analysis was fit to these variables. The greatest contribution to the presence of lead in blood of children was found to be with a multiple coefficient of determination R^2 around 0.5 were: the proximity of the household to a highway; the percentage of food products grown in one's personal garden or kitchen - garden in an annual ration; and the number of cigarettes smoked by the mother. The greatest contribution to the concentration of lead in hair of children with a multiple coefficient of determination R^2 around around 0.4 was explained by: the date of building of the house; availability of a carpet or wall - to - wall carpeting in a child's room; proximity of the house to a highway; location of personal kitchen - garden (within or outside town limits); and the proportion of food products grown in one's personal garden or kitchen - garden in an annual ration.

5.9. TEETH

The lead concentration in teeth of 15 children, aged 5 - 8, living in different districts of town were found to be 1.7 - 4.9 μg/g with an average value of 3.4 μg/g. No significant differences between school and preschool children were observed.

6. Childrens' Health

Changes in the functional state of the central nervous system (CNS) and of the mental and emotional component are typical signs of lead effect on children's health.

Examination of children conducted in 1989 made it possible to detect a longer latent period of response to color and verbal stimulus virtually in all age and sex groups of children and adolescents of Belovo (Table 3) (adapted from V. Yarushkin [5]).

The extension of response latent period is considered as a disturbance of balance between stimulation and inhibition in the cerebral cortex.

TABLE 3. Extension of the latent period of response to verbal stimulus in children and adolescents of Belovo as compared to controls.

Age, years	Sex	Number of children	Response	
			Absolute value, ms	Relative value, %
5.5 – 6.0	Boys	9	7.9	100.6
	Girls	18	49.7	103.6
6.5 – 7.0	Boys	13	83.9	106.2
	Girls	12	155.2	112.1
16.0	Boys	19	36.6	103.9
	Girls	29	28.9	103.1

Mental status of children. Examination of children undertaken in 1996 resulted in the detection of significant changes in their higher mental functions. In Belovo children demonstrated a delayed response, bad bilateral coordination, slow fine finger motility and abnormal logical sequence of performing different tasks. It is indicative of the expressed neurological dynamics, higher regulation and cortical function disturbances.

Anxiety level. The average anxiety level in children of Belovo was significantly higher than in the control group (Table 4). Control indices were estimated in the group of children of similar age (300 people) living in Moscow, Kaluga and Suzdal. The control group includes the greatest number of the examined children (over 70%) having the highest adaptive level of anxiety, but among the children of Belovo one may more frequently observe a high level of anxiety pointing to an evident and manifested alert which appears to be on the verge of a psychological norm. The anxiety level was somewhat different in children living in district with varying levels of environmental pollution with lead (Table 5).

An average level of anxiety in children of Group 2 was slightly lower than for the Group 1, but the differences between the two Groups were statistically insignificant. Statistically significant differences ($p<0.01$) which were even higher in children of Group 1 were found only at the highest level of anxiety. Consequently, the

TABLE 4. Average anxiety level in the examined children of Belovo.

Sex	Number of children (n)	Belovo		Controls	
		Average	Standard deviation	Average	Standard deviation
Girls	84	22.1**	7.4	14.8**	7.6
Boys	50	18.2**	6.1	13.7**	6.2
Total	134	20.7*	7.2	14.2*	7.4

*p < 0.1
**p < 0.01

hypothesis of relationship between the anxiety level in children and the impact of lead-polluted environment is confirmed both by a higher level of anxiety in children living in the most contaminated part of town, and the municipal data as a whole compared to the control indices.

TABLE 5. Anxiety indices in children living in different districts of Belovo: Group 1 - 0.5-2 km from the plant; 130-400 mg/kg of Pb in soil; Group 2 - 3.5 km from the plant; 130 mg/kg of Pb in soil.

Sex	Group 1			Group 2			Total in Belovo		
	n	Average	Standard deviation	n	Average	Standard deviation	n	Average	Standard deviation
Girls	73	22.4	7.7	11	20.6	4.8	84	22.1	7.4
Boys	44	18.5	6.2	6	15.8	4.0	50	18.2	6.1
Total	117	20.9	7.4	17	18.9	5.1	134	20.7	7.2

7. Assessment of Health Risks of Environmental Lead Contamination Faced by the Population of the Town of Belovo

The Integrated Children's Exposure Uptake Biokinetic Model for lead (IEUBK) developed by the US EPA makes it possible to estimate the distribution of the blood lead levels among a hypothetical children's population. The IEUBK Model calculates the uptake of lead by the body on the lead concentration data in different media and intake routes via different exposure routes (inhalation, food, soil, house dust and tap water, ingestion). The risk (probability) of lead concentration in children's blood exceeding a selected allowable level is estimated by means of an assumed (lognormal) probability distribution. Using this Model one can determine the contribution of different exposure ways to the total lead burden as well as evaluating various exposure reduction options aimed at the mitigation of children's exposure to lead.

7.1. ASSESSMENT OF LEAD INGESTION BY THE POPULATION ASSOCIATED WITH DIFFERENT PLANT OPERATING SCENARIOS.

In recent years the Metallurgical Plant in Belovo has been running under three major conditions of its production, i. e. at full capacity before the 1990s, at reduced capacity in 1994 - 1996 and in spring 1996 it was practically at a standstill. Consequently, three major scenarios of exposure to lead under different production conditions of the Plant were considered: at maximum emission (Scenario 1 - lead emissions over 100 tons/year), with reduced amount of emissions down to 15 -20 tons/year (Scenario 2) and when the Plant was virtually at a standstill (Scenario 3).

For each Scenario, the total intake of lead was estimated separately for three different areas of town (Table 6), i. e. near the Plant, downtown at a distance of 1.0 -

2.0 km from it and in the newly - built residential district (3.5 - 4.0 km from the Plant).

Thus, the major pathways leading to the accumulation of lead in the blood of children of Belovo are food products, soil + dust and air. Contaminated tap water seems to make an insignificant contribution: in all scenarios its share is below 1%. The polluted atmospheric air with a fully operating Plant is the major factor effecting the children's blood lead levels in the 1st zone of contamination. For other areas in town its contribution is under 20%. Soil and dust contribution varies between 3 and 13%. In case of a reduced capacity of the Plant (Scenario 1) and when it was closed (Scenario 3) the contribution of inhalation route to the total uptake drops rapidly while the soil and dust lead ingestion pathway up approximately to 20%. In all impact

TABLE 6. Contribution of different exposure routes to lead to the predicted total accumulated daily lead burden in preschool children of the town of Belovo

Scenario	Distance from the Plant, km	Total uptake of Pb, μg/d	Contribution (%)			
			inhalation	tap water	soil and dust	food products
1	0.5	55 - 135	38 - 69	0.2	7 - 13	25 - 49
1	1.0 - 2.0	30 - 43	6 - 20	0.3 - 0.7	6 - 10	73 - 86
1	3.5 - 4.0	26 - 35	6 - 20	0.3 - 0.8	3 - 6	76 - 90
2	0.5	34 – 44	0.9 – 3.0	0.3 – 0.7	19 – 27	70 - 80
2	1.0 - 2.0	31 - 39	1.0 - 3.3	0.3 - 0.8	17 - 24	72 - 82
2	3.5 - 4.0	27 - 34	1.1 - 3.9	0.3 - 0.8	16 - 23	73 - 82
3	0.5	34 – 43	0.1 – 0.2	0.3 – 0.8	20 – 28	72 - 80
3	1.0 - 2.0	31 - 39	0.03 - 0.1	0.3 - 0.8	17 - 25	79 - 82
3	3.5 - 4.0	27 - 33	0.03 - 0.1	0.3 - 0.9	16 - 24	75 - 83

scenarios, the contribution of food ingestion is quite significant (nearly 75-80%), except in the area in the immediate vicinity of the Plant.

7.2. ASSESSMENT OF AVERAGE BLOOD LEAD LEVELS IN CHILDREN.

For the scenarios examined the average blood lead levels in children of Belovo were also calculated. Average predicted blood lead levels in children, aged 6-7, vary over the municipal territory from 11.7 to 9.0 μg/dL. In 1996 an average blood lead level in 7-year old children were measured to be as high as 9.9 μg/dL. This finding is consistent with the results predicted by the IEUBK Model. Predicted data for a fully operating Plant resulted in the range of average levels over the municipal territory from 36.1 to 9.9 μg/dL. Mother's blood lead level of 23.4 μg/dL and the one in a neonate of 25.3 μg/dL were measured in 1989 [5]. These two measurements fall in the middle of the predicted interval by the model, thus indicating a satisfactory agreement between predicted and the observed blood lead levels.

7.3 ASSESSMENT OF PROBABILISTIC DISTRIBUTION OF BLOOD LEAD LEVELS IN CHILDREN AND THE ALLOWABLE LEVEL EXCESS RISK.

Using the probabilistic formulation of the IEUBK Model one can assess the probability of finding excessive blood lead levels in the Belovo children. These probabilities define the magnitude health risks from the lead exposures posed to children, aged 0 - 7 (Table 7).

TABLE 7. Assessment of health risks of environmental lead contamination faced by the children of Belovo under different operating conditions of the Plant (Blood level >10 μg/dL is considered to pose excessive risk)

Fully operating Plant (Scenario 1)			Plant at a standstill (Scenario 3)		
Distance from Plant, km	Blood lead levels μg/dL	Risk %	Distance from Plant, km	Blood lead levels μg/dL	Risk %
0.5	32.8	99.2	0.5	14.3	75.6
1.0 - 2.0	15.1	72.7	1.0 - 2.0	13.0	69.8
3.5 - 4.0	13.0	58.3	3.5 - 4.0	11.2	55.5

These calculations indicate a great risk to children no matter how far they may live from the Plant. The closure of the Plant has led to the reduction of risk nearly by 1/4 in the 1st zone but very small decrease in other distant zones. This result follows from the fact that in the vicinity of the Plant the inhalation component makes a significant contribution to the total risk that is significantly reduced as soon as the Plant was shut down. The US Centers for Disease Control (CDC) provide a number of recommended mitigation measures in case of different blood lead levels. Using this approach, it is possible to find out that even with an idling Plant (Scenario 3), in 70% of children residing in downtown and in 53% living in the newly-built 'sleeping' district, there can be behavioral and educational problems (blood lead level of 10 - 19 μg/dL), and the treatment of 15 - 20% of children should be initiated at once through chelation therapy (at a blood lead level of 20 μg/dL).

A group of children from Belovo with higher lead levels in their blood underwent a coarse of treatment.

8. References

1. Dietrich K., Berger O., Succop P. Lead Exposure and the Motor Developmental Status of Urban 6-Years-Old Children in the Cincinnati Prospective Study // Pediatrics, 91; 1993, pp. 301-307.
2. *CDC Report*: (1991) Preventing Lead Poisoning in Young Children. A Statement By The Centers For Disease Control. U.S.Department of Health & Human Services/Public Health Service.
3. *Iljin V.B.* (1991) Heavy metals in market-garden soils and foods of the city Belovo. Report of the Institute of Soil Science and Agrochemistry of the Siberian Department of Russian Academy of Sciences, Novosibirsk (in Russian).
4. *Iljin V.B.* (1993) Studies of ceolites use for heavy metal inactivation in industrially contaminated soils for ecologically pure food production. Report of the Institute of Soil Science and Agrochemistry of the Siberian Department of Russian Academy of Sciences, Novosibirsk (in Russian).
5. *Yarushkin V.Yu.* (1991) Environmental hygiene and children health in the regions with metallurgical zinc plants. Ph.D. thesis, Irkutsk (in Russian).

AIR POLLUTION IN POLAND

K. SAWICKA-KAPUSTA and M. ZAKRZEWSKA
Department of Animal Ecology,
Institute of Environmental Biology,
Jagiellonian University, Ingardena 6, 30-060 Krakow, POLAND.

For many years Poland belonged to the most polluted countries in Europe. The heavy industry with too much water and energy consuming old technology, caused the contamination of natural environment. Atmospheric pollution in Poland in the eighties was among the heaviest in Europe making this country the third largest polluter after the former Soviet Union and Germany [1,2]. In 1987-1989 Poland produced each year 12.5% of European dust, 10% of sulfur dioxide and 7% of nitrogen oxides. The annual emissions of major air pollutants in 1989 has been estimated to be 3.0 million tons of dust, 4.0 million tons of sulfur dioxide, 1.5 million tons of nitrogen oxide (Tab. 1)

TABLE 1. Estimated emission of main air pollution in Poland in 1987-1989 [mln t yr $^{-1}$] (After [2]).

Substance	Energy	Industry	Transportation	Agriculture	Total
Sulfur dioxide	2.0	1.0	0.1	0.9	4.0
Nitrogen oxides	0.5	0.4	0.4	0.1	1.5
Dust	0.8	1.3	-	0.9	3.0
Carbon monoxide	0.06	1.37	0.92	0.85	3.2
Carbon dioxide	180	130	37	110	457
Aromatic hydrocarbons	0.02	·0.15	0.23	-	0.5

Emissions of carbon monoxide were 3.2 million tons and carbon dioxide about 470 million tons per year [2,3]. The major sources of air pollution in Poland are energy production, industry, transportation and agriculture (Tab. 1). Energy production based mainly on black and brown coal it caused about 65% of sulfur dioxide and 44% of nitrogen dioxide emission [4]. Due to emission of air pollutants from sources located in Poland, and because of trans-boundary transport, the natural environment in the wide area of the country is devastated. In 1980, 27 areas of environmental hazard were identified in Poland. They cover about 10% of Poland's territory (35 000 km) and are inhabited by 13 million people - 30% of the population. Four areas have been declared

I. Linkov and R. Wilson (eds.), Air Pollution in the Ural Mountains, 405–406.

TABLE 2. Average concentration of main air pollutants in Upper Silesia Industrial Region in 1989 [μg m. $^{-3}$] (After [2]).

Substance	Maximum permissible level	Average concentration in Upper Silesia	Maximum concentration in Silesia cities
Sulfur dioxide	32	44	65-83
Nitrogen oxides	50	66	100-120
Dust	50	132	185-214
Carbon monoxide	120	6500	12900-17500
Lead	0.2	0.3	1.3-1.9
Cadmium	0.01	0.10	0.80-1.16
Hydrocarbons	820	5700	11100-13600
Benzo-(a)-pyrene	0.001	0.042	0.070-0.890

as ecological disaster regions: Upper Silesia Industrial Region, Krakow province, Legnica-Glogów mining area and Gdansk area [1,5]. Upper Silesia has been identified as one of the most polluted places in Europe (Tab. 2) [2].

Political changes and economic crisis started in 1990 were the main reasons of pollution decreasing in our country (Tab. 3) [3,6]. Although industrial emissions have rapidly decreased in the recent six years, still the condition of natural environment in Poland is far from acceptable.

TABLE 3. Global emission of main air polluters in 1990-1994 [thousands tons] (After [3]).

Substances	1990	1991	1992	1993	1994
Dust	1950	1680	1580	1495	1400
Sulfur dioxide	3210	2995	2820	2725	2645
Nitrogen oxides	1280	1205	1130	1120	1120
Carbon dioxide	384000	388000	363000	372000	357000

References

1. Nowicki, M., (1993) Environment in Poland, Kluwer Academic Publishers, Dordrecht, Boston, London.
2. Andrzejewski, A. and Baranowski, M. (eds.) (1993) Natural Environment in Poland, PIOŒ, GRID, Warsaw.
3. Environmental Protection (1996), Information and statistical data, General Statistical Office, Warsaw.
4. Oleksyn, J. and Reich, P.B., (1994) Pollution, Habitat Destruction and Biodiversity in Poland, Conversation Biology 8, 943-960.
5. Kassenberg, A. and Rolewicz, Cz., (1985) Land-use Planning Diagnosis of Environmental Protection in Poland, PWE, Warsaw.
6. Statistical Year Book (1995), General Statistical Office, Warsaw.

ATMOSPHERIC POLLUTION IN GREECE

C. PAPASTEFANOU
Nuclear Physics Department
Aristotle University of Thessaloniki, Thessaloniki 54006, GREECE

The problems of atmospheric pollution in Greece have been appreared during the last 30 years and were mainly due to movement of a large part of population into the big cities, in association with the development of the country. Two kinds of problems are existing, those of industrial and those of urban origin.

Industrial pollution is related to areas, where thermal power plants are operating and producing electrical power. In that case, the size of the problems depend on the kind of fuel used, e.g. lignite, a type of coal, or oil fuel and on meteorological conditions. Such problems were risen in the nearby coal-fired power plants areas, at regions of Florina, Kozani and Arkadia.

Different industries, such as oil distilleries (refineries), cement industries, fertilizer industries, steel industries, and minerals mining industries are also making a source of atmospheric pollution. Such problems were risen in the nearby big cities industrial areas of Athens, Thessaloniki, Volos, Kavala, Halkida, Elefsina, Aspropyrgos, Drapetsona, Kordellio (Thessaloniki).

Urban pollution is relating with the traffic and central hot water heating systems of buildings in the big cities, such as Athens, Thessaloniki, Patras, Heraklion,etc.

The first network of stations for controlling atmospheric pollution in Greece started operating in Athens in 1974 with 7 stations of a semi-automatic type. From 1983, this network was completely automatic and completed with new stations recording different pollutants, such as carbon monoxide, CO, nitrogen oxides, (NO,NO_2), ozon, O_3, sulfur dioxide, SO_2, smog (photochemicals). From 1986 similar networks have been established in Thessaloniki, Patras, Heraklion, Larisa, Volos, Kavala, Ioannina, Halkida and in the nearby areas of coal-fired power plants, such as Kozani, Ptolemaida, and Megalopolis. A map showing the stations of atmospheric pollution control in Greece is presented in Figure 1. Annual variation of atmospheric pollutants, such as CO, SO_2, NO_2, O_3, smog, in the city of Athens (the most polluted area), for different stations for the period 1984-93, is shown in Figures 2a-2e.

References

1. Ministry of Environment, Surveying and Public Works of Greece. "Greece: Reserve of Ecology and Civilization. Data, Actions, Programs for Environmental Protection", Athens, Greece, July 1995.

I. Linkov and R. Wilson (eds.), Air Pollution in the Ural Mountains, 407–411.

FIGURE 1. A map showing the stations of atmospheric pollution control in Greece.

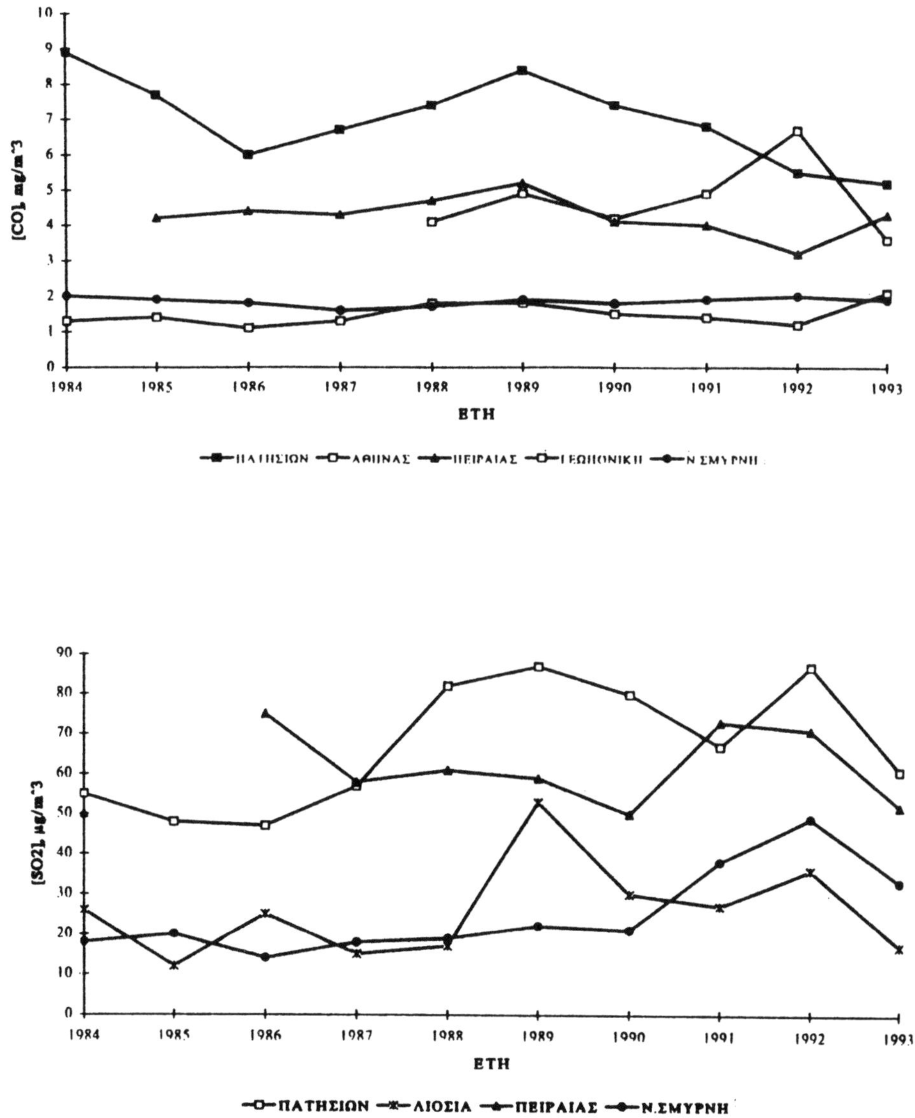

FIGURE 2a-b. Annual variations of carbon monoxide (a) and sulfur dioxide (b) in Athens, Greece.

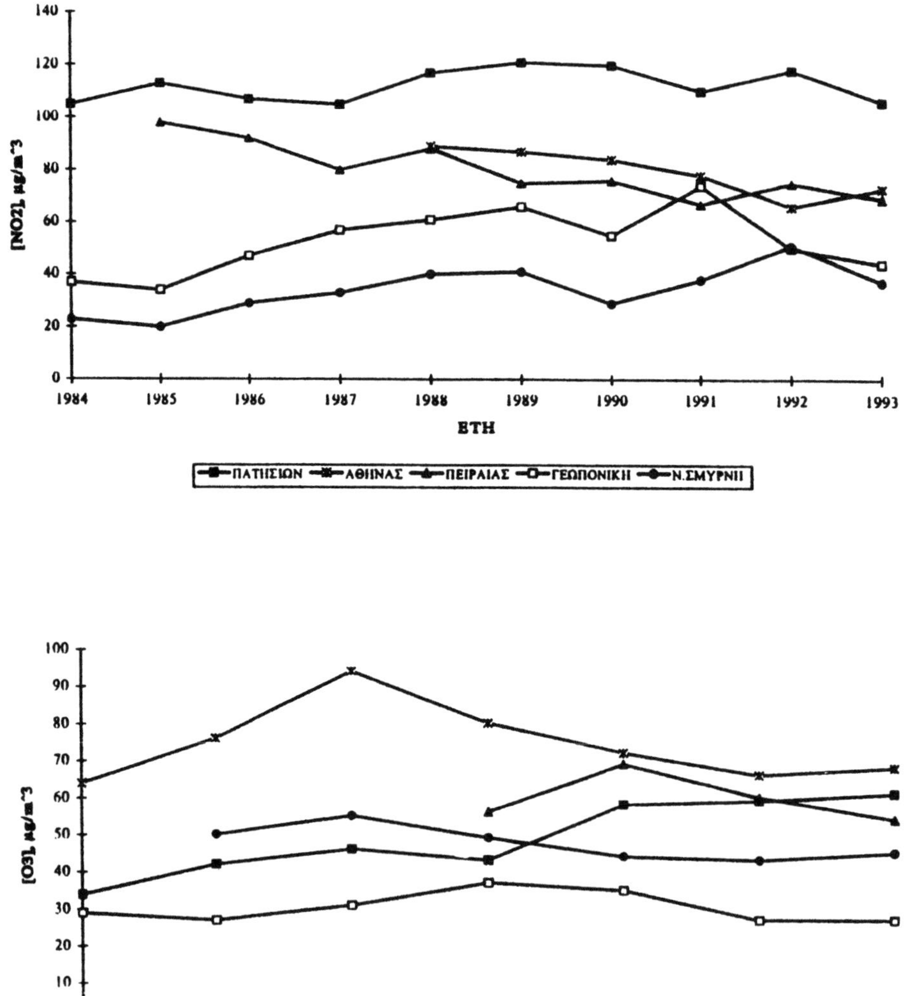

FIGURE 2c-d. Annual variations of nitrogen dioxide (c) and ozone (d) in Athens, Greece.

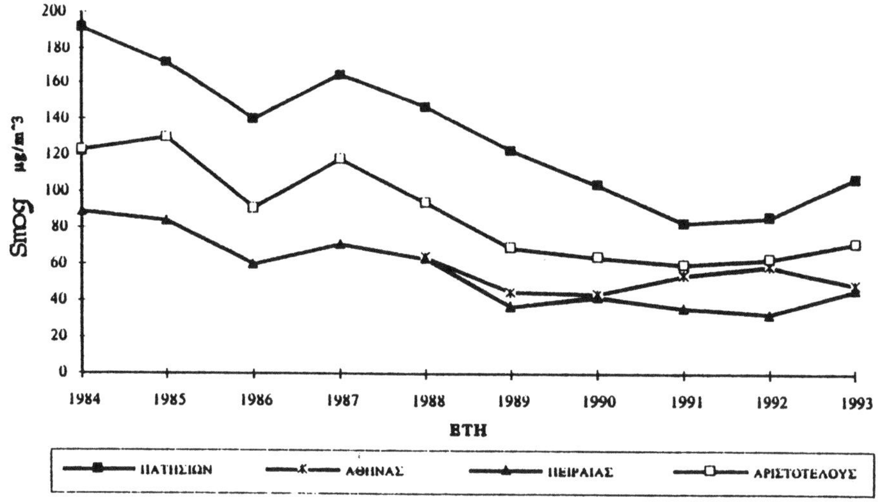

FIGURE 2e. Annual variations of smog in Athens, Greece.

AIR POLLUTION AROUND THE WORLD

Discussion Summary

C.I. DAVIDSON
Departments of Civil & Environmental Engineering
and Engineering & Public Policy,
Carnegie Mellon University, Pittsburgh, Pennsylvania 15213, USA

J.C. WILLSON
Kleinfelder, Inc.
5953 South Iola WayEnglewood, CO 80111, USA

1. Canada

M. Sheena Symington discussed air pollutant concentrations in Canada over the period 1974-1994. The national average annual mean airborne concentration of lead (Pb) has been reduced from 0.69 $\mu g/m^3$ in 1974 to 0.01 $\mu g/m^3$ in 1994. Total suspended particle concentrations have decreased from 78 to 34 $\mu g/m^3$ over this same period, while sulfur dioxide (SO_2) concentrations have been reduced from 13 ppb to 5 ppb. This reduction in SO_2 has been accompanied by a dramatic decrease in cations such as magnesium, calcium, and potassium (Mg, Ca, and K) which has reduced the neutralizing capacities of ecosystems.

The major problem now being addressed in Canada is ozone (O_3). Concentrations have increased from a national average of 15 ppb in 1974 to 21 ppb in 1994, although there is considerable variability from year to year. For example, very high concentrations were observed in the summers of 1983 and 1988 due to hot weather and stable atmospheric conditions.

2. England

Ian Milborrow reported on studies assessing the value of environmental effects of air pollution in England. The studies have considered PM-10 (particles less than 10 μm in diameter), SO_2, and O_3. As with other countries, a high percent of current air pollution is from the transportation sector. The effects have been divided into six categories, including human health, agricultural crops, forests, terrestrial ecosystems, freshwater fisheries, and building materials.

I. Linkov and R. Wilson (eds.), Air Pollution in the Ural Mountains, 413–416.

Results show that there are significant effects occurring from PM-10 in London. There are also significant effects of SO_2 in the industrial areas of the midlands of England. By estimating the costs associated with excess deaths, a total value can be placed on the damage caused by acute and chronic effects of PM-10 and acute effects of SO_2 and O_3. Similarly, a total value can be placed on morbidity associated with these pollutants. The cost of both mortality and morbidity effects is approximately 2% of the GDP of England for the year 1990.

3. Greece

Constantin Papastefanou reported on air pollutants in Greece. Data available for several pollutants in Athens show that smog and ozone concentrations have decreased substantially throughout the city over the past several years; a 50% decrease occurred between 1984 and 1993. However, concentrations of SO_2 and NO_2 (nitrogen dioxide) have increased during this time period. Carbon monoxide (CO) has remained relatively unchanged. Note that Athens has a population of about 5 million, which is more than half of the total population of Greece.

Beryllium-7 activity size distributions have been used as a tool in identifying sites of particularly heavy industrial pollution. Both Athens and Thessaloniki have regions of industrial emissions within the urban area. For these cities, the activity mean aerodynamic diameter (AMAD) for Beryllium-7 has been measured at several sites. It is found that the AMAD at highly polluted sites was shifted toward greater values (1 μm) compared with less polluted locations (0.68 and 0.62 μm in rural and mountain locations, respectively).

4. Kazakhstan

Abay Djabasov summarized studies of morbidity and mortality in Alma-Ata, the capitol of Kazakhstan. A total of 50,000 tonnes of total particles are currently emitted each year from stationary sources in Alma-Ata. This is much less than seven years ago, when pollution levels in Alma-Ata were similar to those in Magnitogorsk as determined via remote sensing from space. However, stationary sources now account for only 10% of particulates in air, while automobile traffic is mainly responsible for the remaining 90%.

The research group in Alma-Ata has developed a model for air pollution that accounts for the peculiar movement of air in the valley where the city is located. They have also collected medical data about morbidity in children and adults due to respiratory diseases and are able to correlate pollution levels with morbidity levels. Furthermore, they have developed methods to estimate the value of life lost due to pollution using several approaches, including data on Kazakhstan's Gross National Product (GNP) per person and earnings lost due to early death. Their final estimate is \$200,000 to \$300,000 per life lost due to air pollution.

5. Poland

Katarzyna Sawicka-Kapusta summarized air pollution in Poland. Both stationary sources and mobile sources (for example, automobile traffic) are important. During 1987-89, emissions of SO_2, NO_x (oxides of nitrogen), CO, and dust were each in the range of 1-4 million tonnes/year. The highest emissions are in the southern part of the country. A total of 27 areas of significant environmental damage have been identified; these areas comprise 10% of the land area and 30% of the population of the country.

Many parts of the country have contamination of air, land, and water above permissible levels. For example, airborne concentrations of lead were 0.3 $\mu g/m^3$ in Upper Silesia in 1989, which exceeded the standard of 0.2 $\mu g/m^3$; in other areas, levels of 1-2 $\mu g/m^3$ were measured. Similarly, concentrations of SO_2 are above 40 μg/m3 in central Cracow, greater than the standard of 32 $\mu g/m^3$.

6. Portugal

Francisco Ferreira discussed air pollution issues in Portugal, including monitoring of airborne concentrations as well as assembling emissions inventories. Concentrations of NO_x, CO, and particles sometimes exceed air quality standards in Lisbon, Porto, and smaller cities in the country. Work is underway to reduce emissions of several pollutants from stationary sources, although CO_2 emissions are increasing and ozone concentrations have increased in recent years. Leaded gasoline is still used throughout Portugal. It is of interest that stationary sources were the principal emitters of many pollutants in 1990, but mobile sources have become more important in recent years.

The emission inventory data show that roughly 50% of CO_2 emissions in the country is from power plants, while about 50% of the NO_x emissions is from mobile sources. Chemical plants, smelters, and refineries contribute significant amounts of other emissions. An inventory of emissions is put together each five years as part of the CORINAIR program of the European Community, which mandates air quality standards.

7. United States

James Wilkinson discussed several current air quality management issues in the U.S. There are a number of steps in the process of making decisions on source controls needed to maintain air quality. First, the impacts of current air quality conditions are determined, especially health and welfare effects. Air quality goals to minimize the effects are then identified. Next, control strategies to reduce emissions are designed in order to achieve the desired air quality goals. Finally, computer models for predicting airborne concentrations are developed and used to determine whether the reductions in emissions will indeed achieve the air quality goals; iteration may be necessary.

Sophisticated models have been developed to relate the emissions to air quality. Some of the models contain enormous numbers of equations (up to several million) that are solved simultaneously. The models incorporate nested grids so that calculations can be done over short distance scales near sources, and over greater distance scales far from sources where changes are taking place only slowly. This provides more reasonable computation times. These methods have been applied to the Southern Appalachian Mountains Initiative (SAMI) to develop a regional air quality strategy.

Part 5

Radioactive Contamination in the Ural Mountains

PROBLEMS OF RADIOACTIVE CONTAMINATION IN THE URALS

V.N.CHUKANOV and B.A.KOROBITSIN
Institute of Industrial Ecology, Ural Branch of Russian Academy of Sciences, Sophya Kovalevskaya Str., 20-a, Ekaterinburg, 620219, RUSSIA

The following natural and man-made sources of radiation are the most significant in the Urals:

(1) A high concentration of natural radionuclides (isotopes of the ^{238}U and ^{232}Th decay chains) in the environment;
(2) Poor management of low-level radioactive wastes;
(3) Testing of nuclear weapons in the atmosphere and underground peaceful explosions;
(4) The production of radioactive materials for military use and the fabrication of nuclear weapons

(1) Because of the geological peculiarities of the Urals the concentration of natural radionuclides in terrestrial materials is in 4-5 times higher than the average in the flat regions of Russia. There are granite areas in Sverdlovsk and Chelyabinsk Oblasts where dose rates in the air reach 500 nGy/h (50 μR/hr). Large surveys of natural radionuclides in groundwater yielded more than one thousand sites with elevated concentration of uranium (up to 0.01 g/l). Exposure to radon is the most significant component of human irradiation by natural sources. The year-average concentration of ^{222}Rn and ^{220}Rn in indoor air of some localities in Sverdlovsk Oblast are 100 Bq/m^3 and 5 Bq/m^3. Thus the total estimated average annual effective dose from inhalation of ^{222}Rn, ^{220}Rn and their decay products may be as high as 10 mSv (1 rem).

(2) Such localities as Ozerniy, Kostousovo, Dvurechensk (Sverdlovsk Oblast) were contaminated due to mining and milling of mineral sands containing monazite in the 1950s and 1960s. Because of poor management of low-level radioactive wastes, monazite sands were used by the population for building construction, wall and ceiling plastering and road strewing. In some dwellings dose rates in the air reached 30,000 nGy/h. In the 1990s most of the dwellings in these localities were sanitized, but there are numerous outdoor sites with elevated dose rates. The similar problems are in the towns Krasnoufimsk (Sverdlovsk Oblast) and Dolmatovo (Kurgan Oblast).

(3) Due to its geographic location and the prevailing wind direction, the Urals suffered from relatively high radioactive fallout after nuclear weapons tests at the

I. Linkov and R. Wilson (eds.), Air Pollution in the Ural Mountains, 419–422.

Semipalatinsk and Novaya Zemlja test sites. The maximum annual deposition rates of fission products during the period of most intensive testing in 1961-1963 were about 0.6 Curies of beta-nuclides per square kilometer. To this day, the level of ^{137}Cs contamination in the Urals is in two times higher than at the same latitude in the other regions of Russia. In 1954 there was a military exercise in Orenburg Oblast with an atmospheric nuclear explosion which caused radioactive contamination of the adjacent territory. In 1965-1987 11 peaceful nuclear explosions were performed in the territory of Orenburg and Perm Oblasts and Republic of Bashkortastan to increase oil output and create underground reservoirs for gas storage. Leakage and venting of gases and aerosols caused the radioactive contamination of near-by territories.

(4) The most significant cause of radioactive contamination and exposure has been the almost fifty-year activity of Mayak facility (the first Soviet plutonium production plant). Routine activities at the Mayak facility since 1948 over 40 years as well as the radiation accidents that occurred have led to an unprecedented radioactive contamination level in a considerable part of the Chelyabinsk, Kurgan, and Sverdlovsk regions. Dumping of radiochemical waste with an activity of 2.8 million curies into the river Techa in 1949-1952 resulted in the radioactive contamination of the Techa-Iset-Tobol river system. Approximately 5500 inhabitants of 22 river localities were resettled. So far, the regime of normal vital activities along the river has not been restored, and 8000 hectares of farming land is still abandoned.

In September of 1957 a storage tank for liquid high activity waste at the Mayak site exploded. Radionuclides with an activity of about twenty million curies were released to the environment, of which around two million curies were discharged outside the territory of the Mayak facility, thereby giving rise to the East-Ural Radioactive Trace (EURT). The northern part of the Chelyabinsk Oblast and the southern part of the Sverdlovsk Oblast suffered radioactive contamination. Some 15,000-23,000 km^2, in a track 300 km in length and 10-30 km wide, were contaminated at a level greater than 0.1 Ci/km^2 of ^{90}Sr. The western part of the town Kamensk-Uralskiy (in 1957 the total population of the town was about 160,000) was contaminated up to 4 Ci/km^2 of ^{90}Sr. Twenty-three rural localities within the radioactive contamination zone were abandoned and 10,400 dwellers were resettled.

The considerable radioactive contamination of the environment and the radiation exposure of the population, especially in the first decades of the activities of the enterprise, are due to routine process discharges of radioactive matter to the atmosphere at the Mayak facility. The effect of these emissions is traceable over a distance of 20 to 30 km westwards and 50 to 70 km eastwards. In spring 1967, radioactive substances of about 600 Ci were scattered by gusty winds from the banks of the Karachai lake to adjacent territories. Some characteristics of these incidents are present in Table 1.

The problems associated with Mayak activities are not restricted to those arising from the previous radioactive contamination of the regional territory. Over the years of operation of the enterprise, radioactive wastes of all categories (about one billion Ci) have piled up, including waste deposition in the open air. Such

TABLE 1. Characteristics of the radiation incidents at Mayak facility

	1949-1956 discharges in the Techa river	1957 Accident	1967 Accident
Total activity, Ci (Bq)	2.75×10^{6} (1.0×10^{17})	2×10^{6} (7.4×10^{16})	600 (2.2×10^{13})
Radionuclide	Contribution to total activity of the mixture, %		
^{89}Sr	8.8	traces	-
^{90}Sr	5.8	2.7	17
^{95}Zr	6.8	12.5	-
^{106}Ru	13	1.8	-
^{137}Cs	12.2	0.036	48
^{144}Ce	-	33	9
Rare-earths	26.8	-	-
Pu	traces	traces	-
Area of the contaminated territory	2000 km^2 of watershed, 80 km^2 were excluded from use	15,000-23,000 km^2 at a level greater than 0.1 Ci/km^2 of ^{90}Sr; about 1,000 km^2 at a level greater than 2.0 Ci/km^2 of ^{90}Sr	1,800 km^2 at a level greater than 0.1 Ci/km^2 of ^{90}Sr; 30 km^2 at a level greater than 2.0 Ci/km^2 of ^{90}Sr
Size of the overexposed population	124,000	300,000	40,000

accumulation of very large amounts of radioactive waste in the densely populated region is fraught with a threat of new large-scale disasters.

In 1991, a Federal program was developed that focused on radiation sanitation of the contaminated territories of the Urals and on medical rehabilitation of the radiation affected population up to the year 1995. In 1996 a new Federal program up to the year 2000 was developed and adopted by the Government.

Two groups of the major problems were defined as follows:

1. Contamination of the environment and human irradiation as the result of routine emission of radionuclides to the atmosphere and three accidents
2. Accumulation of radioactive waste on the territory of Mayak facility

In the period 1992-1995, an urgent treatment of these problems was carried out. The following leading regional organizations participate in this work on radiation sanitation of the contaminated territories and on medical rehabilitation of the radiation affected population, Table 2.

The results of scientific research in 1992-1995 allowed to identify in new way major problems and rank them according to priority.

At present, the following problems seem to be the most severe:

1. Uncontrolled migration of radioactivity from the lake Karachay into the groundwater and hydrographic system;
2. Overfilling of the cascade of artificial reservoirs and the threat of the dam burst;
3. High contamination of the Techa river system within river settlements and continuing leaking of radionuclides from the upper reaches of flood-lands and filtration through the dam of the last reservoir.

TABLE 2. Key organizations in the Federal program focusing on radiation sanitation of the contaminated territories of the Urals and on medical rehabilitation of the radiation affected population

Organization	Person in charge for execution of the Federal program	Field of activity
Mayak facility (Ozersk)	Drozko E.G.	Radioactivity monitoring in the environment and radiation sanitation of the territories of Mayak facility site (including Lake Karachay and Techa Reservoir). Radiation wastes management
Radioecological Research Station of the Mayak facility (Ozersk)	Romanov G.N.	Radioactivity monitoring in the environment and radioecological studies at the territory of EURT (Chelyabinsk Oblast). Agricultural and industrial production at the radioactively contaminated territories
Branch №1 of the Biophysics Institute (Ozersk)	Romanov S.A.	Dose reconstruction for the stuff of Mayak facility and population of Chelyabinsk-65 (Ozersk), assessment of health effects of radiation exposure, medical rehabilitation
Ural Centre for Radiation Medicine (Chelyabinsk)	Akleev A.V.	Dose reconstruction for the inhabitants of the Techa River and the territory of EURT (Chelyabinsk Oblast), assessment of health effects of radiation exposure, medical rehabilitation
Institute of Industrial Ecology (Ekaterinburg)	Chukanov V.N.	Complex analysis of all scientific results of all organizations which participate in the realization of the Federal program, preparation of consolidated reports for Federal Government. Radioactivity monitoring in the environment and radioecological studies at the territory of EURT (Sverdlovsk Oblast), assessment of health effects of radiation exposure

RADIOACTIVE CONTAMINATION OF THE TECHA RIVER

D. BURMISTROV
Urals Research Center for Radiation Medicine, RUSSIA

I. LINKOV
Department of Physics, Harvard University, Cambridge, MA 02138, USA.

Abstract

In radiation and other accidents, it is a major problem to evaluate integrated environmental exposure, since measurements are taken some time after the accident when radiation or other decay has considerably reduced the pollution to be measured. This deficiency in the data must be filled by models. We have developed a compartmental dynamic model for radionuclide fate and transport in Techa River. The database ENVIRONMENT was used for the testing and validation of these models. This database was developed in the Urals Research Center for Radiation Medicine (URCRM, Chelyabinsk, Russia) and contains: (i) archival data on radioactive source term and environmental monitoring provided by the Plutonium Production Association "Mayak" (Chelyabinsk Region), (ii) environmental measurements and whole body counting of exposed population carried out at the URCRM. The model is being successfully applied to describe radionuclide transport in the Techa River and to reconstruct radiation doses for the Techa River populations. This model is modest in requirements and allows a greater flexibility in environmental applications.

1. History of the Techa River Contamination

In the 1950's Techa River included the Koksharov and Metlinsky ponds in the upper part, and the body of the Techa river itself (~240 km) lower the dumb (point 21, Fig.1). The Koksharov pond was created in August of 1951 and had approximate volume of 400000 m^3 with effective width of 230m and length 1,500 m. In May of 1952 the water level in the pond was elevated up to 80-90 cm. The Metlinsky pond was relatively stable, and had volume about $1.5*10^6$ m^3, effective length of 1,700 m, and width of 680 m. The mean width of the river in summer was ~5m in the upper part, and depth: 0.5-1 m. The mean river velocity was about 0.3 m/s. The water discharge had a strong seasonal dependence and varied from 0.6 m^3/s to 15-20 m^3/s (60 m^3/s in the spring of 1951), the mean value was about 1 m^3/s in the upper Techa. Water contained from 1 to 8 mg/l of suspended sediments, with the mean density of about 1.5 g/cm^2 . In 1956 the

I. Linkov and R. Wilson (eds.), Air Pollution in the Ural Mountains, 423–428.

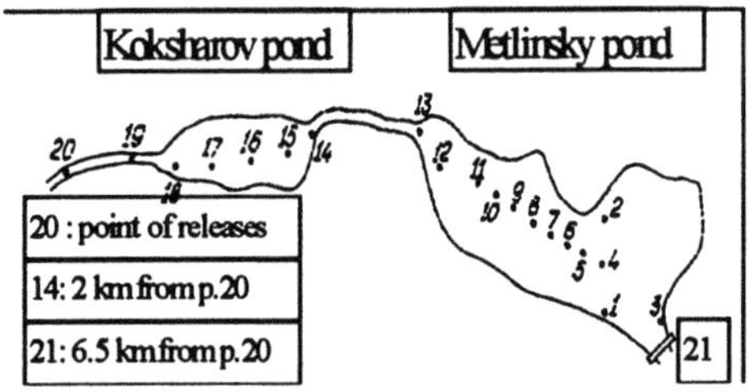

FIGURE 1. Upper Techa River in the 1950's

river was dammed lower the point 21, and a set of water reservoirs as constructed there.

The source of contamination was located at the point 20 (Fig.1). The contamination resulted from routine releases of low- and intermediate-level liquid wastes from the plutonium radiochemical separation facility. In addition, large uncontrolled amounts of radioactive materials were released with cooling water from the waste-storage facility ("wild releases"). About $2.7*10^6$ Ci of beta-emitters were released during the period 1949-1956, with 95% of the releases occurred between March 1949 and October 1952. The mean release during that period was ~4,300 Ci/day. Since November 1951, the releases were redirected into the Lake Karachay, located near the facility. The source and approximate isotope composition of the releases is presented in [1] for several periods of "Mayak" operation. It was estimated that about 40% of releases were solid particles, and other 60% were dissolved nuclides. The approximate isotope composition of routine releases was: 1) 50% ^{144}Ce, 25% $^{95}Zr+^{95}Nb$, 12% $^{103}Ru+^{106}Ru$, 7% ^{90}Sr, 6% $^{140}Ba+^{89}Sr$ (solid phase); 2) 20% ^{137}Cs, 37% $^{103}Ru+^{106}Ru$, 15% ^{90}Sr, 11% ^{144}Ce, 13% $^{140}Ba+^{89}Sr$ ·, 4% $^{95}Zr+^{95}Nb$ (liquid phase). It must be noted, that this composition is approximate. For "wild releases", the solid fraction sometimes reached 90%, and the radianuclide composition varied significantly.

2. Compartmental Model for Radionuclide Fate and Transport in Techa

The compartmental model (Fig.2) was developed to study the dynamics of the radionuclide transport through the Koksharov and Metlinsky ponds. The model takes into account:

1. transport by water flow;
2. sedimentation of solid fraction of releases;
3. exchange between dissolved fraction and solid particles of natural suspended sediments;
4. deposition of natural sediments;
5. effective scoring of bottom sediments by water flow;
6. decay and production of radionuclides.

In the approximation of full mixing of water in the ponds the radionuclide transport is described by 32 systems of ordinary differential equations. Each system relates to dynamics of a radionuclide concentrations in water and bottom sediments.

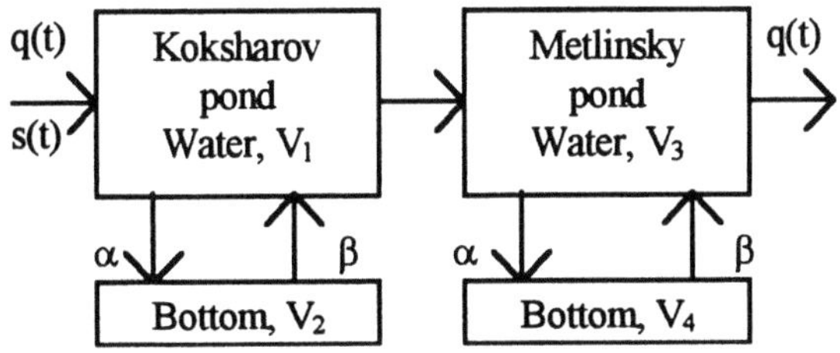

FIGURE 2. Box diagram of the compartment model

The systems are written for each of 8 radionuclides, solid and liquid phases of releases, and two ponds. The equations are solving by Runge-Kutta method, or by specially developed fast numerical method of solution [5]. The last method is based on additional assumption that coefficients of the systems can be considered as constants within intervals of appropriate time scale. Such an assumption allows to use analitical solution of the systems within each time interval with concentrations of radionuclides in the beginning of each interval as initial conditions. This fast method is useful for computation consuming fitting of unknown parameters of the model.

Unknown model parameters of can be determined by fitting model predictions to actual measurements. These parameters are: 1) the effective size of technological and natural particles in the water, 2) the concentration of suspended matter in water and 3) the effective desedimentation rate. We start with fitting using the data on specific activity of beta-emitters in water. We found that logarithms of beta-activity in water is normally distributed. Therefore the mean-square distance in logarithmic scale was choosen as the "distance" between measurements and model predictions. Special software for search of the global and local extrema of multivariate functions were developed and used to fit the parameters mentioned above. To reduce the source term uncertainty we solve a simplified problem: the actual concentration of radionuclides in Koksharov pond was used as the source-term for contamination of the Metlinsky pond. The results of calculation show a good predictive quality of the model (Fig.3a). The agreement of the model calculation with experimental measurements is less satisfactory when the historical source term [1] is used (Fig.3b).

The compartment modelling is one of possible ways to reconstruct external gamma dose-rates on the banks of the Techa River in 1949-50 years [1-4]. There are no actual measurements of dose-rates made during this period. The reconstruction can be done by correlating the dose-rates-in-air with the activity of bottom sediments. The historical source term [1] and the actual beta-activity of water in the Koksharov pond were used as the source-term for calculation. Preliminary results (Figure 4) suggest that significant portion of the total dose comes from 1949-1951 when direct measurements are not available.

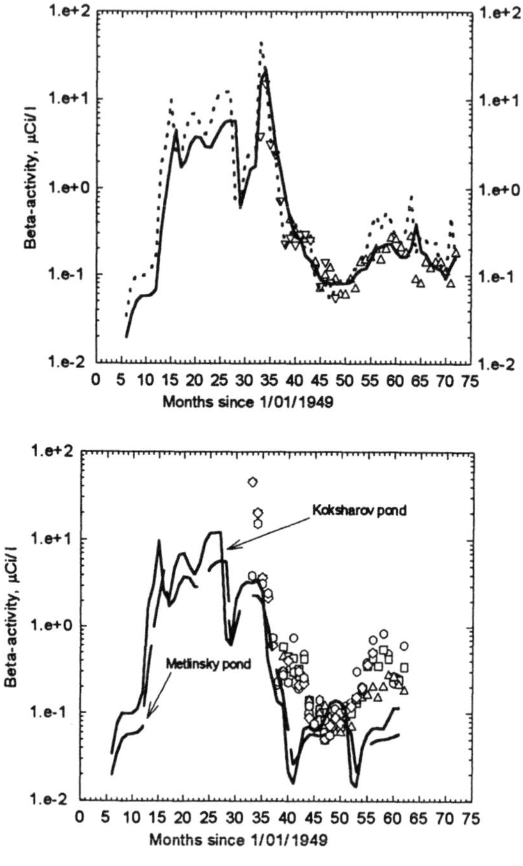

FIGURE 3. Specific activity of beta-emitters in Metlinsky and Koksharov ponds **a)** solid line is calculation for Metlinsky pond (best fit results for monthly averages). Measured beta-activity in the Koksharov pond (dotted line) is taken as the source term, triangles are actual measurements in the Metlinsky pond. **b)** The lines are results of calculations using historical data for radionuclide releases, top pointed triangles and hexagons are measurements in the Metlinsky pond taken from different sources, other notations are measurements in the Koksharov pond from different sources.

3. Acknowledgements

This work was supported by Russian-US Joint Coordinating Committee for Radiation Effects (Project 1.1, "Dose Reconstruction for the Ural Population") and the Nato Collaborative Research Grant SA.12-2-02 (ENVIR.CRG960914) 676(96) LVdc.

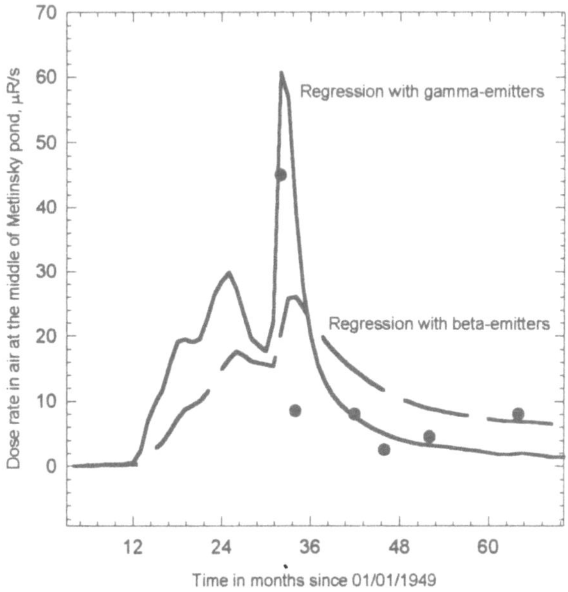

FIGURE 4. An example of the external dose reconstruction in-air. Linear regression between dose-rates and activities of the bottom sediments was done for beta- and gamma-emitters. The points represent actual measurements in the middle of the Koksharov pond. The lines are calculations that were done using historical source term.

4. References

1. M.O.Degteva, V.P.Kozheurov, D.S.Burmistrov, V.V.Valchuk, M.I.Vorobiova, N.G.Bougrov, H.A.Shishkina. (1996). "An Approach to Dose Reconstruction for the Urals Population" Health Physics, **71**:71-76.
2. M.O.Degteva, V.P.Kozheurov, M.I.Vorobiova (1994). "General approach to dose reconstruction in the population exposed as a result of release of radioactive wastes into the Techa River". Sci. Total Environ. **142**:49-61.
3. M.O. Degteva, V.P. Kozheurov, M.I. Vorobiova, D.S. Burmistrov, V.V. Khokhryakov, K.G. Suslova, L.S. Anspaugh, B.A. Napier and A.C. Bouville (1997). "Population Exposure Dose Reconstruction for the Ural Region". In: "Assessing Health and Environmental Risks from Long-Term Radiation Contamination in Chelyabinsk, Russia" (Ed. by S.L.Kellogg and E.J.Kirk). Proceedings from the 1996 AAAS Annual Meeting Symposium, Baltimore, Maryland. AAAS, Washington,DC.

4. M.O. Degteva, E.G. Drozhko, L.S. Anspaugh, B.A. Napier, A.C. Bouville and S.W. Miller. (1996) "Dose reconstruction for the Urals population". Joint Coordinating Commettee on Radiation Effects Research. Project 1.1-Final report, Lawrence Livermore National Laboratory, Livermore, CA, Rept. UCRL-ID-123713.
5. D. Burmistrov, I. Linkov, M. Vorobiova, M. Degteva and R. Wilson (1997) "Radioactive Contamination of the Techa River: Environmental Records and Multimedia Modeling." in: Proceedings of the International Conference on Nuclear Data for Science & Technology, Trieste, Italy (in press).

SIGNIFICANCE OF LONG RANGE ATMOSPHERIC TRANSPORT FOR HEAVY METAL AND RADIONUCLIDE CONTAMINATION OF THE TERRESTRIAL ENVIRONMENT

E. STEINNES
Norwegian Department of Chemistry
University of Science and Technology
N-7034 Trondheim, NORWAY

1. Introduction

The most severe effects of air pollutants are likely to occur in the vicinity of strong point sources. However, at most point sources of air pollution a considerable part of the emissions occur either in gaseous form or as small particles, capable of travelling very long distances in the atmosphere before they eventually reach the ground by wet or dry deposition. Heavy metals or radionuclides released in volatile form from the source are likely to be preferentially concentrated on small particles, which have a large surface-to-mass ratio and are likely to be available for long range atmospheric transport. Thus a substantial fraction of heavy metals or radionuclides emitted from a point source may be deposited in areas very far away from the source, and quite frequently in a different country, resulting in transboundary air pollution.

Sometimes a combination of prevailing winds and precipitation patterns makes the trasnsfer of pollutants from a source region to a recipient area particularly efficient. This situation exists in the southernmost part of Norway, where the natural environment has been exposed to huge amounts of air pollutants originating in other European countries over the years. The most obvious effect of this transboundary pollution is acidification of lakes and subsequent fish-kill [1], but this region has also been subjected to a rather massive "import" of some heavy metals over the years. This has led to extensive contamination of natural surface soils and vegetation, and further transfer of these metals in the foodchains.

If, on the other hand, a pollution problem is caused by an accidental release of pollutants during a short time rather than a continuous pollution situation over a long period of time, the resulting deposition will depend on the meteorological conditions during a period of the order of hours to a few days, and the geographical deposition pattern observed may be very different from that experienced at "normal" conditions. The Chernobyl accident was such a short-time accidental release, and the resulting deposition of radiocaesium in Norway occurred mostly in regions not normally subject to strong impacts of long range transport of pollutants.

In the present paper some selected data illustrating the situation in Norway with respect to long range transport from other countries will be discussed, as a back-

I. Linkov and R. Wilson (eds.), Air Pollution in the Ural Mountains, 429–434.

ground for evaluating the possible impact of industries in the Ural region at places very far from the source region.

2. Heavy metals

The acidification of lakes and rivers in the southernmost part of Norway due to long range atmospheric transport of pollutants from densely populated and heavily industrialized regions elsewhere in Europe was well documented during an extensive research program during the 1970s [1]. During this work it also became apparent that some heavy metals were supplied in large amounts with the same polluted air masses [2]. Following this observation an extensive research and monitoring activity was initiated in order to identify the main source regions of the heavy metal pollution, the geographical deposition pattern of these metals within Norway, and their fate in terrestrial ecosystems.

2.1. SOURCE REGIONS

During two periods in 1978-79 and 1985-86 air concentrations of a number of trace metals were measured daily at Birkenes in southern Norway. Comparison of emission estimates for different European countries with air concentrations at Birkenes using a simple trajectory model for selected days yielded good agreement between measured and calculated concentrations for V, Mn, Cr, Cd, and Pb [3]. Trajectory sector analysis showed that the main contributing source areas for airborne trace elements in southernmost Norway are southeast to southwest of Birkenes [4].

2.2. ATMOSPHERIC DEPOSITION

The atmospheric deposition of trace elements in Norway has been studied regularly since 1977 at about 500 sites all over the country [5-7], using moss analysis. Mosses can be used as biomonitors of atmospheric deposition because they have no root system and therefore absorb nutrient mainly from the air. Moreover they are very efficient collectors of many metals from wet and dry deposition. The moss analysis as such provides only relative deposition patterns, but calibration can be done against bulk deposition values from precipitation samples [8]. In the Norwegian monitoring data the temporal and spatial differences are particularly pronounced in the case of Pb, for which two trends are obvious: a pronounced decline as a function of time, and a consistently higher deposition in the southernmost part of the country where the impact of long-range atmospheric transport is strongest. Other elements showing a very similar geographical deposition pattern include V, Zn, As, Mo, Cd, Sb, Tl, and Bi, indicating that they are all derived predominantly from the same general source region. From the trajectory analysis mentioned above it appears that at least 80 % of the Pb deposition in southernmost Norway is derived from sources out of Norway, and results from principal component analysis of the moss data strongly indicates that the same applies to the other elements showing similar deposition patterns.

2.3. SURFACE SOIL CONTAMINATION

A survey of some selected heavy metals collected in 1977 from all over Norway in the organic-rich surface layer of natural soils, showed that Pb and Cd were concentrated by a factor of 10 in the southernmost part of the country relative to more northerly regions where the contribution from long range atmospheric transport is small [9]. Repeated sampling in 1985 [10] including samples from the underlying mineral soil revealed that a similar geographical trend was not evident in the C-horizon. Hence the excess heavy metal in the south must be the result of integrated atmospheric deposition and subsequent strong retention in the humus layer. Also Zn, As, and Sb were sustantially enriched in the southern topsoils. Analysis of peat cores from ombrotrophic bogs [11] indicates that the deposition of these metals in southern Norway during recent decades was at least 20 times the pre-industrial deposition rate.

Although metals such as Zn and Cd are not normally very strongly bound to humic substances by purely chemical forces, experimental acidification of a mature pine stand in southern Norway showed that these elements were remarkably well retained in the humus layer, probably due to microbial uptake and acidification [12]. Analyses of ectomycorrhiza from coniferous forest soils supported this assumption [13], and indicated that the fungal biomass may account for as much as 30-40% of the soil levels of Zn and Cd.

2.4. UPTAKE IN FOOD-CHAINS

Investigation of some common vascular plant species from similar pine forest ecosystems in 1982 revealed that samples from the southernmost part of Norway consistently exhibited Pb and Cd concentrations about five-fold those observed in the same species growing in central Norway [14], presumably because of differences in atmospheric deposition in the two regions. Repeated analyses in 1992 showed the same geographical difference for both elements, but the Pb concentrations had decreased to about one third of the 1982 values, whereas the Cd levels were the same as 10 years before. These observations together with a decrease in atmospheric Pb deposition rates of about 70% both in southern and central Norway, suggest that direct atmospheric deposition strongly influenced Pb levels in vegetation. The unchanged Cd levels indicate that the excess Cd observed in the plants in southern Norway was due to root uptake from the surface soil contaminated from long range atmospheric transport.

In an investigation of heavy metals in lamb liver from livestocks in ten different places in Norway [15] the Pb concentration showed a very strong positive correlation with atmospheric deposition, whereas for Cd the association, although present, was much weaker. Also for As a connection with atmospheric deposition was evident. In a simultaneous study [16] levels of trace elements in liver from moose, reindeer and red deer in different parts of Norway were investigated. The geographical distribution, particularly in moose, resembled very much that of atmospheric deposition in the case of Pb, As, and Se, whereas for Cd the natural geochemical environment seemed to play a greater role.

3. Radionuclides

Artificial radionuclides are supplied to the terrestrial environment almost exclusively by atmospheric wet or dry deposition. Anthropogenic sources of radioactivity in air have been mainly nuclear weapons testing, release from nuclear installations during regular operation, and accidents such as the Chernobyl disaster. The radionuclides produced from atmospheric nuclear weapons testing were released predominantly to the stratosphere, and hence the radioactive fallout was globally distributed. The radioactivity released from the Chernobyl accident, on the other hand, was to the lower troposphere. Therefore a major part of the fallout was on a local and regional scale. Still some areas much farther away were considerably affected by the accident. The predominant north-westerly movement of air masses from Chernobyl during the first few days after the accident resulted in transport of large amounts of radionuclides to central Scandinavia, and the specific precipitation pattern during April 28 and 29, 1986, produced an irregular, but in some cases substantial, deposition of radioactivity in this region. Based on soil measurements [17] and other evidence we may estimate that in Norway alone about 20,000 km^2 of land received more than 40 kBq m^{-2} (1 Ci km^{-2}) of ^{137}Cs.

The Chernobyl fallout in Norway was initially dominated by ^{131}I, which fortunately decayed to an insignificant level before the main pasture season. The main problem with respect to human exposure, therefore, was associated with radiocesium (^{134}Cs + ^{137}Cs) which constituted about 98% of the long-lived radionuclides supplied to these areas from the Chernobyl fallout. The ^{90}Sr contribution was only of the order of 1-2%, in spite of the fact that the total activities of ^{137}Cs and ^{90}Sr in the reactor core at the time of the accident must have been nearly the same. Apparently ^{137}Cs, being more volatile, was to a great extent released on small particles capable of long range atmospheric transport, whereas the more refractory ^{90}Sr was released predominantly with larger particles which were deposited more locally.

Most studies in radioecology carried out prior to the Chernobyl accident had been concentrated on agricultural food-chains. The fact that most of the contaminated land in Norway after the Chernobyl accident consisted of forest and montain ecosystems provided a unique opportunity to study the behaviour of radionuclides in systems and food-chains not previously investigated in this respect. Exstensive research on these ecosystems has been carried out in Norway and other Nordic countries following the Chernobyl accident, and much of the experience gained may be relevant also to similar areas in the boreal zone, in Russia and other countries . A survey of the results from this research is beyond the scope of this paper, but one particular observation may be mentioned: In most of the natural ecosystems in Norway affected by the Chernobyl accident, radiocesium has shown a very strong accumulation in the upper humus layer of the soil [18]. The apparent strong binding of ^{137}Cs to humus was quite surprising, since cesium does not form strong chemical complexes in general. It is now becoming increasingly evident that the underground fungal biomass plays an important role as a storage reservoir of radiocesium in the humic topsoil [19], in a similar way as described above for heavy metals. This may be a key factor in explaining the observation [20] that the ecological half-life of ^{137}Cs in some food-chains in these ecosystems appears to be very long, possibly approaching the physical half-life of 30 years.

4. References

1. Overrein, L.N., Seip, H.M. and Tollan, A. (1981) *Acid Precipitation - Effects on Forest and Fish.* Final report from the SNSF Project, Report 19/80, Oslo - Ås, Norway.
2. Hanssen, J.E., Rambæk, J.P., Semb, A. and Steinnes, E. (1980) Atmospheric deposition of trace elements in Norway, in D. Drabløs and A. Tollan (eds.) *Ecological Impact of Acid precipitation,* SNSF Project, Oslo - Ås, Norway, pp.116-117.
3. Pacyna, J.M., Semb, A. and Hanssen, J.E. (1984) Emission and long-range transport of trace elements in Europe.*Tellus* **36B**, 163-178.
4. Amundsen, C.E., Hanssen, J.E., Semb, A. and Steinnes, E. (1992) Long-range atmospheric transport of trace elements to southern Norway. *Atmos. Environ.* **26A**, 1309-1324.
5. Steinnes, E., Rambæk, J.P. and Hanssen, J.E. (1992) Large scale multi-element survey of atmospheric deposition using naturally growing moss as a biomonitor. *Chemosphere* **35**, 735-752.
6. Steinnes, E., Hanssen, J.E., Rambæk, J.P and Vogt, N.B. (1994) Atmospheric deposition of trace elements in Norway: Temporal and spatial trends studied by moss analysis. *Water, Air, Soil Pollut.* **74**, 121-140.
7. Berg, T., Steinnes, E., Røyset, O. and Vadset, M. (1995) Atmospheric trace element deposition: Principal component analysis of ICP-MS data from moss samples.*Environ. Pollut.* **88**, 67-77.
8. Berg, T., Røyset, O. and Steinnes, E. (1995) Moss (*Hylocomium splendens*) used as biomonitor of atmospheric trace element deposition: Estimation of uptake efficiencies*Atmos. Environ.* **29**, 353-360.
9. Allen, R.O. and Steinnes, E. (1980) Contribution fronm long-range atmospheric transport to the heavy metal pollution of surface soil, in D. Drabløs and A. Tollan (eds.) *Ecological Impact of Acid precipitation,* SNSF project, Oslo - Ås, Norway, pp. 102-102.
10. Bølviken, B. and Steinnes, E. (1987) Heavy metal contamination of natural surface soils in Norway from long-range atmospheric transport. Further evidence from analysis of different soil horizons, in S.E. Lindberg and T.C. Hutchinson (eds.) *Heavy Metals in the Environment,* CEP Consultants, Edinburgh, Vol. I pp. 201-203.
11. Steinnes, E. (1997) Trace element profiles in ombrogenous peat cores from Norway: Evidence of long range atmospheric transport.*Water, Air, Soil Pollut.,* in press.

12. Berthelsen, B.O., Årdal, L., Steinnes, E., Abrahamsen, G. and Stuanes, A.O. (1994) Mobility of heavy metals in pine forest soils as influenced by experimental acidification. *Water, Air, Soil Pollut.* **73**, 29-48.
13. Berthelsen, B.O., Olsen, R.A. and Steinnes, E. (1995) Ectomycorrhizal heavy metal accumulation as a contributing factor to heavy metal levels in organic surface soils. *Sci. Total Environ.* **170**, 141-149.
14. Berthelsen, B.O., Steinnes, E., Solberg, W. and Jingsen, L. (1995) Heavy metal concentrations in plants in relation to atmospheric heavy metal deposition. *J.Environ. Qual.* **24**, 1018-1026.
15. Frøslie, A., Norheim, G., Rambæk, J.P. and Steinnes, E. (1985) Heavy metals in lamb liver: Contribution from atmospheric fallout. *Bull. Environ. Contam. Toxicol.* **34**, 175-182.
16. Frøslie, A., Norheim, G., Rambæk, J.P. and Steinnes, E. (1984) Levels of trace elements in liver from Norwegian moose, reindeer and red deer in relation to atmospheric deposition. *Acta Vet. Scand.* **25**, 333-345.
17. Bache, S., Bjerke, H., Rudjord, A.L. and Ugletveit, F. (1987) Fall-out pattern in Norway after the Chernobyl accident estimated from soil samples. *Radiat. Protection Dosimetry* **18**, 105-107.
18. Varskog, P., Næumann, R. and Steinnes, E. (1994) Mobility and plant availability of radioactive Cs in natural soil in relation to stable Cs, other alkali elements and soil fertility. *J. Environ. Radioact.* **22**, 43-53.
19. Olsen, R.A. (1994) The transfer of radiocaesium from soil to plants and fungi in seminatural ecosystems, in H. Dahlgaard (ed.) *Nordic Radioecology*, Elsevier, pp. 265-286.
20. Nelin, P. and Nylen, T. (1994) Factors influencing the changes in ^{137}Cs levels with time in boreal-forest plamts in Sweden. *Sci. Total Environ.*, 157, 73-81.

EPIDEMIOLOGIC STUDIES ON THE HEALTH EFFECTS OF IONIZING RADIATION

M.B. GOLDMAN
Department of Epidemiology, Harvard School of Public Health, 677 Huntington Avenue, Boston, MA 02115, USA

1. Background

Exposure to ionizing radiation comes from a variety of natural background sources including cosmic rays, radioactive elements in the soil including uranium, thorium, and radium, internally deposited radionuclides such as ^{14}C and ^{40}K, and from exposure to indoor radon [1-2]. Medical procedures account for the greatest source of man-made exposure, with occupation, nuclear power, consumer products (TV's and smoke detectors), and fallout from weapons testing accounting for relatively little. Health consequences that have been identified include leukemia (except CLL), solid tumors at a number of sites, and heritable effects [3]. Estimates of annual per capita doses vary depending on geographic region and altitude, but are in the range of 1.6 mSv from all sources of radiation except radon [2]. This dose estimate suggests that about 5% of all cancers may be due to radiation exposure (excluding the risk of lung cancer associated with radon exposure) [2,4].

Scientists continue to explore the mechanisms by which radiation initiates cancer and the effect of dose and timing of exposure. Unanswered questions include the nature of dose-rate effects (Are low doses harmful only above a certain threshold level? Do dose rate and type of radiation affect risk? Do high instantaneous doses equal low chronic doses?) and the role of cofactors in modifying risk, particularly the influence of host susceptibility factors such as age, sex, and genetic make-up.

Data concerning the health consequences of radiation exposure come from animal and laboratory studies, case reports, and epidemiologic studies of populations who received medical or occupational exposure or who were exposed as a result of military action. Because of the contributory role of medical exposure, many studies have focused on populations who received diagnostic or therapeutic radiation [5]. These studies have included women who underwent multiple fluoroscopies to monitor air collapse therapy of the lung during tuberculosis treatment [6], patients who underwent radium implants or pelvic radiotherapy for the treatment of cervical cancer [7], and Hodgkin's disease and retinoblastoma patients who were treated with radiotherapy [8-9]. We are currently investigating three populations: patients treated for hyperthyroidism with ^{131}I [10-11], girls irradiated to reduce the size of their thymus gland [12-13], and patients injected with the contrast medium, Thorotrast, during cerebral angiography [14-15]. These studies are designed to examine several

I. Linkov and R. Wilson (eds.), Air Pollution in the Ural Mountains, 435–440.

hypotheses including the influence of age and genetic susceptibility on cancer risk, the mechanism of cancer induction by alpha radiation, and the occurrence of health effects other than cancer.

2. Current Studies

2.1. RADIOACTIVE IODINE THERAPY FOR HYPERTHYROIDISM

In 1961 the United States Public Health Service began the Cooperative Thyrotoxicosis Therapy Follow-up Study to investigate the occurrence of benign and malignant neoplasms in 36,050 hyperthyroid patients treated between 1946 and 1964 [16]. The study was designed to assess the carcinogenic potential of radioactive iodine (^{131}I), which became available as a treatment for hyperthyroidism soon after World War II. Atmospheric releases of ^{131}I occur as a result of nuclear fallout from weapons testing and unscheduled releases. No increase in cancer risk was found, but the patients treated with radiation were followed for only eight years, on average.

We conducted a follow-up study of the subset of 1,762 hyperthyroid women who were treated at the Massachusetts General Hospital Thyroid Unit [10]. The average length of follow-up of the patients was 17.2 years. Ninety-two percent of the cohort was located and eighty-eight percent of 1,058 living patients responded to a mail questionnaire. Eighty percent of the women were treated with radioactive iodine. The average age at treatment with ^{131}I was approximately 50 years.

The standardized mortality ratio (SMR) for all causes of death was 1.3 (95% confidence interval (CI) 1.2-1.4). The standardized mortality ratios for all malignant neoplasms and for breast cancer were 0.9 (95% CI 0.7-1.1) and 1.3 (95% CI 0.8-1.9), respectively. The standardized incidence ratios (SIR) for all malignant neoplasms and for breast cancer were 0.9 (95% CI 0.8-1.1) and 1.2 (95% CI 0.9-1.5), respectively. A nonsignificant excess breast cancer risk was observed 10 years after the onset of thyroid symptoms and was present at the end of thirty years of observation. The excess cases of breast cancer occurred only among those women treated with ^{131}I, but there was no consistent trend with increasing dose. When age at treatment and year of treatment were controlled, women who were ever treated with radioactive iodine had a standardized rate ratio for breast cancer of 1.9 (95% CI 0.9-4.1), compared with those who were never treated with radioactive iodine. A statistically significant excess number of pancreatic cancer cases (SIR = 2.0, 95% CI 1.0-3.7) and a nonsignificant excess of brain cancer cases (SIR = 2.3, 95% CI 0.7-5.3) were also observed.

The Thyrotoxicosis Therapy Study population was recently reassembled and follow-up was extended to an average of 21 years [17]. Radioactive iodine was not linked to cancer deaths overall, but an increase in thyroid cancer was found (SMR = 4.0, 95% CI 2.6-5.9).

2.2. BREAST CANCER AFTER IRRADIATION OF THE IMMATURE BREAST

Few studies have investigated the long-term consequences of childhood irradiation of the breast. Women exposed to ionizing radiation as infants or children may be at even greater risk of developing breast cancer than women who were first exposed at older ages. Radiation exposure of the immature breast may act as a tumor initiator, increasing the susceptibility of the exposed cells to the tumor-promoting effects of a variety of subsequent exposures, including hormones, dietary factors, chemicals, or viruses. A study of female tuberculosis patients who received multiple fluoroscopic examinations of the lungs reported that the greatest absolute excess risk of breast cancer occurred among women who were first exposed between 15 and 19 years of age, while no increased risk of breast cancer was observed among women who were 30 years of age or older at the time of first exposure [6]. Similarly, the atomic bomb survivor studies reported that girls 9 years of age or younger at exposure were at greater risk than those who were 10 to 19, and that these girls were, in turn, at greater risk than those women who were 20 to 29 years of age at the time of the bombing [18]. Furthermore, a recent study of the Japanese atomic bomb survivors suggests that early onset breast cancer is more strongly related to radiation dose than cancer occurring at later ages [19]. This finding suggests that there is a subset of women who are at particular risk of developing radiation-induced breast cancer.

We are conducting a retrospective follow-up study to evaluate the hypothesis that irradiation of the immature breast tissue of prepubertal girls increases their risk of subsequent breast cancer and that these cancers are more likely to develop (or perhaps develop earlier) in those women who have a genetic predisposition to breast cancer. We have identified a population of 700 girls who received prophylactic irradiation for an “enlarged” thymus gland in infancy or early childhood at Massachusetts hospitals between 1924 and 1946 [12-13]. The current age range of this population is 50 to 73 years, so the women are now entering the years of peak breast cancer incidence. A comparison group of sisters and peers is also being studied. One other cohort of similar patients has been followed. These women, whose mean age was 36 years, had an increased risk of breast cancer that was 3.6 times that of their non-irradiated sisters (95% CI 1.8-7.3) [20].

Follow-up of the Massachusetts population provides the opportunity to determine if age at radiation exposure influences the susceptibility of breast tissue to malignant transformation, while being able to assess interaction with reproductive factors such as age at first birth, and radiation variables, such as age at exposure and dose. An integral part of this research project is the opportunity to investigate the interrelationship between age at exposure to radiation and inherited susceptibility. To accomplish this a nested case-control study of the breast cancer cases in the cohort and matched control women, selected from radiation-exposed women in the cohort who did not develop cancer, will be conducted.

2.3. RADIOACTIVE THOROTRAST

We recently conducted a retrospective follow-up study to evaluate the long-term consequences of exposure to the contrast medium, Thorotrast, a colloidal thorium dioxide solution [14-15]. Thorotrast was used during the 1940’s and 50’s as a radiographic contrast agent in an estimated 2.5 million people worldwide. Thorotrast accumulates in the reticulendothelial system (liver, spleen, lymph nodes, and bone marrow) where the blood pool is exposed to the ^{232}Th decay series. The series includes

^{228}Ra, ^{228}Th, ^{224}Ra, and ^{220}Rn. Studies of Thorotrast may increase our understanding of the health effects of these radionuclides. Thorotrast has a biological half-life in humans of 400 years; therefore, once injected, patients are exposed to internal alpha irradiation for life.

The study included 723 patients who underwent cerebral angiography with Thorotrast and 315 patients who underwent cerebral angiography with the non-radioactive iodine-based dye, Diodrast. The average amount of Thorotrast injected was 26 ml, for an estimated annual dose of 25 cGy to liver and 16 cGy to bone marrow. Current vital status has been determined for 92% of the study subjects; 90% of the Thorotrast-exposed and 75% of the Diodrast-exposed patients have died. The median age at angiography was 44 years for both groups. Average follow-up time was 24.5 years for the Thorotrast group and 27 years for the Diodrast group.

The SMR for deaths due to liver cancer for men exposed to Thorotrast was 26.6 (11 observed, 0.4 expected, 95% CI 1.3-47.6) and for men exposed to Diodrast was 4.6 (1 observed, 0.2 expected, 95% CI 0.1-25.6). For women exposed to Thorotrast, the SMR for deaths due to liver cancer was 21.4 (11 observed, 0.5 expected, 95% CI 10.7-38.3). There were no deaths from liver cancer among the women injected with Diodrast (expected number = 0.3). All liver cancer deaths occurred 15 or more years after angiography. Chromosomal studies of lymphocytes and monocytes from living study subjects are underway to elucidate the carcinogenic mechanisms of alpha irradiation.

3. Summary

Continued surveillance of populations exposed to radiation at both high and low doses remains a priority. While ionizing radiation is a known human carcinogen, important questions related to dose, timing, and genetic predisposition remain unanswered. In addition, long-term health outcomes other than cancer have received little attention. Follow-up of populations such as those described here will permit scientists to understand the totality of health consequences resulting from different types and doses of radiation. Such studies will also further our understanding of the molecular basis by which radiation causes cellular damage.

4. References

1. Upton, A.C. (1985) The biological effects of low-level ionizing radiation. Chapter 3 in Cancer Biology. W.H. Freeman.
2. Boice, J.D. Jr, Land, C.E., Preston, D.L. (1996) Ionizing radiation. Chapter 16 in Cancer Epidemiology and Prevention, Second Edition (Schottenfeld D, Fraumeni JF Jr., eds.) Oxford University Press.
3. Doll, R. (1995) Hazards of ionising radiation: 100 years of observations on man. Br. J. Cancer 72, 1339-1349.
4. National Research Council. (1990) Health Effects of Exposure to Low Levels of Ionizing Radiation. BEIR V. National Academy Press. Washington, D.C.

5. Boice, J.D. Jr, Fraumeni, J.F. Jr (eds.). (1984) Radiation Carcinogenesis: Epidemiology and Biological Significance. Raven Press. New York.
6. Boice, J.D. Jr, Monson, R.R. (1977) Breast cancer in women after repeated fluoroscopic examinations of the chest. J. Natl. Cancer Inst. 59, 823-832.
7. Inskip, P.D., Monson, R.R., Wagoner, J.K., Stovall, M., Davis, F.G., Kleinerman, R.A., Boice, J.D. Jr. (1990) Cancer mortality following radium treatment for uterine bleeding. Radiat. Res. 122, 107-119.
8. Aisenberg, A.C., Finkelstein, D.M., Doppke, K.P., Koerner, F.C., Boivin, J.F., Willett, C.G. (1997) High risk of breast carcinoma after irradiation of young women with Hodgkin's disease. Cancer 79, 1203-1210.
9. Wong, F.L., Boice, J.D. Jr, Abramson, D.H., Tarone, R.E., Kleinerman, R.A., Stovall, M., Goldman, M.B., Seddon, J.M., Tarbell, N., Fraumeni, J.F. Jr, Li, F.P. (1997) Cancer incidence after retinoblastoma: radiation dose and sarcoma risk. JAMA 278, 1262-1267.
10. Goldman, M.B., Maloof, F., Monson, R.R., Aschengrau, A., Cooper, D.S., Ridgway, E.C. (1988) Radioactive iodine therapy and breast cancer. Am. J. Epidemiol. 127, 969-980.
11. Goldman, M.B, Monson, R.R., Maloof, F. (1990) Cancer mortality in women with thyroid disease. Cancer Res. 50, 2283-2289.
12. Snegireff, L.S. (1959) The elusiveness of neoplasia following roentgen therapy for thymic enlargement in childhood. Radiology 72, 508-517.
13. Janower, M.L., Miettinen, O.S. (1971) Neoplasms after childhood irradiation of the thymus gland. JAMA 215, 753-756.
14. Janower, M.L., Miettinen, O.S., Flynn, M.J. (1972) Effects of long-term Thorotrast exposure. Radiology 103, 13-20.
15. Goldman, M.B., Janower, M.L., Monson, R.R., Platz, E.A., Schottenfeld, D., Travis, L.B. (1997) Liver cancer after long-term exposure to radioactive Thorotrast (Abstract). Am. J. Epidemiol. 145, S68.
16. Dobyns, B.M., Sheline, G.E., Workman, J.B., Tompkins, E.A., McConahey, W.M., Becker, D.B. (1974) Malignant and benign neoplasms of the thyroid in patients treated for hyperthyroidism: a report of the Cooperative Thyrotoxicosis Therapy Follow-up Study. J. Clin. Endocrinol. Metab. 38, 976-998.
17. Ron, E., Doody, M.M., Becker, D.V., Brill, A.B., Curtis, R.E., Goldman, M.B., Harris, B.S.H., Hoffman, D.A., Maxon, H.R., McConahey, W.M., Preston-Martin, S., Warshauer, M.E., Wong, F.L., Boice, J.D. Jr. Cancer mortality following treatment for hyperthyroidism. In press.
18. Tokunaga, M. Land, C.E., Yamamoto, T., Asano, M., Tokuoka, S, Ezaki, H., Nishimori, I. (1987) Incidence of female breast cancer among atomic bomb survivors, Hiroshima and Nagasaki, 1950-1980. Radiat. Res. 112, 243-272.
19. Tokunaga, M., Land, C.E., Tokuoka, S., Nishimori, I., Soda, M., Akiba, S. (1994) Incidence of female breast cancer among atomic bomb survivors, 1950-1985. Radiat. Res. 138, 209-223.
20. Hildreth, N.G., Shore, R.E., Dvoretsky, P.M. (1989) The risk of breast cancer after irradiation of the thymus in infancy. N. Engl. J. Med. 321, 1281-1284.

RADIOACTIVE AND TRACE ELEMENTS IN THE URAL MOUNTAINS

Report of the Working Group on Environmental and Health Effects of Radiation[1]

W. R. SCHELL
Dept. of Ciencies Ambientals, Universitat de Girona, SPAIN[2]

M. B. GOLDMAN
Department of Epidemiology, Harvard School of Public Health, Boston, MA 02115, USA

I. LINKOV
Department of Physics, Harvard University, Cambridge, MA 02138, USA

1. Introduction

Radionuclides from the Chernobyl NPP accident in 1986 and the Kyshtym nuclear waste accident in 1957 illustrated our lack of preparation for such disasters in Europe. The radionuclides were deposited on forests, agricultural lands, marshlands, rivers and lakes where the population utilized these regions for food, employment, and recreation. We question "How long will these areas remain contaminated and affect the population?" and "How can we document the health risks associated with continued exposure?"

One requirement is to develop a dynamic model that could predict the radiation and chemical dose in the future. Recent developments in the Geographic Information System (GIS) have provided a modeling framework where information can be collected and utilized. However, the application of GIS to ecological models is just beginning. New mathematical tools of linked partial differential equations (PDE) and cellular automata (CA) can be used to great advantage in such complex ecosystems. In addition, simple models that use the minimum number of variables, but explain the generic system, are needed. Recently developed software can provide generic models for ecologists to use and to test concepts. These models can then be used for site-specific conditions. Different degrees of complexity can be incorporated by varying the

---

[1] Members of Working Group: W.R. Schell (Chairman), M.B. Goldman (Rapporteur), I. Linkov (Rapporteur), B.A. Korobitsin, C. Papastefanou, S.A. Romanov, D. Burmistrov, M.N. Kossenko, N.D. Okladnikova, R. Day, E. Steinnes, M. Frontasyeva, R. Wilson

[2] Present address: Department of Environmental and Occupational Health, University of Pittsburgh, Pittsburgh, PA 15261, USA

I. Linkov and R. Wilson (eds.), Air Pollution in the Ural Mountains, 441–444.

compartment number and functional dependence of the rates of intercompartmental pollutant flows.

Radionuclides and toxic trace elements deposited on catchment basins require understanding several ecological units. For example, models can describe the linkage of information on terrestrial lands, marshlands, rivers, and lakes and the toxic material interactions among these environments. We must attempt to integrate the several compartments in contaminant flow since, in trying to describe the natural process, we cannot be restricted to considerations of only parts of the ecosystem. Both field collections and laboratory experiments are needed. In addition, interdisciplinary backgrounds will be required for experimentalists and theorists to drive the model development, testing, use, and validation.

The goals of a collaborative study can be to advance understanding of the evolution of chemical and radiation doses to humans and ecosystems following significant releases to the environment. This study would ultimately provide a systematic evaluation of what is currently known about the history, distribution, and fate of contaminating toxicants and help provide a basis for the design of countermeasures to make the contaminated environment useable for normal activity in the future. Both environmental and human risk assessments should be provided, including identification of the most exposed individuals (i.e., critical groups). This information could then be used in epidemiologic studies on the health consequences of toxic contamination.

In brief, the specific objectives of this type of assessment are to:

1. Quantify the physical, chemical, and biological processes influencing the behavior of ^{137}Cs, ^{90}Sr, Pu, natural radionuclides, and selected toxic trace elements in river catchment basins.
2. Provide a scientific basis for predicting the impact of such contaminants on catchment areas and their consequent inputs into aquatic ecosystems.
3. Obtain detailed information on the factors influencing the levels and transport of elements through the terrestrial/aquatic interface and thereby improve the reliability and accuracy of assessments of radioactive and trace metal exposure to humans and other organisms.
4. Identify the sources of data, reliability of measurements, and extent of interpretation relevant to this program as a major first step.
5. Provide the basis for a practical model, or series of linked models, to be used by regulators and governments to yield an initial prediction of radiation and toxic chemical dose upon which epidemiologic studies and public policy decisions could be based.

2. Prospectus for Collaborative Research

The atmospheric deposition pattern of heavy metals in the Ural region is influenced on a local and regional scale by point sources within the area. It is likely that there is also influence from long range transport from sources up-wind from the selected Magnitogorsk regions. To integrate the sources and distribution of such toxicants can be a formidable task because of the great expanse of territory. The geographical distribution of heavy metals deposition in the entire Ural region has not been systematically studied. A survey sampling technique including collection and analysis of moss samples, as applied in Western and Central Europe, should be a suitable technique for this purpose. In addition, the time history of deposition of toxic chemicals can be determined from core samples collected from appropriate undisturbed repositories where deposition can be dated using ^{210}Pb techniques.

The recently released information about emissions of radioactivity from accidents and from waste disposal related to nuclear facilities in the Ural region has aroused great interest and concern in Russia and abroad. Vast amounts of data on local contamination problems have been collected, but, as far as is known, the regional impact of the release of radionuclides to the air is not known in detail. A sampling and analysis program using selected lake sediments, marshlands, bogs and surface soils could obtain this information.

Concentration data for radionuclides and heavy metals in the terrestrial environment may provide the geographic variation in the integrated exposure of the population to these pollutants that would be relevant for epidemiologic studies. Information on the time resolution of the deposition would be even more interesting from an exposure point of view. Such information can be obtained from analysis of dated cores from peat bogs or from sediment cores collected in lakes.

Epidemiologic studies should be designed using different means to reconstruct doses to specific geographic regions and, where possible, to individuals. Studies that address the interaction between air pollution, heavy metals contamination, and radioactive exposure would be especially useful. The Workshop discussed the scientific importance of the Chelyabinsk and Mayak cohorts. Studies of these cohorts would result in valuable data on several important research questions including the consequences of high dose exposure, the effects of internal exposure, and the effects of low-dose chronic radiation exposure. Scientists agreed that studies of health effects should have high priority.

3. Proposed Program

We propose to develop a modeling concept that integrates the compartments of forest, interface and aqueous zones. The forest is represented as upland areas which can erode toxic elements and transfer them to the interface zone which includes wetlands of rivers and lakes and into the river channels for long range transport to the ocean or to a sink which is the lake sediments. The catchment basins, rivers, and reservoirs impeding the flow would provide samples to develop and test the modeling framework and could include:

1. Collection of moss and surface soil samples e.g. in a 30 x 30 km grid.
2. Collection of vertical cores of peat (or lake sediment) at 5-10 sites along the Ural Mountains.
3. Measurements of selected metals in mosses, bog cores, and soils by appropriate analytical techniques (AAS, ENAA, ICP-MS) following an interlaboratory comparison for quality control.

4. Measurements of long-lived radionuclides (^{137}Cs, ^{90}Sr, Pu, etc.) in sediment and peat cores and in soil profiles.
5. Preparation of maps in the GIS format where the data could be organized.
6. Identification of different sources by atmospheric transport models and relation of the emission inventories to chemical and radiochemical measurements made in the sampling grid.

In addition, sources of funding support for collaborative studies on the health effects of radiation exposure should be sought. These studies should supplement those that are already underway and should take dose modeling and environmental measures of exposure into account.

LIST OF PARTICIPANTS

Name	Address	Contact
Raisa A. Alakayeva	Ufa Research Institute of Occupational Health and Human Ecology, 94 Kuvikina Str., Ufa, 450106, RUSSIA	
Vaciliy A. Bakunin	Chelyabinsk Regional Committee on Ecology and Nature Management, 75, Elkin st., Chelyabinsk, 454092, RUSSIA	phone: 7-351 237-8183
Taicia I. Belotserkovets	Chelyabinsk Institute of Industrial Ecology, 76, Lenin aven., Chelyabinsk, 454089, RUSSIA	phone: 7-351-241-4596
Vyacheslav D. Cherchintsev	Magnitogorsk State Academy of Min ing and Metallurgy, 38,Lenin aven. Magnitogorsk, 455000, RUSSIA	e-mail: post@mgmi.mgn.chel.su phone: 7 351 1328515 FAX: 7 351 1322886
Cliff Davidson	Engineering and Public Policy Carnegie Mellon University Pittsburgh, PA 15213, USA	e-mail: cliff@andrew.cmu.edu phone: 412-268-2951 FAX: 412-268-7813
Richard Day	Department of Biostatistics GSPH, University of Pittsburgh 130 DeSoto Street Pittsburgh, PA 15261, USA	e-mail:rdfac@vms.cis.pitt.edu phone: 412-624-4077 FAX: 412-624-3040
Abay Dzhabasov	Head, Department for Project Design and Planning, KazKredSotsBank, KAZAHSTAN	e-mail: burger@hiidb.almaty.kz phone: 7-3272-64-26-90 FAX: 7-3272-64-67-15,
Michael J. Ellenbecker	University of Massachusetts Lowell Department of Work Environment One University Ave, Lowell, MA, USA	e-mail: ellenbecm@woods.uml.edu phone: 508-934-3272 FAX: 508-934-3050
Francisco Ferreira	Department of Environmental Sciences and Engineering, College of Sciences and Technology, New University of Lisbon Quinta da Torre, Monte Da Garcia 2885 PORTUGAL	e-mail: ff@mail.fct.unl.pt phone: 351-1-2954464 (e0131) FAX: 351-1-2948554

Boris N. Filatov	Volgograd EnvironmentProtection Centre of Harvard University, 12, Zemliatchky, Volgograd, 400048, RUSSIA	e-mail: ucmp@sprint-v.com.ru
Marina V. Frontasyeva	Joint Institute for Nuclear Research Dubna, Moscow Region, 141980 RUSSIA	e-mail: marina@nf.jinr.ru phone: 7-096-216-3653 FAX: 7-096-216-5891
Evgeniy L. Genihovich	Voeykov Geophysical Observatory, 40 - 86, Engels aven., St.Petersburg, 194156, RUSSIA	e-mail: mgo@glas.apc.org.ru phone: 7-812-247-4390
Marlene Goldman	Department of Epidemiology, Harvard School of Public Health, 677 Huntington Ave Boston, MA 02115. USA	e-mail:goldman@episun1.harvard.edu phone: 1-617-432-4586 FAX: 1-617-566-7805
Isaak V. Gyelxfyenbuym	Perm Regional Environmetal Protec tion Committee, 51, Lenin st., Perm, 614600, RUSSIA	e-mail: comm@ecol.perm.ru phone: 7-342-234-9422
Tom C. Hutchinson	Trent University, Environmental and Resource Studies, P.O. Box 4800 Peterborough, Ontario K9J 7B8, CANADA	e-mail: thutchinson@trentu.ca phone: 1- 705 748-1634 FAX: 1- 705 748-1634
Jouni Jaakkolla	Kimmeltie 11 C 31 Espoo, FINLAND	e-mail: jjaakkol@jhsph.edu jojaakkola@medicum.helsinki.fi
Asilkhan J. Kaliev	Orenburg State Agrarian University 100 - 58, Karagandinskay, Orenburg 460352, RUSSIA	
Boris A. Korobitsin	Institute of Industrial Ecology, 20-A, S.Kovalevskoy, Ekaterinburg, 620219, RUSSIA	e-mail: kba@ecko.rcupi.e-burg.su phone: 7-3432-74-9394 FAX: 7-3432-74-3771
Valentina S. Koshkina	Institute of Biomedical Problems, 145 -10 K.Marks Ave., Magnitogorsk, 455043, RUSSIA	phone: 7-351-134-0440
Anatoliy K. Kostukov	Magnitogorsk Iron, Inc., 92,Kirov str., Magnitogorsk, 455002,RUSSIA	phone: 7-351-133-7983
Alexander P. Krasavin	Research Institute of Industrial Ecology “MSRIIEKO - TEK“, 60, N. Ostrovsky st., Perm,614070, RUSSIA	phone: 7-342-232-9529 FAX: 7-342-232-7443

Igor Linkov	Department of Physics Harvard University Cambridge, MA 02138, USA	e-mail: linkov@huhepl.harvard.edu phone 1- 617 495 4208 FAX: 1 -617 495 0416
Alexander B. Magid	Institute of Problem for Applied Ecology and Natural Resources 12\18 Marta str., Ufa, 450005, Bashkortostan, RUSSIA	e-mail: ROOT@ipeprb.bashkiria.su phone: 7-347-228-6892
Ian P. Milborrow	School of Social Sciences, Univeristy of Bath, Claverton Down, Bath, BA2 7AY, UK	e-mail: hssipm@bath.ac.uk phone: 44 1225 826954 FAX: 44 1225 826381
Nail S. Minigazimov	Institute of Problem for Applied Ecology and Natural Resources, 12\1, 8 Marta str., Ufa, 450005, Bashkortostan, RUSSIA	e-mail: root@ippeprb.bashkiria.su phone: 7-347-228-6870 FAX: 7-347-228-3913
Benoit Morel	Engineering and Public Policy Carnegie Mellon University Pittsburgh, PA 15213, USA	e-mail: bm1v@andrew.cmu.edu phone: 1-412-268-3758 FAX: 1-412-268-3757
Nikolay A. Motovilov	Sterlitamak Integrated Works, Inc. "Soda", 7, Babushkina, Sterlitamak, Bashkortostan, 453122, RUSSIA	phone: 7-347-112-7609
Boris A. Nikiforov	Magnitogorsk State Academy of Mining and Metallurgy, 38, Lenin aven, Magnitogorsk, 455000, RUSSIA	e-mail: post@mgmi.mgn. chel.su phone: 7-351-132-1287 FAX: 7-351-132-2886
Jerome Nriagu	Environmental and Industrial Health, School of Public Health, 109 Observatory Street Ann Arbor, MI 48109, USA	e-mail: jnriagu@sph.umich.edu phone: 1-313-936-0706 FAX: 1-313-764-9424
Ahmet Haluk Ozkaynak	Harvard School of Public Health 665 Huntington Avenue Boston, Massachusetts, 02115, USA	e-mail: haluk@sparc6b.harvard.edu phone: 1-617-432-4112 Fax: 1-617-432-4122
Connstantin Papastefanou	Aristotle University of Thessaloniki Nuclear Physics Department Thessaloniki 54006, GREECE	e-mail: papastefanou@physics.auth.gr phone: 30-31-998005 FAX: 30-31-998058
Sergey V. Pavlov	IPPER, 12\1, 8 Marta str., 455005, Ufa, RB, RUSSIA	e-mail: ipendy@mts.bashkiria.su phone: 7-347-228-6870
Gerhard Petersen	GKSS Research Centre, Institute of Hydrophysics, Max-Planck-Strasse Geesthacht D-21502, GERMANY	e-mail: petersen@gkss.de phone: 49-4152-871847 FAX: 49-4152-871888

Olga V. Ponomareva	International Institute for Health Risk Assessment, 1-5-7, Bolotnikovska, Moscow, 113556, RUSSIA	phone: 7-095-955-1386 FAX: 7-095-954-0310
Larissa I. Privalova	Ural Regional Center for Environmental Epidemiology, 3 Otdelny per, Ekaterinburg, 620219, RUSSIA	phone: 7-343-274-1725 FAX: 7-343-260-4138
Sergey K. Puntus	Environment Protection Centre Tagilecoprom", 15, Frunze, Nizhny Tagil, 622002, RUSSIA	e-mail: serg@tecopr.tagil.ru
Boris A.Revich	Russian Academy of Sciences Institute for Economic Forecasting, 32 Krasikova str., Moscov, 117418, RUSSIA	e-mail: revich@ector.msk. su phone: 7-095-332-4391 FAX: 7-095-310-7071
Naum A. Samoylov	Ufa State Oil Technical University 1, Kosmonavtov st., Ufa, Bashkiria 450062, RUSSIA	phone: 7-347-242-0713
Sergey V. Samoyiov	Committee on Ecology and Nature Management, 5-a, Kadyry str., Tashkent, 700078, UZBEKISTAN	phone: 7-377-272-3771 FAX: 7-377-241-3990
Katarzyna Sawicka-Kapusta	Institute of Environmental Biology Jagiellonian University U1. Ingardena, 30-060 Cracow, POLAND	e-mail: sawicka@eko.uj.edu.pl phone: 33 63 77, ext. 448 Fax: 48 12 34 07 85
William R. Schell	University of Girona Placa Hospital 6, Girona 17071, SPAIN	e-mail: cawrs@stark.udg.es phone: 34 72 41 82 92 , FAX: 34 72 41 83 61
Azat M. Shagiakhmetov	Intitute of Problem of Apply Ecology and Natural Resouces Use, 10/1, 8-March Str., Ufa, RUSSIA	e-mail: ipendy@diaspro.com phone: 28-87-75 FAX: 28-87-90
MaryBeth Smuts	US EPA -Region 1 JFK Building, Boston, MA 02067 USA	e-mail: smuts.marybeth@epamail.epa.gov phone: 617/565-3232 FAX: 617/5654940
Tom Spittler	US EPA Regional Laboratory 60 Westview St. Lexington, MA 02173 - USA	e-mail: spittler-tom@epamail.epa.gov phone 1- 602-546-1798 Fax. 1- 617-860-4397

Eiliv Steinnes	Norwegian University of Science and Technology, Department of Chemistry N-7034 Trondheim, NORWAY	e-mail: eilivs@alfa.avh.unit.no phone: 47-73596237 FAX: 47-73596940
Sheena Symington	Trent University, Environmental and Resource Studies, P.O. Box 4800 Peterborough, Ontario K9J 7B8, CANADA	e-mail: ssymington@trentu.ca phone: 1- 705 748-1634 FAX: 1- 705 748-1634
Svetlana N. Teplova	Chelyabinsk Medical Academy, 64, Vorovskogo, Chelyabinsk, 454092, RUSSIA	
Alexander M. Volkov	Ministry of Emergency Situations and Environmental Safety of Republic of Bashkortostan, 12\1, 8 Marta, Ufa, 450005, Bashkortostan, RUSSIA	phone: 7-347-228-9400
James Wilkinson	Georgia Institute of Technology 200 Bobby Dodd Way Atlanta, GA 30332, USA	e-mail: jwilkins@isis.ce.gatech.edu phone: 404-894-3089 FAX: 404-894-9724
Jeane Willson	Kleinfelder, Inc. 5953 South Iola Way Englewood, CO, 80111, USA	e-mail: jwillson@mindspring.com phone: 1-303-220-5758 FAX: 1-303-742-8232
Richard Wilson	Department of Physics Harvard University Cambridge, MA 02138, USA	e-mail: wilson@huhepl.harvard.edu phone: 1- 617 495 3387 FAX: 1 - 617 495 0416
Nina V. Zayitseva	Research Institute of Ecopathology 3-81, Puchkin st., Perm, 614000, RUSSIA	e-mail: srci@pstu.ac.ru phone: 7-342-233-5025 FAX: 7-342-234-8676

AUTHOR INDEX

R

S

V

W

X

Z

SUBJECT INDEX

MIX
Papier aus verantwortungsvollen Quellen
Paper from responsible sources
FSC® C105338

If you have any concerns about our products,
you can contact us on
ProductSafety@springernature.com

In case Publisher is established outside the EU,
the EU authorized representative is:
Springer Nature Customer Service Center GmbH
Europaplatz 3, 69115 Heidelberg, Germany

Printed by Libri Plureos GmbH
in Hamburg, Germany